TOPICS IN SURFACE CHEMISTRY

THE IBM RESEARCH SYMPOSIA SERIES

TOPICS IN SURFACE CHEMISTRY

Edited by

Eric Kay

and

Paul S. Bagus

International Business Machines Corporation
San Jose, California

PLENUM PRESS · NEW YORK AND LONDON

Library of Congress Cataloging in Publication Data

Main entry under title:

Topics in surface chemistry.

(IBM research symposia series)
Proceedings of an international symposium held Sept. 7—9, 1977, in the Federal Republic of Germany.
Includes index.
1. Surface chemistry—Congresses. I. Kay, Eric. II. Bagus, Paul. III. Series: International Business Machines Corporation. IBM research symposia series.

QD506.A1T66	541'.3453	78-4888

ISBN-13: 978-1-4613-4005-8   e-ISBN-13: 978-1-4613-4003-4
DOI: 10.1007/ 978-1-4613-4003-4

Proceedings of an International Symposium
on Topics in Surface Chemistry
held in the Federal Republic of Germany, September 7—9, 1977
and sponsored by IBM

© 1978 Plenum Press, New York
Softcover reprint of the hardcover 1st edition 1978
A Division of Plenum Publishing Corporation
227 West 17th Street, New York, N.Y. 10011

Preface

The papers in this volume were presented at an international symposium on Topics in Surface Chemistry which was held in Bad Neuenahr, West Germany, September 7-9, 1977. The symposium was sponsored by IBM Germany.

It has been recognized for many years that our understanding of bulk phenomena and their subsequent exploitation depends largely on our ability to define correlations between microscopic structure and the physical and chemical phenomena of interest.

The role played by surface phenomena in the overall behavior of a material has been a subject for speculation for a long time, but only during the last decade or so have experimental and theoretical tools been developed which make it possible to investigate surface structure and related surface phenomena uniquely.

Numerous surface spectroscopies have been developed in recent years intended to describe the geometric, vibrational and electronic structure of a surface. Our present understanding of surface, thin film and interfacial phenomena in solid state physics owes much to these developments. In chemistry much of the interest in surface science has come from the obvious implications to such important and diverse fields as catalysis and corrosion. It takes little imagination to recognize that there are many other areas where advances in surface science can be brought to bear.

It was the purpose of this IBM sponsored conference to bring together key scientists, particularly from Europe, who, though active in quite diverse fields, appear to be asking related questions about the role of surface structure and phenomena as encountered in their particular fields of interest.

The motivation for the conference was to explore common ground, especially in chemical aspects of surface and interfacial phenomena. A conscious effort was made to intersect but also go beyond topics covered by other chemically oriented surface confer-

ences, most of which had been motivated historically by the wide
interest in catalysis.

Five distinct fields were represented at the conference. The
sessions on "Fundamental Aspects of Surface Chemical Bonding" and
"Optical Excitations at Surfaces" examined recent progress in
understanding surface electronic and vibrational structure, including structure of sorbed species. Photoelectron Spectroscopy,
High Resolution Energy Loss Spectroscopy, Surface Raman Spectroscopy, as well as Attenuated Total Reflection and Surface Photovoltage Spectroscopy, were discussed. Both inorganic and organic
systems were considered. The session on "Atomic and Molecular
Scattering from Surfaces" took cognizance of the fact that ultimately any real understanding of surface chemistry at a gas-
surface interface must include a detailed description of energy
partitioning and the dynamics of surface scattering processes.
The session on "Surface Studies in Electrochemical Systems" dealt
with interfacial electrochemical phenomena and explored contemporary
chemical and physical approaches designed to study and control
electron transfer at such solid-liquid interfaces. The session
on "Ordered Array of Organic Molecules at Surfaces and Interfaces"
explored some very exciting chemical and physical characteristics
of complex organic molecules cast into well-defined monolayer
assemblies and also encountered in micellar structures.

Eric Kay   Paul Schweitzer
IBM San Jose Research Laboratory   IBM Germany, Sindelfingen
Symposium Chairman   Symposium Manager

Contents

ASPECTS OF SURFACE CHEMICAL BONDING

OPTICAL EXCITATIONS AT SURFACES

REACTIONS ON SEMICONDUCTOR ELECTRODES

R. Memming

Philips GmbH Forschungslaboratorium Hamburg

D2000 Hamburg 54, Germany

ABSTRACT

The essential charge transfer processes in reactions at semiconductor electrodes can be described on the basis of an energy band model. The basic mechanisms are described and examples for typical processes are given. The main emphasis is put on photoeffects and various processes induced by light absorption in the semiconductor or in the electrolyte are discussed in detail.

INTRODUCTION

Problems of the interface semiconductor-electrolyte were approached originally from two different fields, from electrochemistry and from solid state physics. Semidonducting properties played a certain role already in an early stage of electrochemistry, for instance in studies of oxide films on electrodes. Since the electrical properties of semiconductors were not known at that time systematic investigations of electrode processes were not possible. Basic studies in this field started shortly after single crystals of semiconductors were available with well defined electrical and optical properties.

In solid state physics the main interest was in studies of semiconductor surfaces since surface effects influence the electrical properties of many semiconductor devices.

From electrochemical investigations not only information about the properties of semiconductor electrodes was obtained but also about the basic mechanisms
of charge transfer processes between electrodes and
molecules in the electrolyte which are not accessable
by using metals as electrode material. The studies in
semiconductor electrochemistry during the last 15 years
were not restricted to elemental semiconductors such as
germanium and silicon. Especially investigation with
large-band-gap semiconductors such as GaP, ZnO, SiC
and SnO_2 (band gaps between 2.3 and 3.8 eV) have contributed to a better understanding of the energy parameters and kinetics of electrochemical reactions. This
report is mainly focused on these processes. The principles of pure charge transfer processes are discussed
and illustrated by various examples.

POTENTIAL AND CHARGE DISTRIBUTION

The potential distribution at the semiconductor-
electrolyte interface differs from that of a metal
electrode in so far that a potential drop does not
only occur across the Helmholtz double layer (U_H) but
also across the space charge (U_{sc}) below the semiconductor surface (Fig. 1). The formation of such a space
charge layer is due to the fact that the density of
charge carriers in the semiconductor is much smaller
than in a metal. Accordingly, the energy bands at the

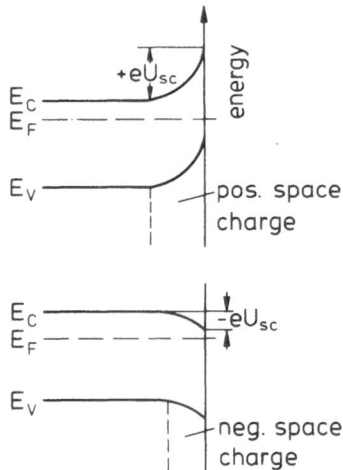

Fig. 1: Potential distribution at interface.

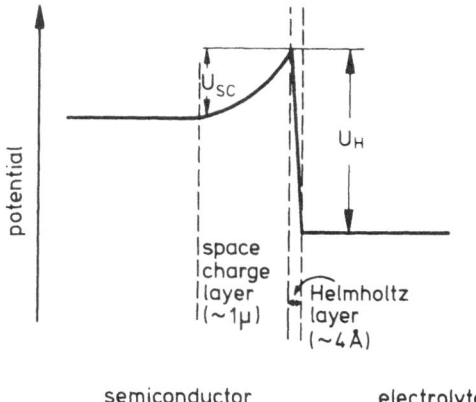

Fig. 2: Energy bands below the surface

semiconductor surface are bent upwards or downwards depending on whether the space charge is positive or negative (Fig. 2). The Fermi level is assumed to be constant. Measuring the electrode potential versus a reference electrode one obtains

$$U_E = U_{sc} + U_H + const, \qquad (1)$$

in which U_E, U_{sc} and U_H are the potentials of the electrode, across the space charge layer and across the Helmholtz layer, respectively, whereas the constant contains all other potentials occuring at the reference electrode.

Varying the electrode potential by applying an external voltage the question arises which potential drop, U_{sc}, U_H or both, are changed. This problem can be solved by capacity measurements. The capacity of a space charge is defined as

$$C_{sc} = \frac{dQ_{sc}}{dU_{sc}} . \qquad (2)$$

The space charge (Q_{sc}) - potential (U_{sc}) dependence can be derived from the Poisson equation

$$\frac{d^2 U_{sc}}{dx^2} = - \frac{\rho(x)}{\varepsilon \varepsilon_o} \tag{3}$$

in which the total charge $\rho(x)$ is given by

$$\rho(x) = e[N_D - N_A - n(x) + p(x)]. \tag{4}$$

The charge is determined by all mobile carriers (electrons, $n(x)$, and holes $p(x)$) and by ionized donors N_D and acceptors N_A. The electron and hole densities n_s and p_s at the interface are related to the carrier densities n_o and p_o in the bulk of the material by the Boltzmann equations:

$$n_s = n_o \exp\left(- \frac{eU_{sc}}{kT}\right), \tag{5a}$$

$$p_s = p_o \exp\left(+ \frac{eU_{sc}}{kT}\right). \tag{5b}$$

If equilibrium between electrons and holes exists throughout the whole semiconductor then

$$n_o p_o = n_s p_s = n_i^2 ,$$

in which n_i is the inversion concentration, mostly a very low value for large band gap semiconductors decreasing exponentially with the band gap.

Integrating the Poisson equation and using equations (2), (4) and (5) one obtains an exact relation between space charge capacity C_{sc} and U_{sc}. This equation is rather complex [1] and will not be discussed here. In the exhaustion region (i.e. $n_s < n_o$ for n-type and $p_s < p_o$ for p-type) this equation simplifies to

$$\frac{1}{C_{sc}^2} = \left(\frac{2}{\varepsilon \varepsilon_o N_D e}\right) \left(U_{sc} - \frac{kT}{e}\right) \quad \begin{array}{l}\text{(Mott-Schottky} \\ \text{equation)}\end{array} . \tag{6}$$

In the case of large band gap semiconductors ($E_g > 2$ eV) the exhaustion region is relatively large so that Eq. (6) can be applied for determining the potential distribution.

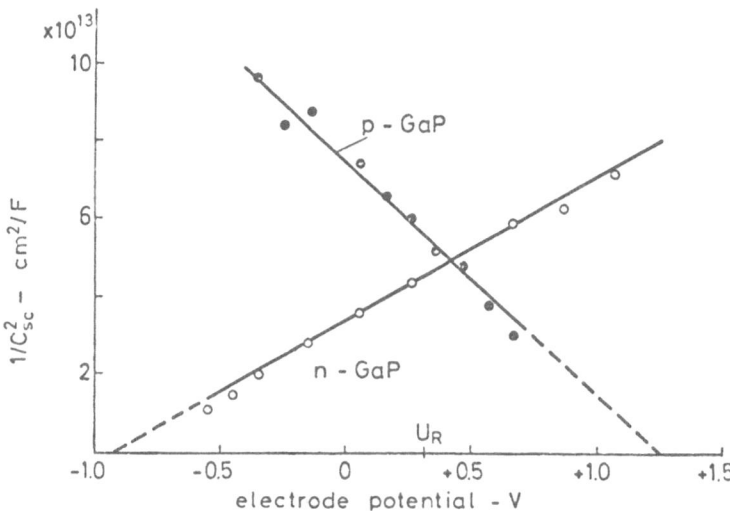

Fig. 3: Capacity curves for n- and p-GaP.

A typical result as obtained with n- and p-type GaP-electrodes [2] is given in Fig. 3. According to this figure $1/C_{sc}^2$ varies linearly with the electrode potential U_E. Since the slope is identical to the theoretical value in Eq. (6) one has to conclude that any variation of the electrode potential occurs across space charge layer, i.e.

$$\Delta U_E = \Delta U_{sc} \qquad\qquad (7)$$

whereas the potential across the Helmholtz layer remains unchanged. This result has been obtained with all semiconductors provided that the surface or the bulk was not degenerated (Fermi level in the conduction or valence band). For further details it must be referred to the literature [1, 3].

From the capacity curves (Fig. 3) some further important results can be derived as follows:

According to Eq. (6)

$$U_{sc} \rightarrow \frac{kT}{e} \approx 0 \text{ for } \frac{1}{C_{sc}^2} \rightarrow 0, \text{ i.e.}$$

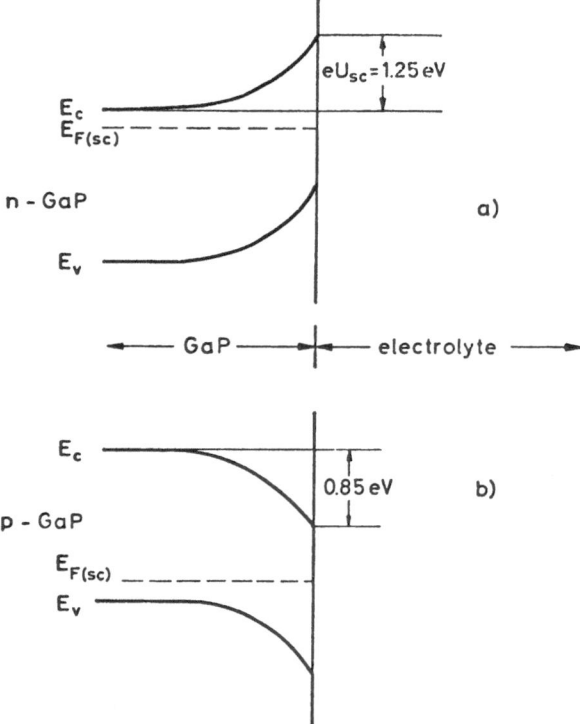

Fig. 4: Band bending of GaP at rest potential.

the extrapolation of the capacity curves in Fig. 3
yields an electrode potential, at which the energy
bands must be flat up to the surface ($U_{SC} = 0$). These
so-called flat band potentials U_{fb} differ obviously
very much for n- and p-type GaP. Reconstructing now
the band bending for these two dopings from Fig. 3 for
a certain electrode potential (e.g. at $U_E = U_R$) then
one obtains a band picture as given in Fig. 4. According to this model the distance of the Fermi level to
the band edges at the surface are equal for n- and
p-type, i.e. the position of the energy bands at the
surface is fixed independent of the electrode potential or
the doping of the electrode. At equilibrium the electrochemical potential (Fermi level) must be equal
throughout the whole system. Accordingly, varying the
doping U_{SC} also changes. In order to keep the Fermi
level constant a corresponding change of the potential
drop at the metal-semiconductor contact on the back

side of the semiconductor crystal must change in exactly the opposite direction. At the flat band potential at which $U_{sc} = 0$, one also knows the position of
Fermi level (equal now to the chemical potential of the
semiconductor) with respect to a reference electrode.
Since the Fermi level of a semiconductor is given by

$$E_F = E_c + kT \ln \frac{n_o}{N_c} \quad \text{for n-type}$$

and   (8)

$$E_F = E_v - kT \ln \frac{n_o}{N_v} \quad \text{for p-type}$$

(N_c and N_v density of states in conduction and valence
band) one can also determine the position of the conduction and valence band at the surface (E_c^s and E_v^s).

Investigations with a large number of semiconductors have supported this result. According to such an
analysis the energy position of energy bands of various
semiconductors are given in Fig. 5 using the normal
hydrogen electrode (NHE) as a reference [1]. It should
be mentioned that these energy values can also be related to the vacuum level as it is usually done in
solid state physics (left hand scale in Fig. 5). According to calculations by Lohmann [4] the energy of
NHE versus vacuum is of the order of

$$E_{NHE} = -4.5 \text{ eV}. \tag{9}$$

The values given in Fig. 5 depend on the pH-value of
the aqueous electrolyte. This result is due to the
fact that the potential drop across the Helmholtz-
layer (U_H) depends on pH. In one case this dependence
has been studied quantitatively [5].

The position of energy bands of various semiconductors can be considerably different which is of great
importance for selecting a proper semiconductor for
studying a certain electrochemical process. As far as
electron transfer reactions are concerned another important result can be derived from measurements of the
potential distribution: Since any externally applied
potential occurs completely across the space charge
layer only the surface concentrations of electrons
and holes are varied according to Eqs. (5a) and (5b).

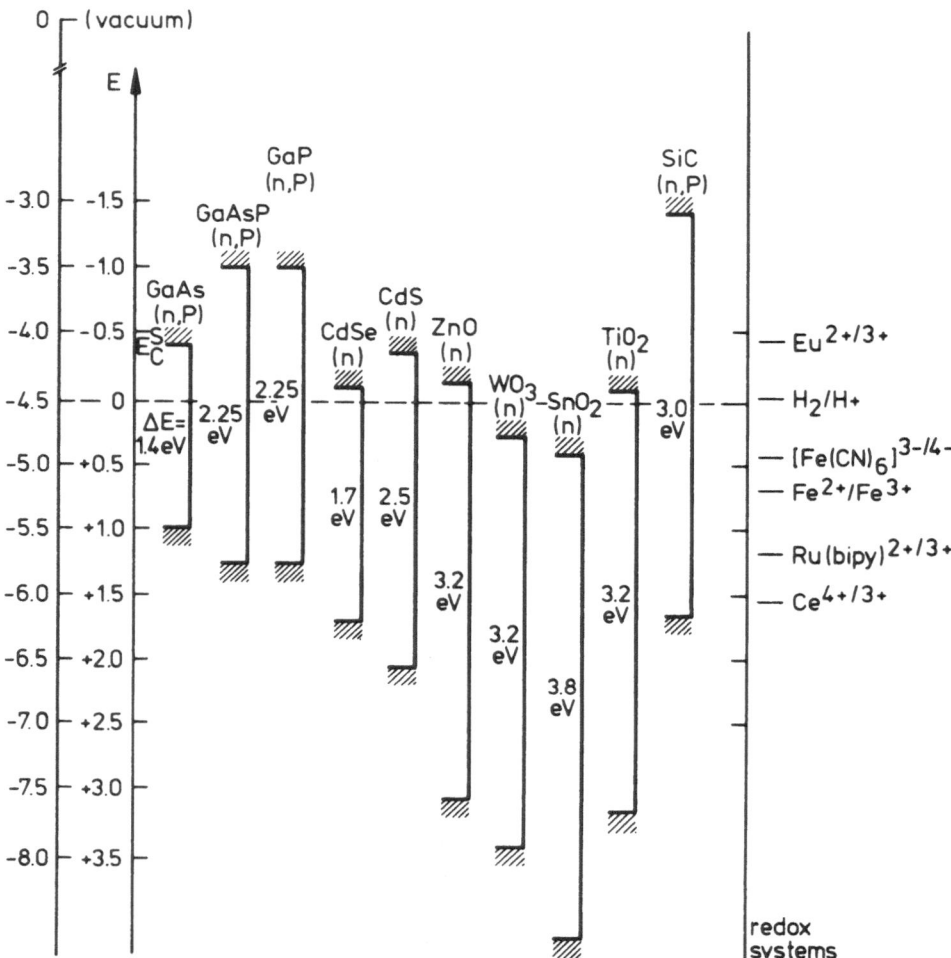

Fig. 5: Position of energy bands at the surface of
various semiconductors at pH 1.

CURRENT-POTENTIAL DEPENDENCES

The simplest way of getting information about reactions at semiconductor electrodes is the measurement
of a current-potential curve, an example given in Fig. 6
for n- and p-type GaP [6]. In this case the anodic process corresponds to the anodic dissolution of the material whereas the cathodic currents are due to hydrogen evolution. The cathodic current increases with potential for n-type whereas it is limited to a very low
value in the case of p-type. This result can only be
interpreted by assuming an electron transfer from the
conduction band to protons in the electrolyte, because

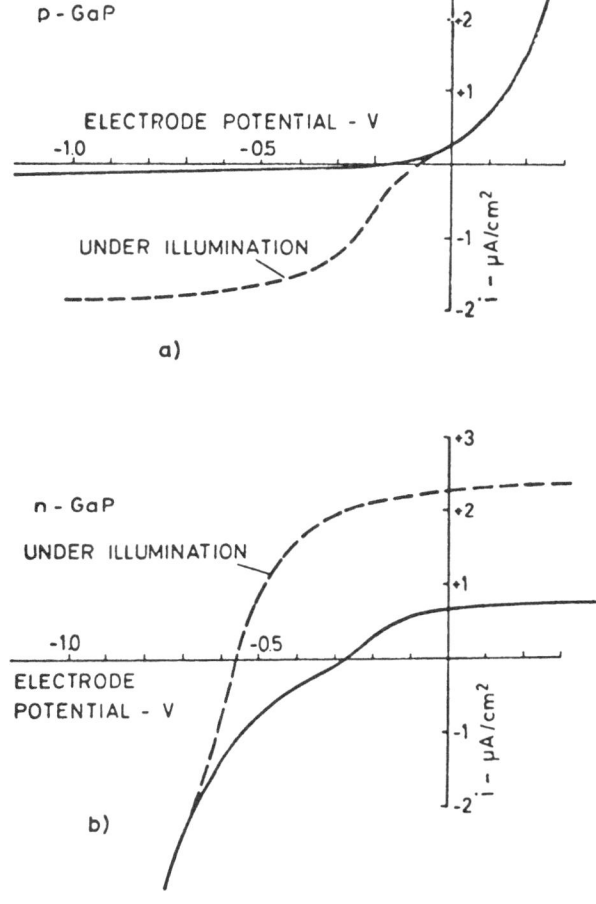

Fig. 6: Current potential curves for GaP in 1 N KCl.

only very few electrons are available in the conduction
band of p-type. This is supported by the fact that the
cathodic current at p-type is increased by light absorbed by the semiconductor. The same arguments are
valid for the anodic dissolution. In this case, however, the process occurs via the valence band because
electrons can only be injected into the valence band
if holes are present. The small currents found with
n-type in the anodic range and with type during
cathodic polarization are limited by diffusion of the
corresponding minority carriers. In some cases these
currents remain at low levels even for very large po-
larizations. For instance, even for small band gap
semiconductors such as n-GaAs no current increase has
been observed up to 25 V [7]. The diffusion current
values and the potentials at which they suddenly increase is very much determined by the quality of the
crystal and the doping. In some cases avalanche breakdown (GaAs) [7], in others tunneling processes through
the space charge (ZnO [8], SnO_2 [9]) have been observed

According to Fig. 6 charge transfer process may
either occur via the conduction or the valence band.
The question arises, which factors determine whether
a certain process occurs via the conduction or the
valence band. More insight can be obtained by studying rather simple processes such as the electron transfer between semiconductor and redox system.

ENERGY LEVELS AND FERMI LEVEL IN ELECTROLYTES

Electrolytes containing an oxidation-reductor system
are characterized by the redox potential. Values of
this potential are usually given in the conventional
scale using the normal hydrogen electrode as a reference
point. Taking, however, the vacuum level as a reference
as it is common for solids, then the redox potential
corresponds to an energy required for transferring an
electron from a redox system in the electrolyte into
the vacuum. Accordingly, the electron energy is defined in the same way as in solids and one can define
a Fermi level of the redox system, $E_{F,el}$. At equilibrium
at the semiconductor-redox system interface the Fermi
levels of the solid (E_F) and of the redox system
($E_{F,el}$) must be equal [1, 3]:

$$E_F = E_{F,el} \cdot$$

Values are also given in Fig. 5.

The question arises now whether one can define or derive also energy states for redox systems in a similar way as in semiconductors or metals. Certainly occupied and empty states are represented by the reduced and oxidized species of a redox-system, respectively. However, the energy position of these states can be considerably different due to the strong interaction of the redox system with the solvent. Each ion is surrounded by a solvation shell. The interaction depends on the size and charge of the ion. If an electron is transferred from the reduced species (Red), e.g. a Fe^{2+}-ion, into the vacuum then this process is followed by a rearrangement or reorientation of the solvent molecules in the solvation shell. A similar rearrangement is required for the reverse process. A certain energy, the so-called rearrangement energy λ is involved. The complete cycle is given by

$$
\begin{array}{ccc}
Red_{solv,ox} & \xrightarrow[-A]{+e^-} & Ox_{solv,ox} \\
-\lambda \downarrow & & \uparrow -\lambda \\
Red'_{solv,red} & \xrightarrow[I]{-e^-} & Ox_{solv,red}
\end{array}
\qquad (11)
$$

in which the indices represent the state of the solvation shell, I the ionization energy and A the electron affinity. Accordingly

$$ I - A = 2\lambda . \qquad (12) $$

It is assumed in this model that the rearrangement is slow compared to the electron transfer (Frank-Condon principle). Consequently the energy levels of the reduced (occupied) E^0_{Red} and oxidized species (empty) E^0_{Ox} are not equal. They differ from the Fermi level $E_{F,el}$ by λ as schematically shown in Fig. 7a.

The energy levels are not pure discrete levels but are distributed over a certain energy range due to the fluctuation of the solvation shell. The corresponding distribution functions of the density of states are given by

$$
D_{red} = \exp - \frac{(E-E_{F,el}-\lambda)^2}{4kT\lambda} \qquad D_{ox} = \exp - \frac{(E-E_{F,el}+\lambda)^2}{4kT\lambda}
$$

$$ (13) $$

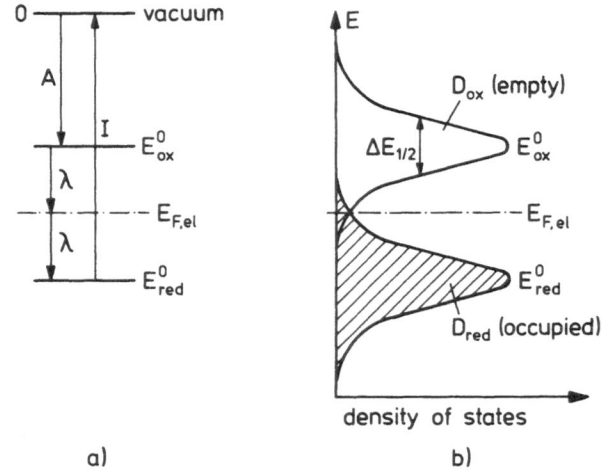

Fig. 7: Energy levels (a) and distribution (b) in redox systems.

and are represented in Fig. 7. This energy picture for redox systems has been proposed at first by Gerischer [10]. The half width $\Delta E_{1/2}$ of the distribution curves are also determined by λ:

$$\Delta E_{1/2} = 0.53 \, \lambda^{1/2} eV.$$

Typical values for inorganic redox systems are of the order of 1 eV as determined by electrochemical methods [11, 12, 13], i.e. the energy levels of such a system are spread over a considerable energy range.

The relative position of energy levels at the interface can now be easily obtained using the condition that the Fermi levels are equal on both sides of the interface. Examples for two redox systems of different normal potentials are given in Figs. 8a and c. Redox systems of a large positive normal potential exhibit rather low lying energy levels in this energy scheme, as also can be derived from Fig. 5 in which various redox potentials are given.

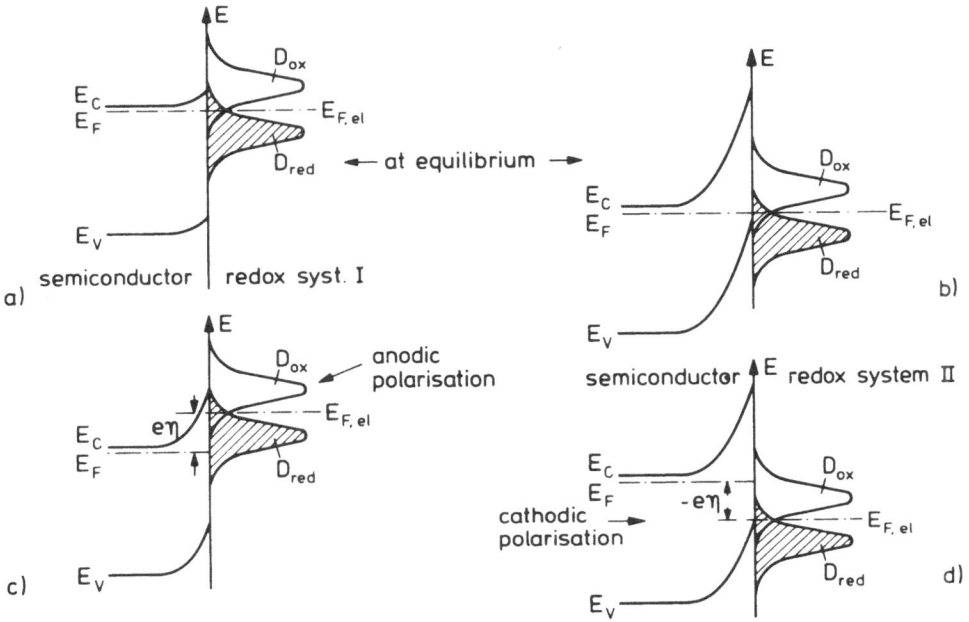

Fig. 8: Relative position of energy levels at both
        sides of the interface
        a) and b) at equilibrium,
        c) at anodic polarization,
        d) at cathodic polarization.

CHARGE TRANSFER PROCESSES

        Electron transfer and a corresponding current can
occur according to the energy picture if levels of equal energy exist on both sides of the interface. One can distinguish between various cases: In Fig. 9a, for instance,
the occupied and empty states of the redox system overlap with the conduction band. Applying anodic or cathodic
potentials corresponding currents can be observed. It
should be emphasized again that any externally applied
potential leads only to a variation of the band bending
whereas the relative position of the energy bands and
levels on both sides of the interface remain unchanged.
This is demonstrated for two cases in Figs. 8b and d.
The Fermi levels are not equal anymore, they differ now
by the externally applied potential η, i.e.

$$E_F - E_{F,el} = e\eta \; .$$ (14)

One can distinguish now between various cases: In Fig. 8a the empty (D_{ox}) and the occupied states (D_{red}) overlap with the conduction band. The anodic and cathodic currents are then given by

$$i_c^+ \approx \int_{E_c}^{\infty} N_c D_{red} dE \quad \text{(potential independent)}, \quad (15a)$$

$$i_c^- \approx \int_{E_c}^{\infty} n_s D_{ox} dE \quad \text{(potential dependent)}. \quad (15b)$$

In the first case electrons are transferred from occupied levels D_{red} to empty states in the conduction band, the latter being nearly identical to the total density of states N_c at the lower edge of the conduction band since only few are occupied by electrons even in n-type material. In this case the corresponding current is independent of the applied potential. The situation is different for a cathodic process, i.e. for an electron transfer from the conduction band into empty

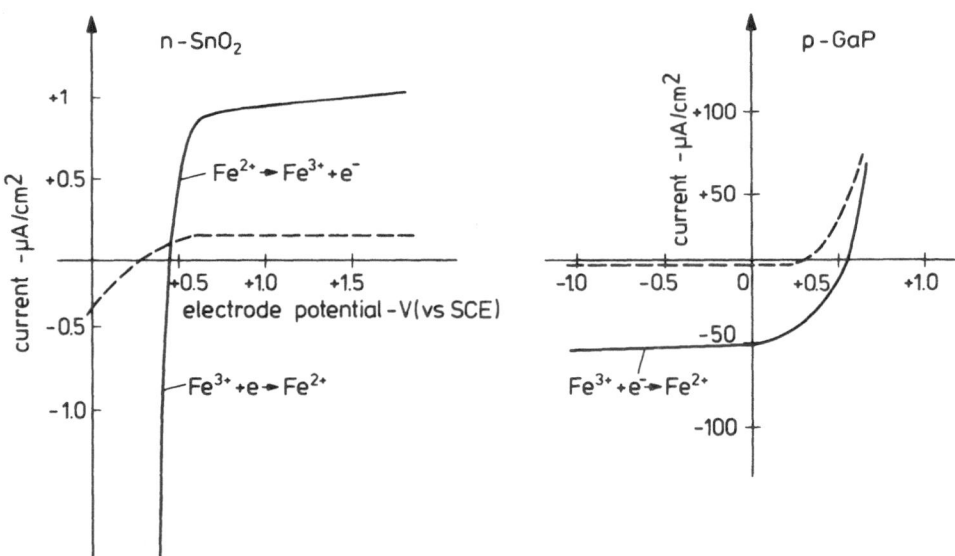

Fig. 9: Current potential dependence for $n-SnO_2$ and p-GaP in $5 \cdot 10^{-2}$ M Fe^{2+}/Fe^{3+} in 0.1 $N_s SO_4$.

states D_{ox} of a redox system. In this case the current depends on the density of electrons n_s at the surface which is potential dependent according Eq. (5a); i.e. the cathodic current increases exponentially with η. An example is given for the system $SnO_2/(Fe^{2+/3+})$ in Fig. 9a [11].

Similar equations are valid for a charge transfer via the valence band. In this case the energy levels of the redox system should overlap with the valence band (Figs. 8c and d). The current equations are given by:

$$i_v^+ \approx \int_\infty^{E_v} p_s D_{red} dE \quad \text{(potential dependent)}, \quad (16a)$$

$$i_v^- \approx \int_\infty^{E_v} N_v D_{ox} dE \quad \text{(potential independent)} . \quad (16b)$$

Here an anodic current, i.e. electron transfer from D_{red} into the valence band, can only occur if holes are present at the surface. The hole density is potential dependent according to Eq. (5b). For cathodic polarization, electrons are transferred from the valence band to the empty states of the redox system (D_{ox}), the current is independent of the applied potential, an example is given in Fig. 9b [6].

It should be emphasized again, that currents determined by the surface concentrations of electrons (n_S) or holes (p_S) only increase with potential if the corresponding bulk concentrations (n_O and p_O) are sufficiently large. Otherwise the currents will be diffusion limited.

As far as the current equations (Eqs. (15) and (16)) are concerned it should be mentioned that also other theories have been developed such as by Marcus [14] and Levich [15] and Dogonadze [16]. In all these cases the reorientation energy λ plays a dominant role and all of these authors obtained the same dependence for $i = f(\lambda)$. This is due to the fact that a harmonic oscillator picture has been used for the fluctuation of the solvation shell. The description of a possible charge transfer by the energy scheme discussed above has been introduced by Gerischer [10] and is now well accepted in semiconductor electrochemistry and proved by many results. On the basis of this model one can predict very well which energy band

will be involved if the position of energy levels in the
semiconductor and the redox is known (see Fig. 5). In
the case of redox systems in which two electron steps
are involved (e.g. H_2O_2 [2]) a prediction is more difficult because such a system has to be described by two
normal potentials [1, 2].

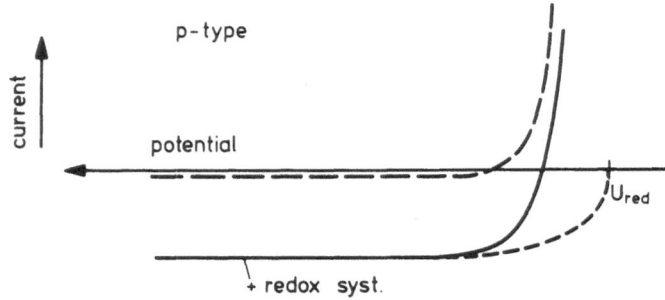

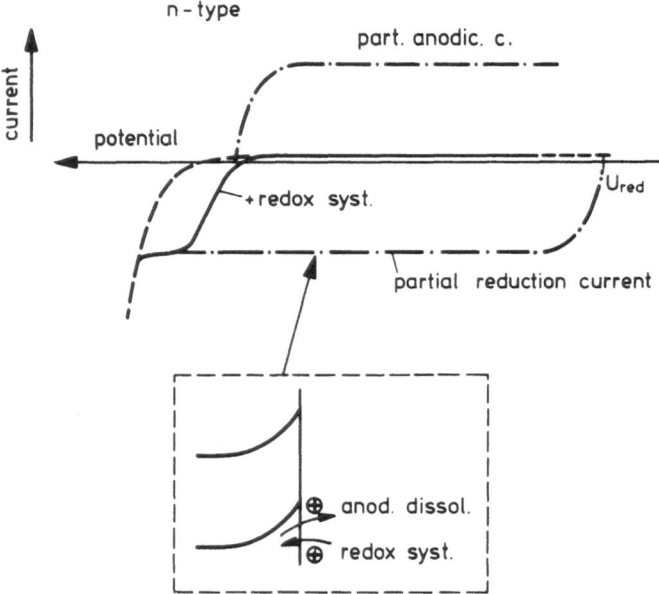

Fig. 10: Current potential dependence for n- and p-type
         semiconductor electrodes in the presence of
         oxidizing agents (schematic).

Finally it should be mentioned that the current-potential curves due to a charge exchange between semiconductor and redox system may be distorted by the anodic dissolution process occuring with many semiconductor electrodes. An example is schematically shown in Figs. 10a and b. Assuming a redox process occuring via the valence band, the cathodic currents (electron transfer from E_v to D_{ox}) are expected to be equal for n- and p-type materials. In the case of n-type, however, the cathodic current occurs at much more negative potentials than with p-type. This difference can be interpreted by the different band bending found for n- and p-type at a given electrode potential (see insert of Fig. 10). Electron transfer from the valence band to the redox system means hole injection. According to the sign of the field within the space charge of n-type these holes are pushed back towards the surface and are consumed for the anodic dissolution. Consequently, the small current found with n-type is actually composed of two partial currents; i.e. corrosion occurs now in a much larger potential range than for p-type which is important in etching processes.

PHOTOEFFECTS

One of the most interesting phenomena in semiconductor electrochemistry are the photoeffects. One photoeffect has already been mentioned in the discussion of the basic current-potential curves in Fig. 6. If minority carrier are involved in an electrode process then currents can be enhanced by light absorbed by the semiconductur as e.g. the anodic dissolution current for n-type GaP (Fig. 6). It is interesting to note, however, that the anodic current at n-type occurs during illumination already at much more negative potentials that at p-type in the dark. This problem can immediately be solved after having discussed the origin of the photoeffect.

Light excitation in a semiconductor leads to the formation of electron-hole pairs as indicated in Fig. 11. They are separated by the electric field across the space charge region. In the case of an upwards band bending the holes are pushed towards the surface and the electrons towards the bulk of the crystal. Under open circuit conditions this leads to a decrease of the electric field and consequently of the band bending. This change can be detected as a corresponding shift of the electrode potential (open circuit) towards cathodic potentials. At sufficiently large light intensities

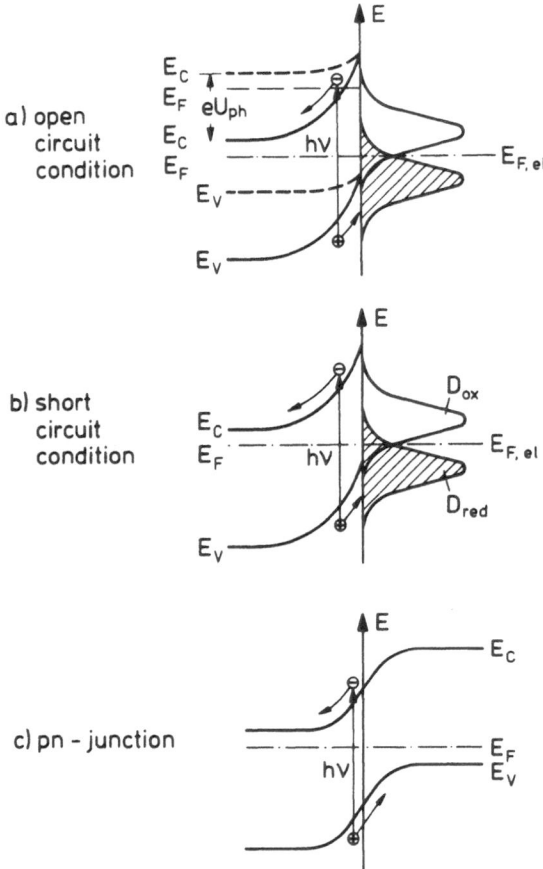

Fig. 11: Energy distribution at the interface semicon-
         ductor-redox system (a) and b)) and for pn-
         junction (c).

even the flatband potential can be reached. Under short
circuit or potentiostatic conditions, however, the
field across the space charge remains constant (Fig. 11b)
and the holes are now forced to cross the interface
leading to a corresponding photocurrent in a potential
range more positive than the flat band potential.

     The same kind of arguments are also valid for a
p-type semiconductor and a negative space charge.
Actually the origin of the photoeffect is identical
to that known already for solid-solid interfaces such
as pn-junctions and semiconductor/metal interfaces and
the same relationship between photopotential U_{ph} or

photocurrent i_{ph} and light intensity I can be used:

$$U_{ph} \approx \frac{kT}{e} \ln (1 + \gamma I)$$

$$i_{ph} \approx I \, .$$

(17)

Applications of Photoeffects

These photoeffects are not only of importance in the analysis of charge transfer processes at semiconductor electrodes but have also found some applications such as photoselective etching, in few photographic systems [17, 18] and for solar energy conversion. Because of its great interest the latter example will be discussed in more detail.

Solar Energy Conversion

The similarity of semiconductor/electrolyte interfaces to pn-junctions implies to use these systems also for conversion of solar energy into electrical energy. Actually a relatively large number of research groups started to study this possibility during the last couple of years. Because of the importance of this application the basic principles for constructing such a cell will be given as follows:

In order to obtain a large photopotential a combination of semiconductor and redox system should be selected that a large band bending exists at equilibrium (without illumination). This can be achieved by combining e.g. an n-type semiconductor with a redox system of which its Fermi level $E_{F,el}$ (redox potential) is quite close to the valence band of the semiconductor. According to Fig. 5 a suitable combination would be n-type CdSe and Ce^{4+}/Ce^{3+}. At equilibrium the Fermi level must be equal across the whole system leading to a band bending as shown in Fig. 12. An inert metal electrode is used as a counter electrode. Then no potential difference occurs in the dark. During illumination a corresponding photopotential can be built up. Under short circuit conditions, shown in Fig. 12 the holes (p^+) created by light excitation are pushed towards the surface whereas the electrons flow through the external circuit to the metal electrode. The corresponding charge transfer reactions are given by

$$Red + p^+ \rightarrow Ox \quad \text{anodic process at} \atop \text{semiconductor} \tag{18}$$

and

$$Ox + e^- \rightarrow Red \quad \text{cathodic process at} \atop \text{metal} \tag{19}$$

as indicated in the lower part of Fig. 12.

In this case electric energy is produced by a re-
generative photoelectrochemical process. The question
arises, however, whether the anodic process in Eq.
(19) really takes place or whether the holes are consumed for the anodic dissolution of the semiconductor
crystal. This is indeed a difficult problem because only
some semiconductors are stable and these have too large
band gaps (> 3 eV) as e.g. TiO_2 and SnO_2. A large

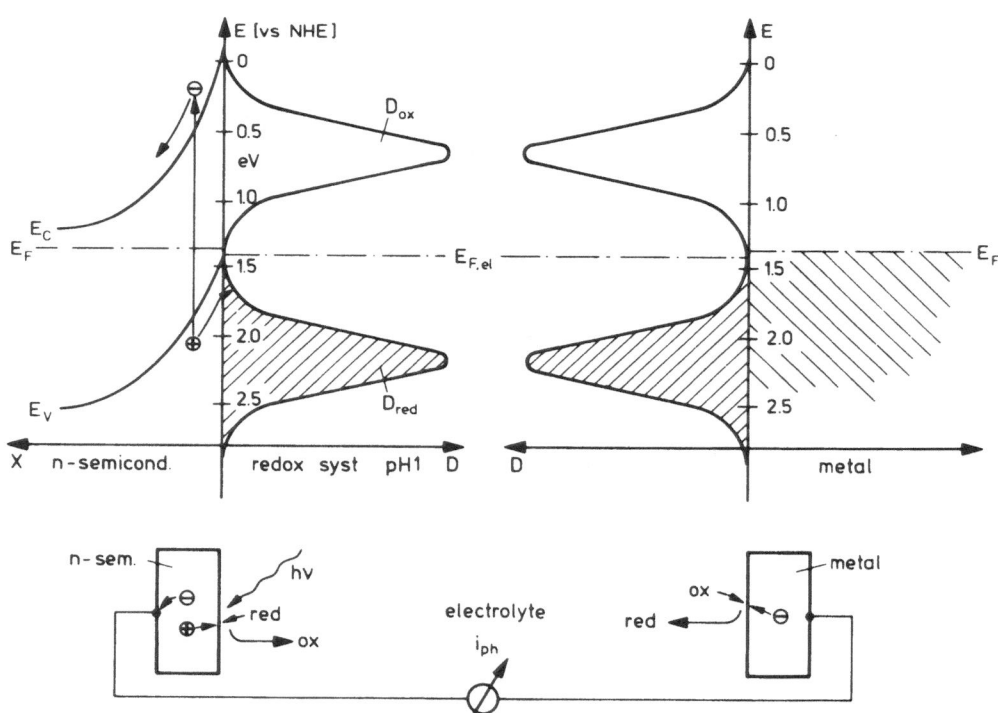

Fig. 12: Solar cell under short circuit conditions.

efficiency can only be expected for semiconductors of smaller band gaps, the optimum band gap being around 1.5 eV. Although no semiconductor of such a band gap is known which does not show anodic dissolution, A few cells have been made in which the redox process competes sufficiently well with the anodic dissolution. Examples are n-CdS/(S/S^{2-}) [19], n-CdS/([Fe(CN)$_6$]$^{4-}/^{3-}$) [20], n-CdSe/(S/S^{2-}) [21] and n-GaAs/(Se/Se^{2-}) [21]. In the two latter cases conversion efficiencies of about 8% have been obtained (theoretical value for GaAs: 25%). In all systems the quantum efficiency for the photo-current is around 80%. The photopotentials, however, are relatively low due to the normal potentials of the redox system selected for these cells [22].

It has also been suggested to use the semiconduc-tor-electrolyte systems for conversions of solar energy into chemical energy [23]. This is in principle possible in the same kind of cell as presented in Fig. 12, using an electrolyte without a redox system. In this case electrons reaching the metal electrode via the external circuit are consumed for H_2-evolution. The anodic pro-cess at the semiconductor electrode should then be O_2-evolution and not anodic dissolution. Although such a system would be of great interest it has not been re-alized for band gaps < 3 eV.

Electrode Processes with Excited Molecules

In the previous chapters photoeffects have been discussed initiated by light absorbed in the semiconduc-tor. In this chapter phenomena will be described occuring during excitation by light absorbed by mole-cules dissolved in the electrolyte. Most of these stud-ies were performed with dye solutions (see Ref. [1] and [24] and literature cited there). In the following the essential effects will be discussed taking Ru(bipy)$_3^{2+}$ as an example. The advantage of this com-plex is that the ruthenium can occur in different oxida-tion states and corresponding well defined redox po-tentials exist.

Principally such a redox system can also be characterized by a Fermi level $E_{F,el}$, by occupied and empty energy levels and their distribution and the re-orientation energy λ. In the case of a dye or Ru(bipy)$_3^{2+}$ energy levels (π-orbitals) exist which are empty or occupied by one or two electrons. For instance in the redox reaction

$$Ru(bipy)_3^{2+} \rightleftarrows Ru(bipy)_3^{2+} + e^-$$

only electrons of the highest occupied molecular level
are involved whereas for

$$Ru(bipy)_3^{2+} + e^- \rightleftarrows Ru(bipy)_3^{1+}$$

electron transfer can only occur via the lowest empty
molecular level. The redox potentials of these two reactions should differ roughly by the difference of the
two molecular levels. The average value of the two redox
potentials, determines now the equilibrium condition if
this system is in contact with a semiconductor electrode.
Accordingly one has also two sets of distribution functions as indicated in Fig. 13, which are reduced to two
distributions if only Ru^{2+} is present. The relative

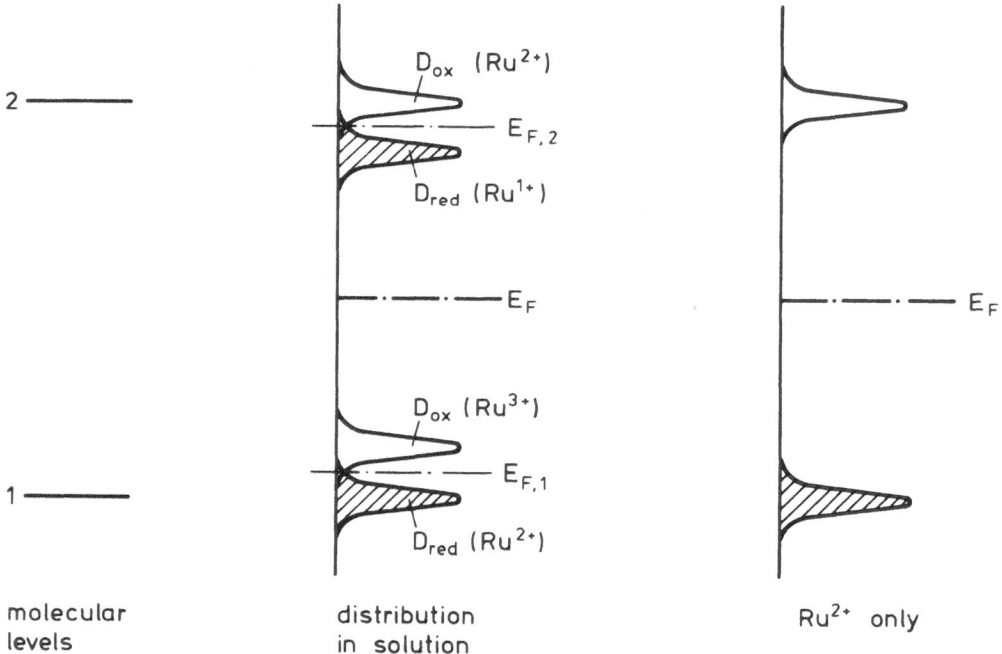

Fig. 13: Energy levels in organic redox systems.

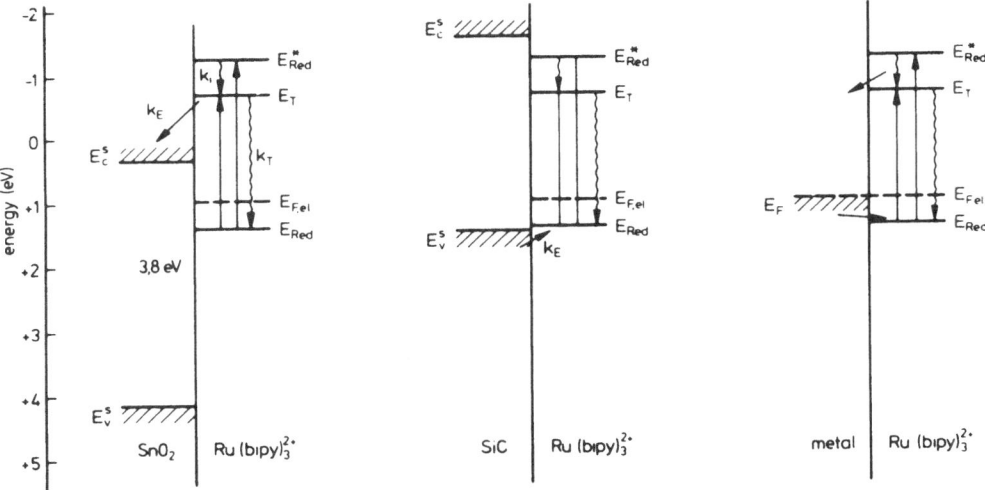

Fig. 14: Relative position of energy levels at semicon-
ductor/Ru(bipy)$_3^{2+}$-interface.

position of energy levels, for semiconductor/electro-
lyte (Ru(bipy)$_3^{2+}$)-interface is shown for two examples
in Fig. 14.

Electron transfer from the Ru-complex to SnO$_2$ is
possible if electrons are excited by light to an upper
level as indicated by arrows in Fig. 14. A corresponding photocurrent has been observed as given in Fig.
15a [25]. The photocurrent spectrum resembles nearly
the absorption spectrum of Ru(bipy)$_3^{2+}$. The electron
transfer occurs actually via the triplet state E_t
because the yield of the intersystem crossing is close
to unity [26]. The electron transfer (represented by
the rate constant k_E) has to compete with the recombination process (k_T) being a radiation transition (k_T =
$2 \cdot 10^6$ s^{-1}).

According to the relative position of energy levels
the reverse process should occur at SiC (Fig. 14):
During light excitation an electron is transferred from
the valence band to the half occupied level E_{Red} of an
excited molecule. A corresponding cathodic current has
been observed (Fig. 15). In the case of metals such processes cannot be detected because currents in both directions occur simultaneously (see arrows in Fig. 15).

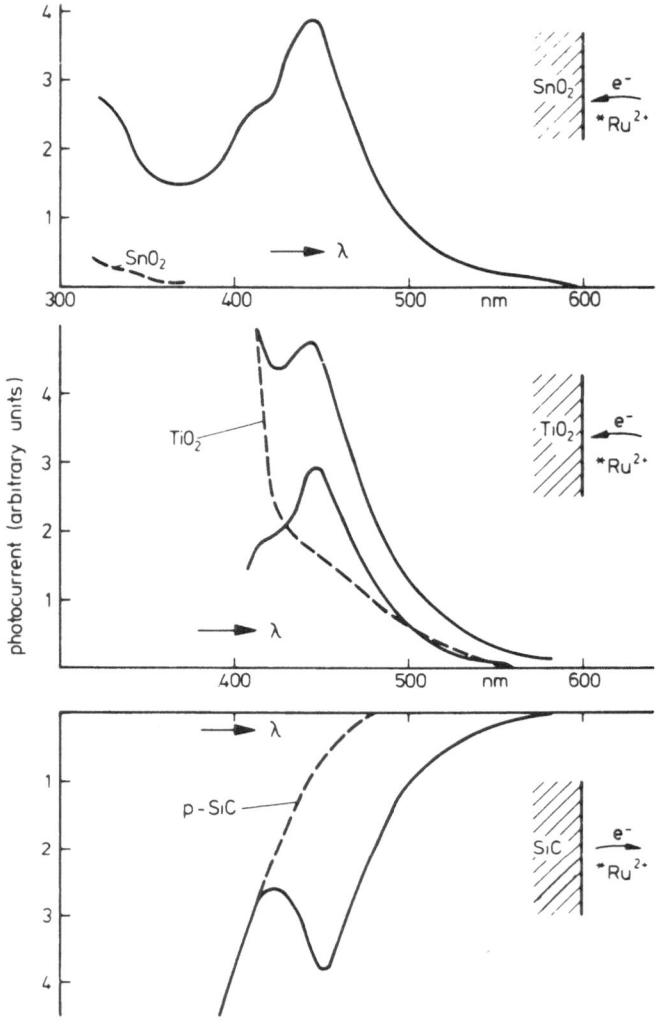

Fig. 15: Sensitized photocurrents for 10^{-3} M
         Ru(bipy)$_3^{2+}$ pH 1.

It must be emphasized that only excited molecules
being close to the electrode surface are involved in the
transfer. According to the short lifetime of an excited
molecule its diffusion length is small. In the case of
the Ru-complex the diffusion length L $\approx$ 250 Å for a
lifetime of τ = 5·10^{-7} s. For dyes the diffusion length
would only be few Å-units because of the short lifetime

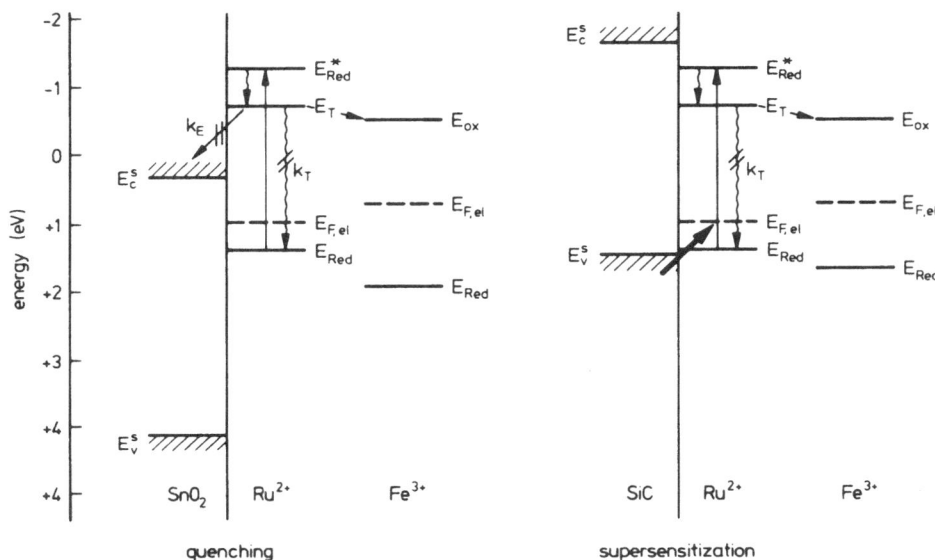

Fig. 16: Quenching and supersensitization by Fe^{3+}-ions.

($\approx 10^{-9}$ s), i.e. in this case only adsorbed molecules participate in the process.

It is interesting to note that the sensitization currents discussed above can be quenched or supersensitized by addition of other redox systems to the electrolyte. An example is given for Fe^{2+}/Fe^{3+} in Fig. 16. The unoccupied levels E_{ox} of this redox system overlap with the triplet of $Ru(bipy)_3^{2+}$ so that an electron transfer is now also possible from the excited molecule to Fe^{3+}. This process competes with the electron transfer (k_E) to SnO_2 and the photocurrent is quenched as proved experimentally [25]. Using SiC instead of SnO_2 the same process (electron transfer from E_t to E_{ox}) leads, however, to an increase of the electron transfer from SiC to the Ru-complex (indicated by thick arrow in Fig. 16) because also the recombination process (k_T) in $Ru(bipy)_3^{2+}$ is quenched [27]. The opposite effect, quenching of the cathodic current at SiC and supersensitization of the anodic current at SnO_2, has

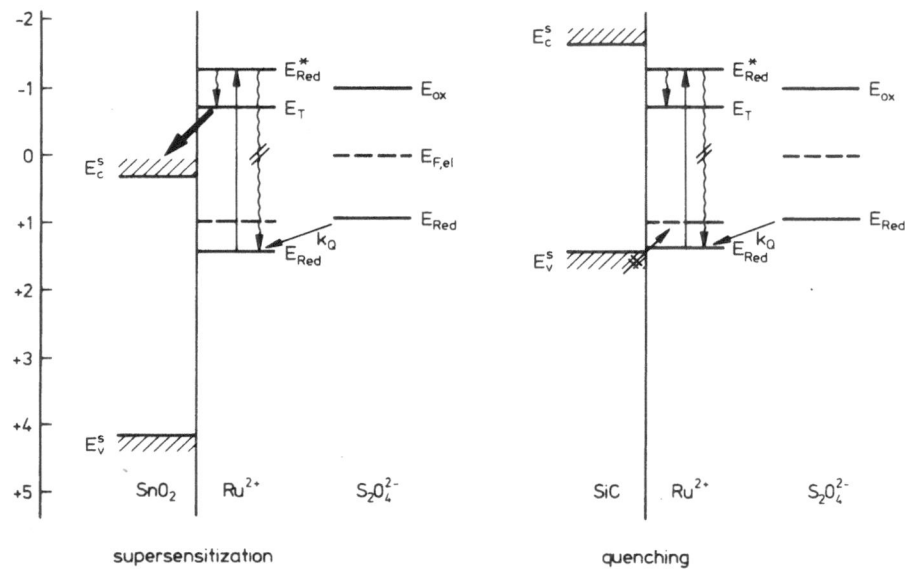

Fig. 17: Quenching and supersensitization by $S_2O_4^{2-}$-ions.

also been observed using $S_2O_4^{2-}$ as a reducing agent [27]. The mechanism is given in Fig. 17.

Luminiscence

        In the previous chapters photoeffects have been discussed in which light excitation has led to an enhanced electron transfer. The reverse process, luminescence caused by charge transfer across the interface, is also possible. The simplest case is the injection of holes into the valence band of an n-type semiconductor followed by a recombination of electron-hole pairs as schematically shown in Fig. 18a. In some cases this is a radiative process as it has been found with GaP [28] and ZnO [8]. Another example is the luminescence created by electron transfer from n-SiC to $Ru(bipy)_3^{2+}$. Here the luminescence process occurs in the Ru-complex as shown in Fig. 18b [29]. The same kind of luminescence has recently been observed with rubrene radical cations at ZnO-electrodes [30].

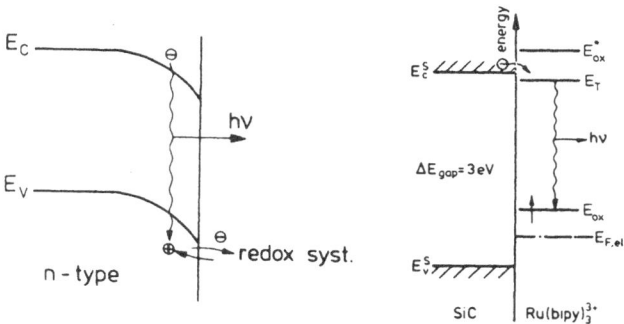

Fig. 18: Luminescence
         a) by hole injection into the valence band,
         b) by electron transfer to $Ru(bipy)_3^{3+}$.

CONCLUSION

The basic principles of charge transfer processes
at semiconductor electrodes are now quite well understood. Interesting results have also been obtained with
organic crystals as electrode materials which were not
discussed here. In corresponding studies basically the
same models have been used. For further information it
must be referred to the literature [24, 31].

REFERENCES

1. R. Memming, in Electroanalytical Chemistry, ed. by
   A.I. Bard, Marcel Dekker, New York, in press.
2. R. Memming, J. Electrochem. Soc. 116, 785 (1969).
3. H. Gerischer, in Physical Chemistry, Vol. 9A, ed. by
   M. Eyring, D. Henderson and W. Jost, Acad, Press,
   New York 1970.
4. F. Lohmann, Z. Naturforsch. 22a, 813 (1967).
5. M. Hofmann and H. Gerischer, Z. Elektrochem. 65,
   771 (1961).
6. R. Memming and G. Schwandt, Electrochim. Acta 13,
   1299 (1968).
7. Hollan, Tranchart, R. Memming, to be published.
8. B. Pettinger, H.R. Schöppel, T. Yokoyama and
   H. Gerischer, Ber. Bunsenges. Phys. Chem. 78,
   450 (1974).
9. F. Möllers and R. Memming, Ber. Bunsenges. Phys.
   Chem. 76, 469 (1972).

10. H. Gerischer, Z. Phys. Chem. N.F. 26, 223 (1960);
    26, 326 (1960); 27, 48 (1961).
11. R. Memming and F. Möllers, Ber. Bunsenges. Phys.
    Chem. 76, 475 (1972).
12. R.A.L. van den Berghe, F. Cardon and W.P. Gomes,
    Surface Sci. 39, 368 (1973).
13. I.M. Hale, in Reactions of Molecules at Electrodes,
    ed. by V.S. Hush, Wiley + Sons, New York 1971.
14. R.A. Marcus, J. Chem. Phys. 28, 962 (1965).
15. V.G. Levich, Advances in Electrochemistry and Elec-
    chem. Eng. Vol. 4, ed. by P. Delahay, Interscience
    Publ. Inc., New York 1966.
16. R.R. Dogonadze and A.M. Kuznetsov, Izv. Nauk. 10,
    1787 (1964).
17. H. Jonker and C.J.G.F. Janssen, Phot. Sci. Eng. 13,
    45 (1969).
18. F. Möllers, H.J. Tolle and R. Memming, J. Electro-
    chem. Soc. 121, 1160 (1974).
19. H. Gerischer and I. Gobrecht, Ber. Bunsenges. Phys.
    Chem. 80, 327 (1976).
20. M.W. Wrighton, A.B. Ellis, P.T. Wolczanski, D.L.
    Morse, H.B. Abrahamson and D.S. Ginley, J. Am.
    Chem. Soc. 98, 2774 (1976).
21. A. Heller, K.C. Chang and B. Miller, J. Electro-
    chem. Soc. 124, 697 (1977).
22. R. Memming, J. Electrochem. Soc. in press.
23. A. Fujishima and K. Honda, Bull. Chem. Soc. Jap.
    44, 1148 (1971).
24. H. Gerischer and Willig, in Topics of Current
    Chemistry, 61, Springer-Verlag, Berlin 1976.
25. M. Gleria and R. Memming, Z. f. Phys. Chem. N.F.
    98, 303 (1975).
26. J.N. Demas and A.W. Adamson, J. Am. Chem. Soc. 93,
    1800 (1971).
27. M. Gleria and R. Memming, unpublished results.
28. K.H. Beckmann and R. Memming, J. Electrochem. Soc.
    116, 368 (1969).
29. M. Gleria and R. Memming, Z. Phys. Chem. N.F. 101,
    171 (1976).
30. L.S.R. Yeh and A.I. Bard, Chem. Phys. Lett. 44,
    339 (1976).
31. W. Mehl, in Reactions of Molecules at Electrodes,
    ed. by N.S. Hush, Wiley + Sons, New York 1971.

PHOTOELECTRON EMISSION INTO ELECTROLYTES

J. K. Sass

Fritz-Haber-Institut der Max-Planck-Gesellschaft

Faradayweg 4-6, 1000 Berlin 33, Germany

1. INTRODUCTION

The interfacial region between a metal and an electrolyte represents one of the most important environments in surface chemistry. The present paper is concerned with photoemission processes at this interface, but with the emphasis not only on chemical but also on physical aspects. The need for such an interdisciplinary approach in surface studies is increasingly recognized and is indeed exemplified in these Symposium Series. The selection of experimental topics and their discussion in this article are intended to reflect this attitude.

The emission of a photoelectron from a metal into an electrolyte and its manifestation as photocurrent in a galvanic cell is a much more complicated phenomenon than a photoemission event in a vacuum. Our present understanding, though fragmentary, seems to indicate the following chain of subsequent processes:

1. Absorption of a photon in the metal, creation of an exited electron.

2. Transport to the emitter surface; possibility of energy-degrading and momentum-altering scattering processes.

3. Escape from the electrode if electron has sufficient energy and parallel component of momentum can be conserved.

4. Rapid dissipation of excess kinetic energy in the solution, with detailed mechanism unknown; solvent is more or less

"frozen" during this fast process and only electronic interaction is possible.

5. Upon thermalization at the lowest non-bonding level (1),
   for example of water, polar solvent molecules rearrange
   and form solvated electron, resulting in further energy
   degradation.

6. Solvated electrons diffuse back to the emitter and are re-
   captured by the electrode, unless solution contains species
   with scavenging properties which react irreversibly with
   solvated electrons.

Obviously, this concept of subsequent steps is highly schematic and considerable overlap of these processes is to be expected. In steps 1. through 3. which apply equally to photoemission in
a vacuum this problem has frequently been addressed and alternative
"one-step" theoretical models (2-4) have been developed. In the
present context this simple step-by-step overview may suffice, however, and for a detailed discussion the reader is referred to more
general reviews on the subject (5-7).
     A very useful and important property of the metal-electrolyte
contact is its low threshold for electron emission. The lowest non-
bonding level in water, for example, is located energetically
~ 1.3 eV below the vacuum level. The Fermi-level of the emitter
electrode, on the other hand, can be shifted considerably within
the potential region of polarizibility and as a result the threshold may be as low as 2 eV. This aspect has also been described previously in considerable detail (5-7) and is merely recalled here.
It opens up the possibility of investigating photoemission processes at very low energies which are not accessible in the vacuum.
Typical examples of such studies and the new information they provide are given in Sec. 2.

     An intriguing interfacial phenomenon is discussed in Sec. 3:
surface plasmon excitation on a rough silver surface. Photoelectron
excitation may occur as a result of the plasmon decay and a strong
dependence of this effect on crystallographic orientation and surface morphology is observed. Here again, the low-energy capabilities of the photoemission-into-electrolyte technique prove extremely useful.

     The dissipation of the electron kinetic energy in polar li-
quids is the topic in Sec. 4. The information on the scattering
mechanism is obtained by applying the anisotropic and polarization-
dependent model of the photoemission process, developed and experimentally documented in Sec. 2, to this problem. The results indicate that the initial angular distribution of emitted electrons is
essentially randomized during the energy dissipation.

In addition to photoelectron emission a different but compli-
mentary photoeffect is observed at the metal-electrolyte interface. It consists of the decomposition of water molecules by excited holes in the metal and is described in Sec. 5. On semiconductor electrodes this effect is well known and the possibility
of a real emission process of the photoholes is tentatively inferred from the experimental results.

2. PHOTELECTRON EXCITATION STUDIES WITH POLARIZED LIGHT

The considerable value of polarized light in photoemission
studies is now firmly established. This section demonstrates its
particular usefulness in experimental situations when only the integral yield, that is the total number of emitted electrons, can
be measured. The results described in this and the following section were performed with solutions containing a high scavenger
concentration such that the collection efficiency was essentially
unity (see Sec. 4). It is then possible to monitor the photoemission properties of the metal in a fashion undistorted by the subsequent processes in the electrolyte.

A typical example of the polarization dependence of photoemis-

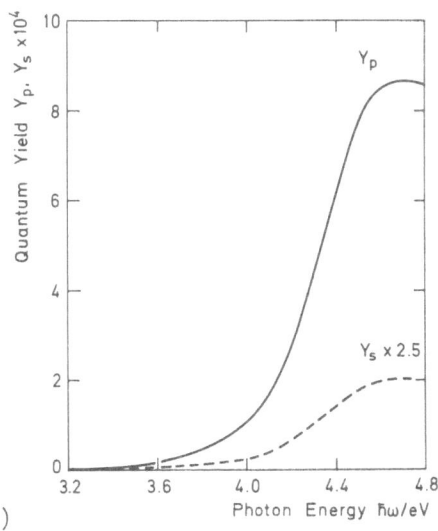
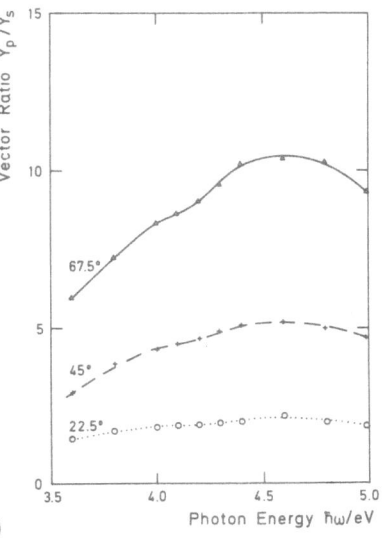

a)   b)

Figure 1. a) Quantum yields Y_p, Y_s for p-polarized and s-polarized
light from Cu(111); angle of incidence $\alpha = 67.5°$, threshold
$\hbar\omega_o = 3$ eV. b) Vector ratios Y_p/Y_s for three incident angles.

sion is shown in Fig. 1. In this experiment the photoyields Y_p and Y_s from a Cu(111) single-crystal surface were measured with both p- and s-polarized light (8). These yields differ considerably (Fig. 1a), with Y_p being much larger than Y_s. In the vector ratio Y_p/Y_s this anisotropy manifests itself even more clearly and in Fig. 1b the experimental values of Y_p/Y_s for three angles of incidence are shown. The increase of the vector ratios with increasing angle clearly suggests that the normal component of the photon field is most effective in producing photoelectrons.

A possible explanation of this effect in terms of direct optical transitions in the bulk of the metals is schematically illustrated in Fig. 2a. For excitations between nearly free-electron bands, a situation which prevails in low-energy photoemission from copper (9), electrons are predominantly excited in the direction of the photon field (9-11). The conduction-to-conduction band transitions which are of relevance here occur in the vicinity of the L-point. Due to the energetically degenerate star of wave vectors which reflects the crystal symmetry, equivalent transitions occur in eight regions of momentum space. The relative population in these regions is, however, strongly dependent on their orientation with respect to the direction of the electric field vector $\vec{e}$.

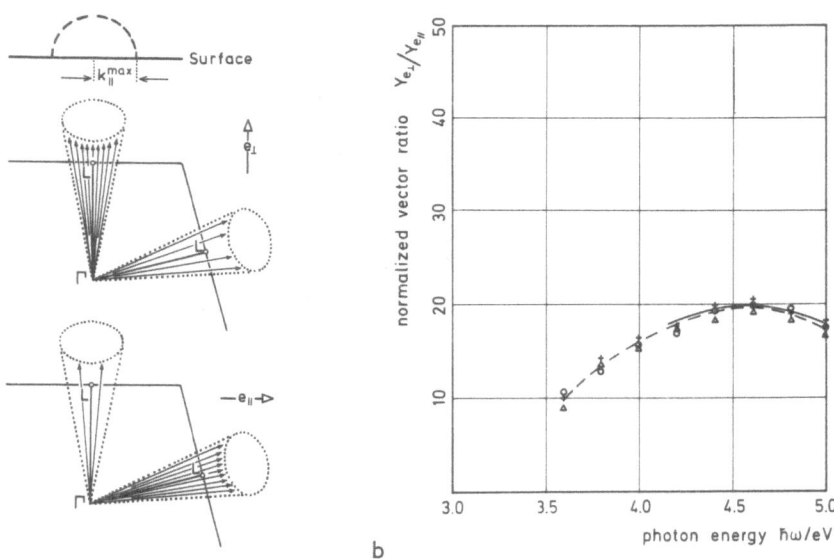

a   b

Figure 2. a) Schematic illustration of the anisotropic photoelectron mechanism between nearly-free electron bands. Electric field vector components $e_\perp$ and $e_\parallel$ of the light, perpendicular and parallel to the surface are indicated. b) Comparison of calculated (——) and experimental (- - -) vector ratios on Cu(111). Experimental values from Fig. 1b are reduced to an angle-independent form (see text).

In an optical absorption experiment this highly anisotropic distribution of photoelectrons does not manifest itself: the sampling occurs over all equivalent regions and as a result a cubic crystal is known to behave optically isotropic. In photoemission, however, there is a crucial restriction which limits the parallel component of the crystal momentum of the excited electron to a maximum value (12). Only electrons moving more or less perpendicular to the surface can escape. Consequently, only one of the equivalent regions usually contributes to the photoemission current, with the number of photoelectrons varying with polarization and incident angle. It seems therefore that the results shown in Fig. 1b are readily explained in the framework of this model, at least qualitatively.

A more quantitative evaluation of the results shown in Fig. 1b has also been performed (8). Taking into account the band structure of copper and the polarization dependence of the transition probability vector ratios $Y_{e\perp}/Y_{e\parallel}$ have been calculated and they are shown in Fig. 2b. The subscripts $e\perp$ and $e\parallel$ refer here to electric fields perpendicular and parallel to the surface, respectively. By using Fresnel's equations the experimental vector ratios can also be reduced to this angle-independent form (8), and the results in Fig. 2b indicate that the experimental $Y_{e\perp}/Y_{e\parallel}$ do indeed coincide for the three angles. Similar results have also been obtained for a Au(111) surface (8). The agreement of the calculated $Y_{e\perp}/Y_{e\parallel}$ with these normalized experimental vector ratios seems also very satisfactory.

The successful use of Fresnel's equations for the reduction of the experimental data has an important additional implication. Recently, the optical properties of metal surfaces have been the subject of considerable theoretical advancement and controversy (13-15). The discontinuous concept of the surface region has been shown to be unrealistic and the occurrence of longitudinal excitations is now generally recognized and has been confirmed experimentally (16-18). The approach which led to the experimental vector ratios $Y_{e\perp}/Y_{e\parallel}$ in Fig. 2b, on the other hand, seems to permit a check of the validity of Fresnel's equations: optical measurements on a metal surface should be performed over a wide range of incident angles. If one observes consistency as in Fig. 2b it can be safely assumed that longitudinal field effects are of minor importance in that particular experimental situation.

A direct-transition analysis of photoemission from copper is restricted to an energy range above 4.15 eV where the onset of these excitations occurs. The large vector ratios in Fig. 1b or Fig. 2b persist, however, at lower energies and the photoelectron excitation mechanism responsible for emission below the direct-transition threshold has not yet been identified with sufficient unambiguity. In a recent experiment (19) this problem has been

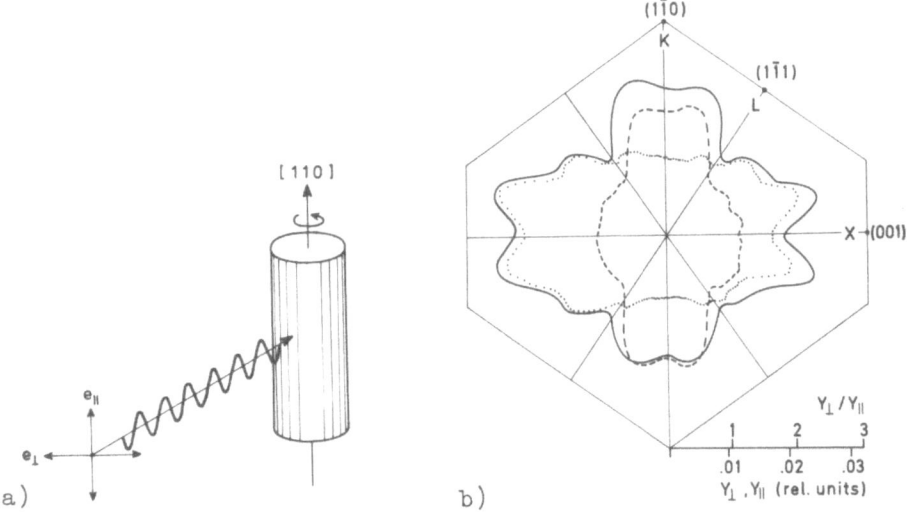

<u>Figure 3.</u> a) Schematics of the experimental geometry for photo-
emission studies on a single-crystal metal cylinder at normal
incidence. The subscripts ∥ and ⊥ are here referred to the
axis of the cylinder. b) Photoyields from a copper single-
crystal cylinder in the experimental configuration of Fig. 3a;
normal incidence of the light, $\hbar\omega_o$ = 3 eV.

studied using a somewhat unusual experimental configuration. A copper single-crystal cylinder was illuminated on its curved surface
(see Fig. 3a) <u>at normal incidence</u>[+) with polarized light. Upon rotation of the cylinder around its axis many crystallographic orientations are exposed to the light beam.

The experimental photoyields obtained in this study are shown
in Fig. 3b. They lead to the interesting and novel conclusion that
indirect transitions strongly reflect the bulk crystal symmetry,
with a polarization dependence which appears to resemble that of the
direct transitions. The results are not yet fully understood, but
they should provide an impetus for a more detailed theoretical
study of these low-energy, Drude-like transitions (20).

<u>Thin Metallic Adsorbate Layers</u> - This subsection deals with
photoemission studies of metal adsorbates and thin films. In an
electrochemical cell such overlayers are readily prepared <u>in situ</u>
and the deposited amount can be determined with high accuracy. A

[+) This experimental geometry deemphasizes surface-potential and
surface-field effects which can only be excited with p-polarized
light.

particularly convenient experimental arrangement for this purpose
is sketched in Fig. 4a. During the deposition process a source electrode of the desired material faces the working electrode in a thin-
layer arrangement. By dissolving this source electrode electrochemically metal ions are created which may diffuse to the working electrode where they are in turn deposited. The faradaic current accompanying this deposition is an accurate measure of the amount of material adsorbed on the substrate electrode. For the optical measurements the source electrode can be moved out of optical path as indicated in Fig. 4a.

With this arrangement the optical plasma resonance in very
thin silver films deposited onto a Cu(111)-substrate has recently
been studied (21). This collective oscillation can only be excited
with p-polarized light and shows up as a large peak at the bulk
plasma frequency in the photoemission yield (22). The experimental
vector ratios Y_p/Y_s in Fig. 4b clearly show the appearance of the
plasma resonance with increasing film thickness. Surprisingly, this
collective effect manifests itself already at a very early stage
of the deposition. It should be mentioned here that silver is known
to grow on copper in a layer-by-layer fashion (23).

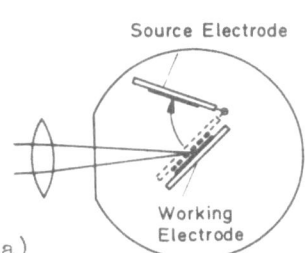

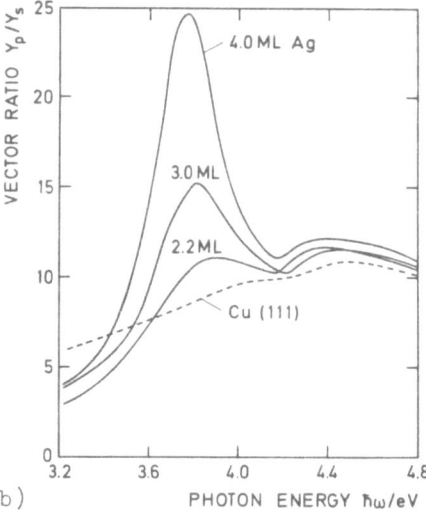

Figure 4 . a) Thin-layer-cell experimental arrangement for the
electrochemical deposition of metal adsorbates from a source
electrode onto the substrate working electrode. b) Vector ratios
Y_p/Y_s for thin silver films of a few monolayers (ML) deposited
electrochemically onto a Cu(111)-substrate; $\alpha = 67,5^{\circ}$,
$\hbar\omega_o = 3$ eV.

The effect of submonolayer deposits of copper and silver on the photoemission properties of a Au(111)-surface has also been studied (24). Drastic changes of the photoyields and a substantial decrease of the vector ratio Y_p/Y_s was observed in that investigation. A considerable advantage of such adsorbate studies in an electrochemical cell is the fixed threshold for electron emission (5, 7).

3. INTERFACE PLASMONS IN PHOTOEMISSION

The manifestation of surface-plasmon excitation in a photoemission experiment was first demonstrated on polycristalline aluminium films which were deposited onto substrates of suitable roughness (25). Either a rough surface or an internal reflection geometry (26) is required to optically excite surface plasmons.

In a photoemission-into-electrolyte study a considerable enhancement of the yield around 3.5 eV due to surface-plasmon excitation was observed on polycristalline silver films (27). Deposited onto quartz substrates at room temperature, these films showed the effect quite clearly without any intentional roughening. With cleaved mica as a substrate, however, surface-plasmon excitation is extremely weak (18), indicating an almost perfectly smooth film (25).

The surface-plasmon decay into single-particle excitations, which in turn can produce "photoelectrons", has recently been studied in more detail on the (100) and (111) faces of silver (28). In the electrochemical cell initially smooth surfaces of both orientations were successively roughened in situ (29) and both the photoemission and the electroreflectance (30) from these surfaces were monitored.

The photoemission results of this work are shown in Fig. 5. Only on the (111) face of silver does the surface-plasmon excitation contribute additional photocurrent. The photoyield on (100) remains structureless during the roughening although the electroreflectance measurements, not shown here, indicate considerable roughening also on this surface. The difference between the two faces is probably due to the varying crystallographic availability of decay channels which can be matched to states outside the emitter. The concept of surface-evanescent states seems to be of particular importance for the noble metals in this low-energy situation (4, 8, 31).

An interesting high-energy shoulder of the surface-plasmon structure is observed on the (111) surface when it is only slightly roughened (see Fig. 5). It can be tentatively attributed to the excitation of surface plasmons with large vectors since in this region of the dispersion relation the energy is expected to increase (32). Upon further roughening of the surface this shoulder more or less

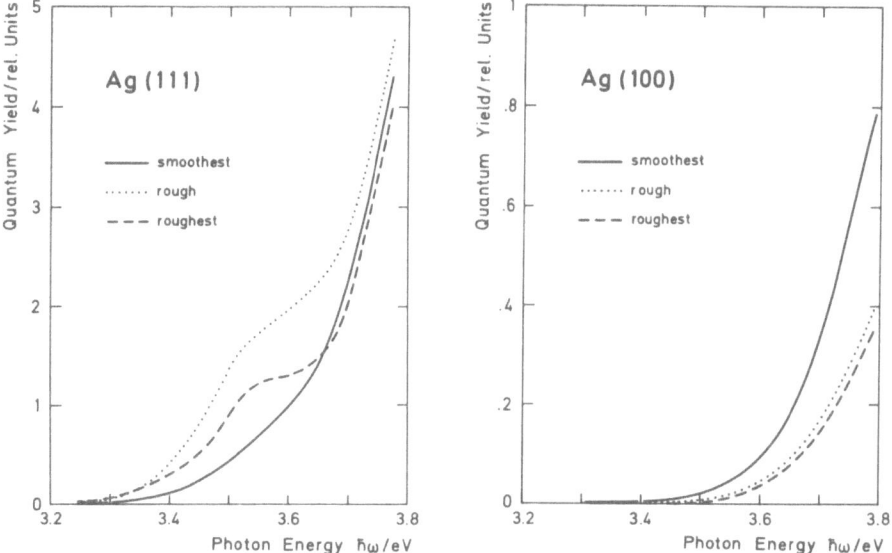

Figure 5. Photoyields at near-normal incidence of the light from smooth and electrochemically roughened (111) and (100) surfaces of silver in the vicinity of the surface-plasmon energy.

diminishes indicating that the roughness spectrum is shifting to a distribution centered on lower momentum values.

The results presented in this section again demonstrate the low-energy capabilities of the photoemission-into-electrolyte technique. With the work function of silver being slightly above 4 eV this study could not have been performed in a vacuum.

4. ENERGY DISSIPATION IN POLAR LIQUIDS

Although the thermalization of the emitted photoelectrons is an important part of the overall photoemission-into-electrolyte process very little is as yet known about the detailed mechanism by which the considerable excess kinetic energy is rapidly dissipated. The present section is concerned with the possibility of obtaining information about these scattering processes by varying the scavenger concentration in the electrolyte.

As outlined in Sec. 1 the presence of scavengers in the solution is an essential prerequisite for observing a photocurrent because the solvated electrons would otherwise diffuse back to the emitter elec-

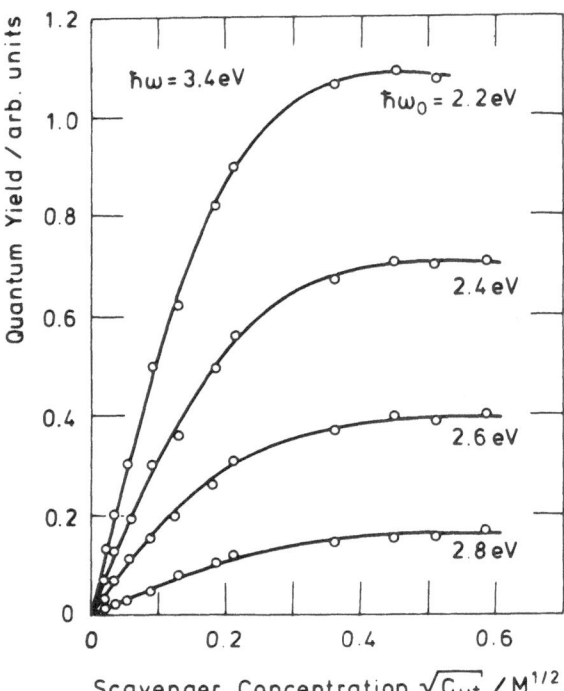

Figure 6. Influence of scavenger concentration on the photoyields
from a polycrystalline lead electrode at different emission
threshold in solutions of KCl, total electrolyte concentration
1M (after Rotenberg and Gurevich[6]).

trode. An experimental example of the influence of the scavenger concentration on the photoyield is shown in Fig. 6 for a lead electrode
in solutions of different acidity (6). Protons appear to be very effective scavengers and they allow a wide range of concentrations to
be covered.

The regions of very low and of very high scavenger concentration
are of particular interest in Fig. 6. At low concentrations the yield
increases with the square root of the proton concentration and at
high concentrations a saturation behaviour is observed. This latter
result indicates that it is possible to collect essentially all of
the emitted photoelectrons. The transition behaviour from low to high
concentrations in Fig. 6 depends on the "excess" photon energy
($\hbar\omega - \hbar\omega_0$) which corresponds to the highest possible kinetic energy
of photoelectrons in the electrolyte.

Rotenberg and Gurevich (6) have shown how to obtain information
on the mean thermalization length of the emitted electrons. The main

advantage of their approach is to avoid the introduction of a specific source function into the theoretical treatment. This source function characterizes the spatial dependence of the appearance of solvated electrons in the vicinity of the emitter surface. Obviously,
the further away from the surface the solvated electrons are formed
the more likely is their encounter with a scavenger species, particularly at low concentrations.

We refrain here from a description of the Rotenberg-Gurevich
model and merely present the results of their analysis of different
measurements, including those of Fig. 6. In Fig. 7 the mean thermalization distance for different metals is shown as a function of the
excess energy ($\hbar\omega - \hbar\omega_0$). A linear relation is obtained by this analysis which does not extrapolate to zero, but coincides for the three
metals.

A serious complication arises when one tries to analyze in more
detail the information which is contained in Fig. 7. The experimental
results were obtained on polycristalline samples and neither the angle
of incidence nor the polarization of the light is specified by the
authors. It seems difficult under these conditions to assess the influence of the energy as well as the angular distribution of the excited electrons. In addition to the specific metal and energy difference ($\hbar\omega - \hbar\omega_0$), this distribution depends decisively also on the
polarization and the incident angle (2, 8, 9-11, 13, 18, 31, 33).

In order to get a deeper insight into this problem the concen-

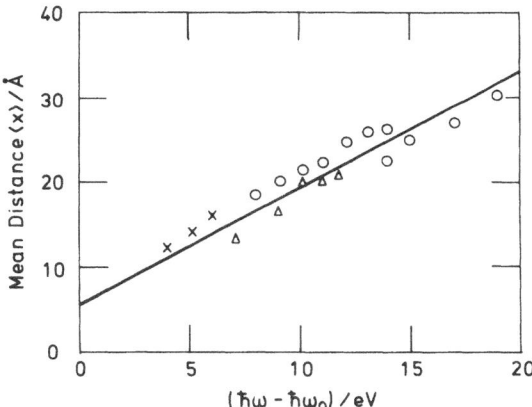

Figure 7. Mean thermalization range of electrons as a function
of the energy difference ($\hbar\omega - \hbar\omega_0$) obtained from measurements
as are shown in Fig. 6 (after Rotenberg and Gurevich[6]).

tration dependence of the photocurrent has recently been investigated (34) on the Au(111) surface with polarized light at oblique incidence (see Sec. 2). The photoemission from this low-index face is dominated by direct transitions down to about 3 eV and the angular as well as energy distribution of photoelectrons can be specified, at least qualitatively, in considerable detail (8).

The most important features are as follows: quite generally, electrons with high kinetic energy escape from the emitter at oblique angles (10); for p-polarized light the emission intensity is essentially uniform with respect to escape angle; for s-polarized light, however, emission normal to the surface is totally suppressed and there is appreciable intensity only at oblique angles. The question then arises to what extent these two different situations manifest themselves in the concentration dependence of the photocurrent; that is, do they give rise to distinctly different source functions which characterize the spatial distribution of thermalized electrons.

The experimental vector ratios shown in Fig. 8 are seen to be independent of the scavenger concentration. This result indicates that the mean thermalization range, with respect to the emitter surface, does not depend to any significant extent on the initial angular distribution of the photoelectrons. We may conclude therefore that the energy-degrading scattering processes are associated

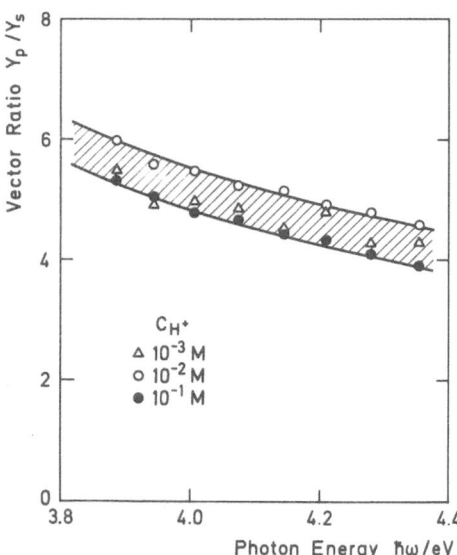

Figure 8. Vector ratios Y_p/Y_s on Au(111) in solutions of LiCl (1M) containing different scavenger concentrations; $\alpha = 45°$, $\hbar\omega_0 = 3$ eV.

with substantial momentum transfer to the electrons, and that the
initially anisotropic distribution is rapidly randomized in the
course of the energy dissipation.

5. SOLVENT DECOMPOSITION BY PHOTOHOLES

This section briefly describes a photoeffect at the metal-electrolyte interface which is very similar to photoelectron emission
in some respects and different in others. It is caused by the excitation of photoholes and is observed experimentally as a photocurrent
of opposite sign compared to electron emission (35). A very similar
phenomenon is observed in semiconductor electrodes (36) and in Fig.
9a the process is schematically illustrated for a metal surface. In
aqueous electrolytes the photoholes have been found to oxidatively
decompose the solvent, with the final reaction product being oxygen
(35, 37).

Experimental results on a Au(111) surface are shown in Fig. 9b.
When plotting the square root of the photocurrent as a function of
photon energy two linear sections are observed in the curve. The
threshold for the process as obtained by extrapolation is seen to shift

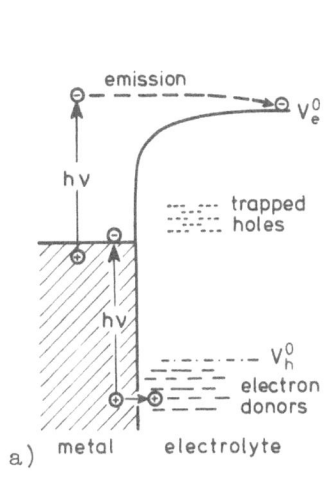

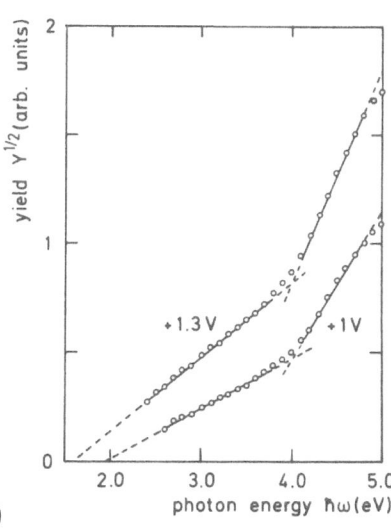

Figure 9. a) Schematic illustration of photohole excitation and
subsequent "emission" as compared to photoelectron emission.
b) Photohole currents from Au(111) at two different electrode
potentials which corresponds to two energetic positions of the
Fermi-level.

with electrode potential, i. e. with the energetic position of the Fermi level. This result and the quadratic increase of the photocurrent with photon energy suggest an interesting implication. It seems possible that photohole emission takes place into a valence band structure in water clusters. The confirmation of this concept of a delocalized positive charge in water requires further studies, however.

ACKNOWLEDGEMENTS

The author would like to thank Prof. H. Gerischer and H.J. Lewerenz for many fruitful discussions and ideas. The outstanding capabilities of E. Piltz who performed many of the experiments described in this paper are also gratefully acknowledged.

REFERENCES

1. J. Jortner, Ber. Bunsenges. Physik. Chem. 75, 696 (1971).

2. C. Caroli, D. Lederer-Rozenblatt, B.Roulet and D. Saint-James, Phys. Rev. B 8, 4552 (1973).

3. W. L. Schaich and N. W. Ashcroft, Phys. Rev. B 3, 2452 (1971).

4. P. J. Feibelman and D. E. Eastman, Phys. Rev. B 10, 4932 (1974).

5. A. M. Brodskii and Yu. V. Pleskov, in Progress in Surface Science, Ed. S. G. Davison (Pergamon Press, New York, 1972).

6. Z. A. Rotenberg and Yu. Ya. Gurevich, J. Electroanal. Chem. 66, 165 (1975).

7. J. K. Sass and H. Gerischer, Interfacial Photoemission, Chpt. 16 in Photoemission and the Electronic Properties of Surfaces, Ed. B. Feuerbacher et al. (Wiley-Interscience, 1977).

8. H. Laucht, J. K. Sass, H. J. Lewerenz and K. L. Kliewer, Surf. Sci. 62, 106 (1977).

9. H. Becker, E. Dietz, U. Gerhardt and H. Angermüller, Phys. Rev. B 12, 2084 (1975).

10. G. D. Mahan, Phys. Rev. B 2, 4334 (1970).

11. W. L. Schaich, Phys. Stat. Sol. 66, 527 (1974).

12. B. Feuerbacher and R. F. Willis, J. Phys. C 9, 169 (1976).

13. K. L. Kliewer, Phys. Rev. B 14, 1413 (1976).

14. P. J. Feibelman, Phys. Rev. Lett. 12, 1092 (1975).

15. F. Forstmann and H. Stenschke, Phys. Rev. Lett. 38, 1365 (1977).

16. I. Lindau and P. O. Nilsson, Phys. Scripta 3, 87 (1971).

17. M. Anderegg, B. Feuerbacher and B. Fitton, Phys. Rev. Lett. 27,
     1565 (1971).

18. J. K. Sass, H. Laucht and K. L. Kliewer, Phys. Rev. Lett. 35,
     1461 (1975).

19. J. K. Sass, H. J. Lewerenz, E. Piltz and K. Horn, J. Phys. C,
     in press.

20. K. L. Kliewer and K.-H. Bennemann, Phys. Rev. B 15,
     3731 (1977).

21. J. K. Sass, S. Stucki and H. J. Lewerenz, Surf. Sci., in press.

22. H. Raether, in Physics of Thin Films, Vol. 7 (Academic Press,
     1977).

23. M. J. Gibson and P. J. Dobson, J. Phys. F., 5, 1828 (1975);
     H. Neddermeyer, Habilitationsschrift, Univ. München (1976).

24. J. K. Sass, H. Laucht and S. Stucki, in Proceedings of an Inter-
     national Symposium on Photoemission, Noordwijk, The Netherlands,
     Eds. R. F. Willis et al. (European Space Agency, Paris, 1976).

25. J. G. Endriz and W. E. Spicer, Phys. Rev. Lett. 27, 570 (1971).

26. A. Otto, Z. Physik 216, 398 (1968); E. Kretschmann, Z. Phys.
     241, 313 (1971); R. Kötz, D. M. Kolb and J. K. Sass, Surf. Sci.
     69, 359 (1977).

27. J. K. Sass, R. K. Sen, E. Meyer and H. Gerischer, Surf. Sci.
     44, 515 (1974).

28. T. Furtak and J. K. Sass, to be published.

29. R. Kötz and D. M. Kolb, Surf. Sci., 64, 96 (1977).

30. J. D. E. McIntyre and D. E. Aspnes, Surf. Sci 24, 417 (1971).

31. T. Furtak and K. L. Kliewer, to be published.

32. C. B. Duke and U. Landmann, Phys. Rev. B 8, 505 (1971).

33. J. K. Sass, Surf. Sci. 51, 199 (1975).

34. H. J. Lewerenz, H. Neff and J. K. Sass, to be published.

35. H. Gerischer, E. Meyer and J. K. Sass, Ber. Bunsenges. Physik.
     Chem. 76, 1191 (1972).

36. A. Fujishima and K. Honda, Nature 238, 37 (1972).

37. E. Meyer, Dissertation, Technische Universität München, 1973.

FORMATION OF SURFACE COMPOUNDS ON ELECTRODES

A. Bewick and M. Fleischmann

Chemistry Department, University of Southampton

Southampton SO9 5NH, Great Britain

The formation of new phases on electrodes is observed under many conditions which include: the cathodic deposition of bulk deposits of metals; the anodic deposition of bulk deposits of insulators (e.g. metal salts and oxides) and semi-conductors (e.g. oxides); the cathodic deposition of two-dimensional phases of metals; the anodic deposition of two-dimensional phases of insulators and of materials which are semi-conducting in the bulk state. In addition, certain transitions in the structures of adsorbed layers are related to phase formation.

In section 1 we review and illustrate the electrochemical kinetics of phase formation which has been a basic tool in the investigation of these phenomena; in section 2 we refer to structural measurements and to recent extensions in the methodology; the final section deals specifically with recent investigations of transformations in two-dimensional systems.

1. THE KINETICS OF PHASE FORMATION

The driving force of electrochemical reactions, the electro-chemical free energy, is determined by the electrode potential. This potential can be accurately controlled and rapidly programmed and the kinetics of phase formation (as determined from the current-time transients) can therefore be examined over a wide time range. The simplest conditions from the point of view of analysis are found when a potential step is applied to an electrode so as to initiate the formation of a new phase. At the simplest level this is a two-stage process consisting of nucleation and crystal growth and, at constant potential, the rate constants for both steps are independent of time.

Nucleation requires the formation of a critical cluster the size of which depends on the overpotential η (the electrode potential measured with respect to the equilibrium potential of the bulk phase). For three-dimensional nucleation the rate constant has the form[1,2]

$$A = K \exp -(\kappa\sigma^3/\eta^2) \tag{1}$$

where the surface free energy σ may be a composite quantity (determined by the nucleus/substrate, substrate/electrolyte and nucleus/electrolyte surface energies) K is a frequency factor and the form of κ is dependent on the geometry of the cluster. For the simplest case of totally irreversible reactions on uniformly growing surfaces the form of the crystal growth rate constant will be

$$k = k_o \, \underset{i}{\pi} \, c_i^{\nu_i} \exp -(\frac{\alpha n FE}{Rt}) \tag{2}$$

where the nature of the product is determined by the kinetics of lattice formation. In the initial stages of crystal growth, where the centres grow independently, the current-time transient may be predicted for any given growth geometry[2]. For example, for three-dimensional growth of a single hemisphere

$$i = nFkS = \frac{2\pi nFM^2k^3t^2}{\rho^2} \tag{3}$$

Nucleation normally takes place at preferred sites on the substrate so that the number of nuclei usually follows

$$N = N_o (1 - \exp -(A't)) \tag{4}$$

(3) combined with (4) gives the expression for the current in the initial stages

$$i = \frac{2\pi nFM^2k^3N_o}{\rho^2} \left[t^2 - \frac{2t}{A'} + \frac{2}{(A')^2} - \frac{2}{(A')^2} \exp -(A't) \right] \tag{5}$$

This in turn leads to two limiting forms: if $1/A'$ is large compared to the time scale over which centres grow independently then nucleation appears to be linear with time (progressive nucleation) and

$$i = \frac{2\pi nFM^2k^3At^3}{3\rho^2} \tag{6}$$

(where $A = A'N_o$). On the other hand, if $1/A'$ is small compared to this time scale, then N_o centres are converted into nuclei in the very initial stages (instantaneous nucleation) and

$$i = \frac{2\pi n F M^2 k^3 N_o t^2}{\rho^2} \tag{7}$$

In view of the very rapid variation of A with η (equation (1)), it is also possible to change the kinetics from (6) to (7) at will by using a short pulse at a higher overpotential to form the nuclei preceding crystal growth at a lower overpotential[3]. Fig. 1 gives an example of the deposition of an oxide to illustrate this behaviour[4].

A model of particular significance in the field of crystal growth is that of the formation of a single layer plane by the two-dimensional growth of centres. In the initial stages we observe

$$i = \frac{\pi n F M h k^2 A t^2}{\rho} \tag{8}$$

and

$$i = \frac{2\pi n F M h N_o k^2 t}{\rho} \tag{9}$$

for progressive and instantaneous nucleation and assuming cylindrical growth. In the later stages centres coalesce ("overlap") and the effect of this can be taken into account by appropriate statistical analysis[5,2]. It is found that for random nucleation the area of substrate covered is given by

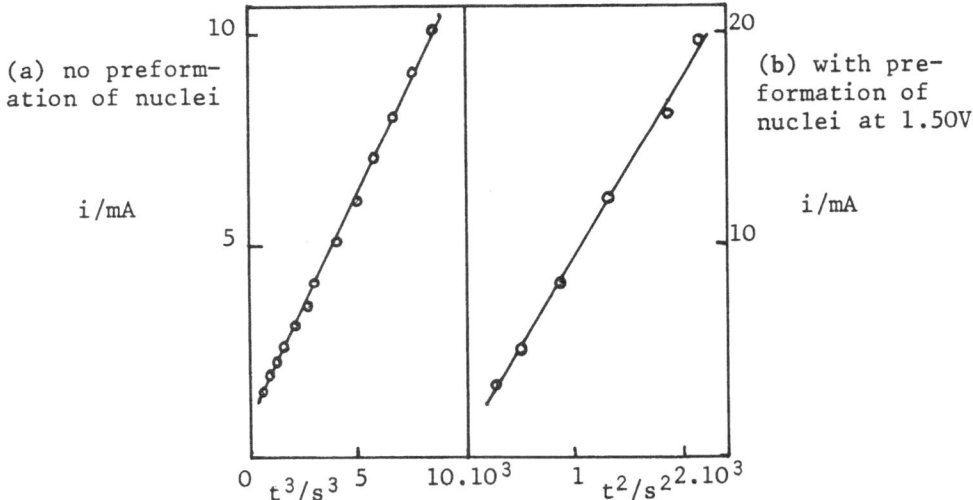

(a) no preformation of nuclei

(b) with preformation of nuclei at 1.50V

i/mA

i/mA

Fig.1 Anodic deposition of NiOOH from 0.25M Ni(Ac)$_2$ at 1.40V, pH = 4.8.

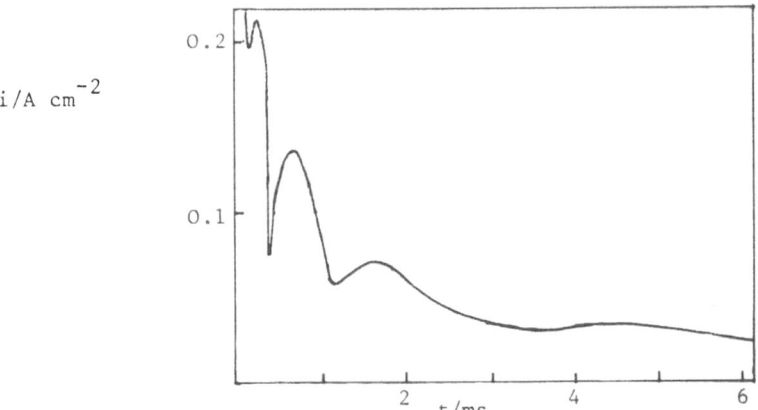

Fig 2A. Formation of Hg_2Cl_2 on mercury in 0.1M HCl; overpotential
36 mV.

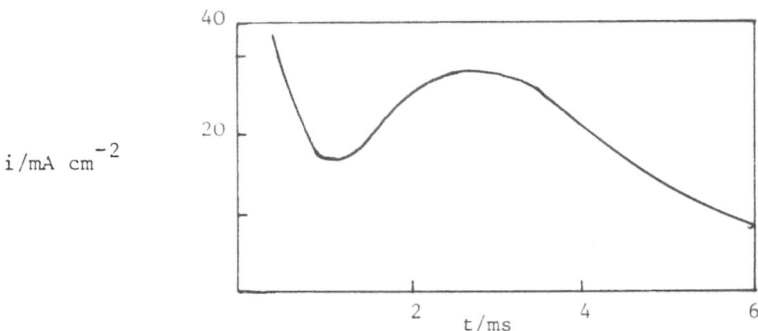

Fig 2B. Formation of HgO on mercury in 1M KOH at an overpotential
of 20 mV.

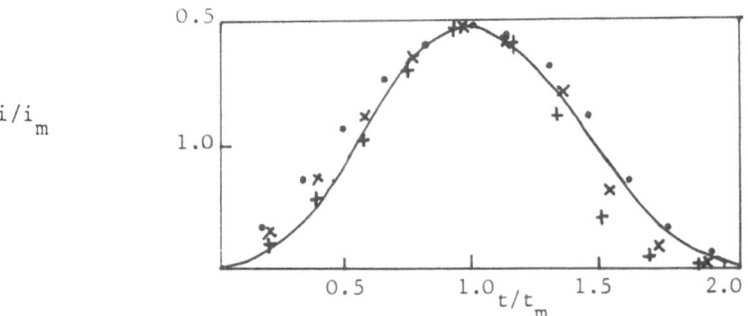

Fig 2C. Reduction of In_2O_3 on indium amalgam at -130 mV (Hg/HgO)
Oxide formed at -400mV for • 0.025, x 0.25, + 2s

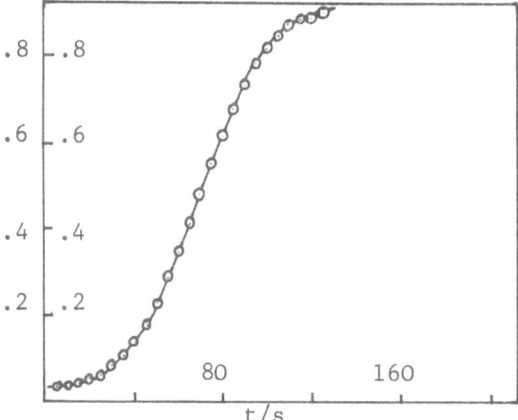

Fig.3. Deposition of nickel onto vitreous carbon, Watt's bath 20°C, −0.825V (SCE). Full line: computer fit by non−linear regression to eq. (15) but allowing for the initial current 36.5±1.97μA and an induction time of 3±0.6 s. Parameters zFk'=872.5±3.48μA, $\pi M^2 k^2 A/3\rho^2 = (2.35\pm0.07) \times 10^{-6}$ s^{-3}.

$$S = 1 - \exp - S_{ext} \tag{10}$$

where S_{ext} is the area predicted if all the centres had grown without overlap. In this way we can deduce[2,6] the full current−time curves for the growth of a single layer for progressive and instantaneous nucleation

$$i = \frac{\pi n F M h k^2 A t^2}{\rho} \exp - \frac{\pi M^2 k^2 A t^3}{3\rho^2} \tag{11}$$

and

$$i = \frac{2\pi n F M h N_o k^2 t}{\rho} \exp - \frac{N_o M^2 k^2 t^2}{\rho^2} \tag{12}$$

The general expression for nucleation according to (4) is

$$i = \frac{\pi n F M h k^2 N_o}{\rho} \left[2t - \frac{2}{A'} + \frac{2}{A'} \exp{-A't} \right]$$
$$\exp - \frac{\pi M^2 k^2 N_o}{\rho^2} \left[t^2 - \frac{2t}{A'} + \frac{2}{A'^2} - \frac{2}{A'^2} \exp{-A't} \right] \tag{13}$$

Many examples of growth according to this pattern have been observed. Fig.2 illustrates the formation of successive monolayers of calomel on mercury[6], the formation of a first layer of mercuric oxide on mercury[7] and the reduction of a monolayer of indium oxide

on indium amalgam[8].   The model discussed above is a modification
of the classical mechanism of crystal growth in which a single
layer is formed from a single nucleus.

Growth according to this classical mechanism has also been
observed for the deposition of silver on dislocation free single
crystal silver substrates grown electrochemically in capillaries[9].
The single nucleus can be formed by applying an initial pulse at a
higher overpotential before observing growth at a lower overpotential.   The detailed shape of the current-time transient depends
upon the shape and orientation of the growing centre relative
to the walls of the capillary and, for capillaries of rectangular
cross section, regions of constant current may be observed in
the current-time transient due to the propagation of a step of constant length[10];   by applying a sufficiently high first pulse to such
electrodes, a polyatomic step is sometimes formed and may be observed directly by Nomarski interference contrast microscopy[11].

The effect of overlap in three-dimensional growth can also be
taken into account by assuming the repeated formation of layers with
a delay proportional to the layer number.   Simple expressions are
obtained if the growth forms are right circular cones.   If the rate
constants parallel and perpendicular to the surface are k and k',
then for instantaneous nucleation

$$i = nFk' \left[1 - \exp - \frac{\pi N_o M^2 k^2 t^2}{\rho^2} \right] \tag{14}$$

while for progressive nucleation

$$i = nFk' \left[1 - \exp - \frac{\pi M^2 k^2 A t^3}{3\rho^2} \right] \tag{15}$$

The general expression for nucleation according to (4) is

$$i = nFk' \left[1 - \exp - \frac{\pi M^2 k^2 N_o}{\rho^2} \left(t^2 - \frac{2t}{A'} + \frac{2}{A'^2} - \frac{2}{A'^2} \exp - A't \right) \right] \tag{16}$$

Fig.3 gives an example of the fit of equation (15) to the initial
stages of the deposition of nickel on a vitreous carbon substrate[12].

It can be seen from the examples given above that many electrocrystallisation reactions follow the growth laws deduced from
simple geometric and statistical postulates.   The behaviour in more
complex cases can often still be interpreted in terms of a succession of events based on these laws.   For example, the deposition of
a first layer according to (11) or (12) is frequently followed by
the formation of further monolayers[6] (see fig.2).   In the region
of each maximum the layers grow essentially independently of each
other and the kinetics can be shown to be in accord with (11).   In

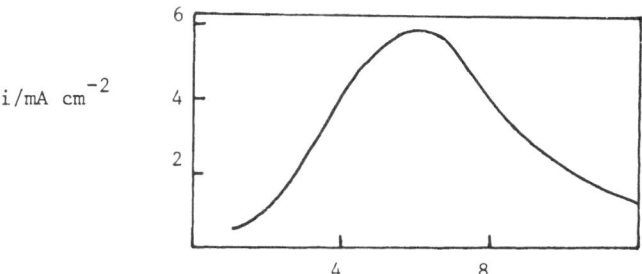

Fig. 4. Formation of multimolecular layer of HgO on mercury in 1M KOH at an overpotential of 25 mV.

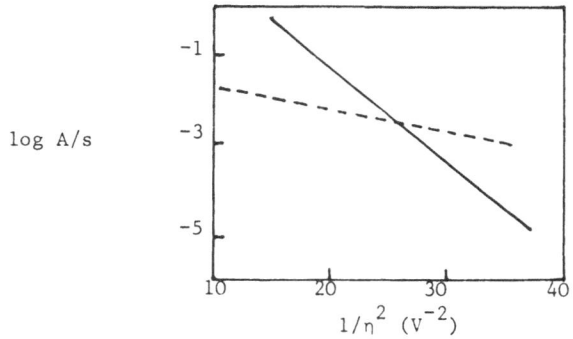

Fig. 5. Variation of nucleation rate constant with overpotential for formation of β-lead dioxide. ──── deposition onto platinum from 0.25M $PbClO_4$ + 0.1M $HClO_4$; ─ ─ ─ Oxidation of $PbSO_4$ in 1M H_2SO_4

other cases the formation of a monolayer is followed by the formation of a three-dimensional deposit according to (14) or (15). Thus the first layer of mercuric oxide, fig 2, is followed by the deposition of the three-dimensional layer[7], fig 4. In this case, however, the supply of mercuric ions becomes restricted by the progressive coverage of the substrate (passivation) and the rate therefore passes through a maximum according to

$$i = nFk' \left[1 - \exp - \frac{\pi M^2 k^2 A t^3}{3\rho^2} \right] \exp - \frac{\pi M^2 k^2 A t^3}{3\rho^2} \tag{17}$$

Nucleation rate constants can be derived by combining measurements at a single and at two potentials with optical or electron microscopic counts of the number of growth centres, equations (6) and (7). The kinetics often follow expressions such as (1)[13,14], fig 5; for the particular case of the electrodeposition of lead

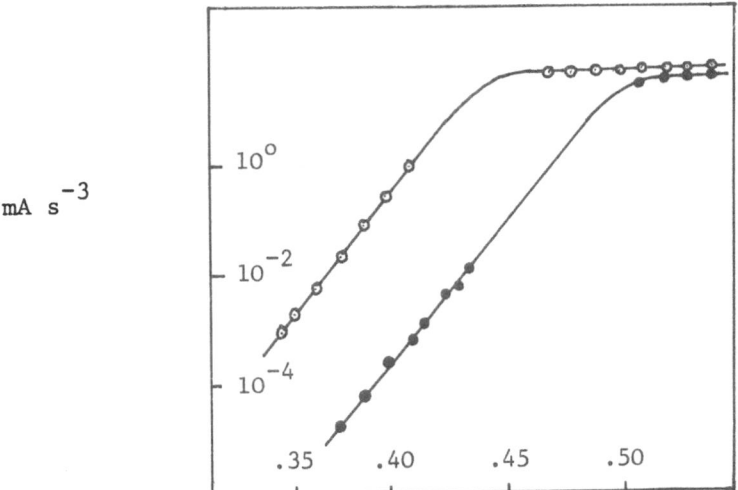

Fig. 6. Variation with electrode potential and pH of the rate
constant of crystal growth for the formation of AgO.   o N KOH
• 0.1 N KOH.

dioxide, data for an extended set of conditions have been shown to
fit this equation provided the variation of σ with electrode potential is taken into account[13].   The frequency factors K are, however,
usually much larger than the theoretical estimates[13] while the sizes
of the critical clusters (deduced from the parameter K) are unrea-
sonably small[13,14].   It has therefore been proposed[15,16] that it is
more correct to apply the atomistic model of nucleation[17] to electrocrystallisation.   The atomistic and continuum models however
lead to essentially similar results and criteria which permit a
clear distinction between the models have not so far been derived.

It will be apparent that the rate constant k governing crystal
growth is a composite quantity, being determined by all the processes leading to lattice formation.   It is found, however, that
the kinetics of crystal growth frequently follow simple laws such as
the "Tafel" equation (2), at least at intermediate overpotentials[3,11].
In some examples such as the growth of argentic oxide, fig 6, the
slopes of the logarithmic plots and the concentration dependence can
be simply interpreted in terms of the slow formation of the lattice
from adsorbed intermediates generated in a fast electrochemical pre-
equilibrium step, in this case according to[18,19]

$$OH^- \rightleftharpoons OH_{ads} + e$$
$$OH_{ads} + Ag^+ \xrightarrow{slow} AgO + H^+$$

At high overpotentials lattice forming sites are covered by a monolayer of adsorbed intermediates and the overall rate of crystal growth in fact follows a potential dependent Langmuir isotherm.

The simplicity of these kinetics stands in sharp distinction to the theoretically predicted patterns of behaviour which have been extensively discussed[20,21,22]. It is generally predicted that lattice formation at Kossel kink sites at the edges of layer planes will be fast in view of the high density of these sites. The slow stages would therefore be either the diffusion of species through the solution to the edges or the diffusion of adatoms (possibly bearing a partial charge) or of adions over the surface, the adsorbed species being formed in preceding electrochemical reactions. The steps themselves have usually been predicted to be formed by the "rotation" of emergent screw dislocations. It can be seen, however, that in those cases where the rate of lattice formation can be related to unit length of an edge of a layer plane (fig.2 and equations (11) and (12)), lattice formation is still the slow step. This suggests that there is in fact a low density of kink sites at the edges of the layer planes and that the slow step is the direct discharge of the lattice forming species at these kink sites. Direct evidence for such a mechanism has been obtained recently for the dissolution of {211} planes of iron by gold decoration of the steps along the <311> edges and by measuring the extent of the misorientation with this direction[23].

The participation of surface diffusion can only be directly demonstrated in a few examples such as the deposition of a monolayer of nickel on mercury[24], fig 7, where the form of the transients can be simulated provided the growth centres are surrounded by diffusion zones. Moreover, the generation of steps by two-dimensional nucleation appears to be rapid in general and it is likely that the frequency factors for this process will also be found to be high as compared to those predicted by current theories. The participation of dislocations has so far only been suggested by a few examples such as the growth of mercuric oxide[7], fig 4. In this case it is found that

$$k' \propto k^2 \qquad (18)$$

as would be predicted by such a model. Direct observation of the pyramidal growth forms of silver on previously levelled single crystal faces having a low density of dislocation has been shown to be consistent with growth determined by emergent screw dislocations[25]. The additional retardation in the rate of electrocrystallisation over and above that due to the discharge reaction, equation (2), has also been ascribed to step generation by dislocations[2,26]: the low step density ensures that each step line grows independently of adjacent steps and the variation in step line density with overpotential (the curvature of the emergent spiral changes with overpotential) determines the additional retardation. The generation of steps by two dimensional nucleation, however, would cause a com-

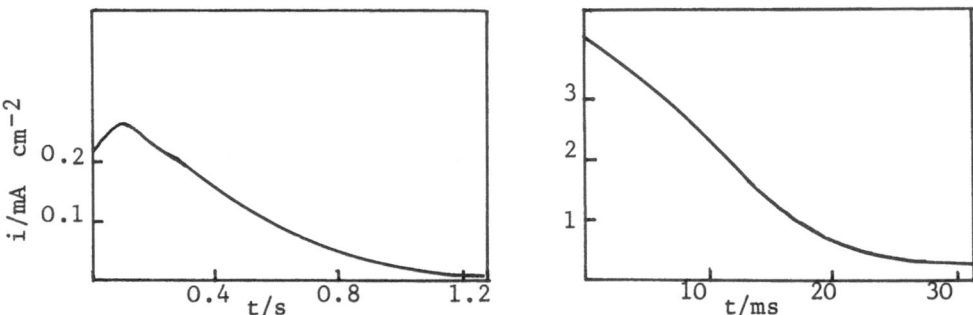

Fig 7a. Current-time transients for the deposition of a monolayer
of Ni at potentials of -600 and -670mV SCE. The effects of surface
diffusion become more dominant with increasing negative potential.

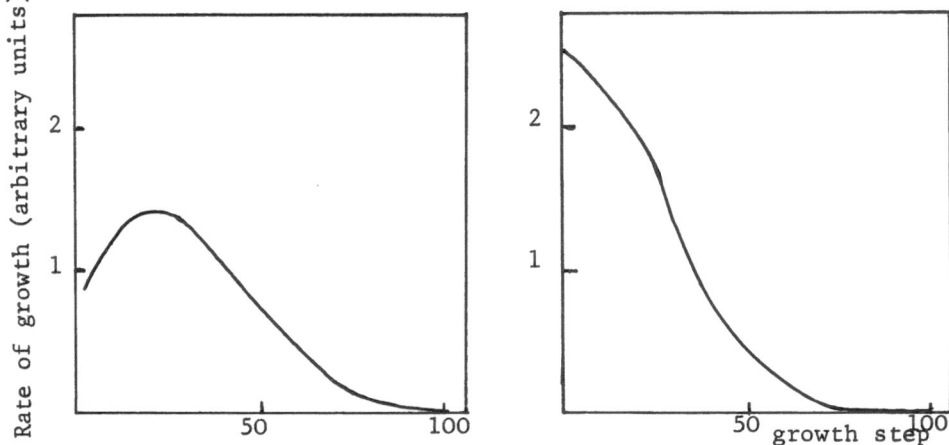

Fig 7b. Simulation of the growth kinetics allowing for surface
diffusion of adatoms at a lower and higher overpotential. The
conditions are comparable to the measurements in fig 7a.

parable retardation. It should also be noted that in other cases
such as the growth of lead dioxide[13] or of nickel[12]

$$k' \propto k \qquad\qquad\qquad (19)$$

and in these cases there is therefore no evidence for a special
mechanism for the generation of lattice growth sites. On the other
hand, in the case of dissolution of iron[23] referred to above, steps
are produced by the dissolution of an atom at the apex of a pyramid
and kinks at the intersection of two steps in a plane. It seems
certain that many specialised mechanisms not predicted by current
theories will in due course be found to be operative.

The kinetics of electrocrystallisation reactions have also been examined using a number of alternative electrochemical techniques. In galvanostatic experiments the overall rate of reaction is pre-scribed and, in deposition on inert substrates, a maximum in the potential with time is observed due to slow nucleation followed by expansion of the reacting interface.   The detailed form of the curves has so far not been interpreted.   Potential-time curves measured in the initial stages of the displacement from equilibrium on a substrate of the material to be deposited[27,28] (as well as A.C. impedance measurements[21]) have been interpreted in terms of the de-position and diffusion of adatoms.   Successive galvanostatic-poten-tiostatic measurements have also been used for this purpose[29].   How-ever, it is not clear to what extent concentration changes in the solution surrounding the growth sites contribute to the effects ob-served[30].

A particularly convenient method which has been developed in recent years for simple electrode reactions relies on the applica-tion of a voltage ramp linear with time and the measurement of the resultant current response.   An example of the application of this method to electrocrystallisation is illustrated in fig 8.[31].   The analysis of the complex patterns of interaction of the various fac-tors is now becoming feasible and is referred to in the next sec-tion.

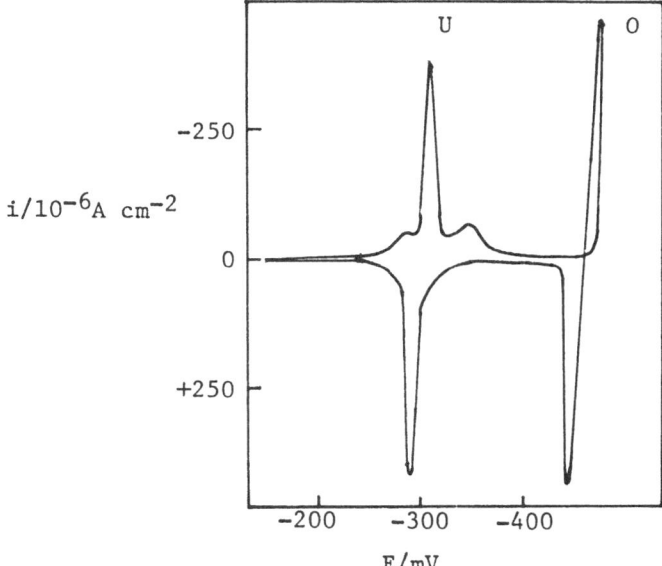

Fig.8.   Linear sweep voltammogram for the deposition of lead on silver in the underpotential region (U) and the overpotential region (O).   Sweep speed 30mV s^{-1}.

RECENT DEVELOPMENTS IN METHODS OF INVESTIGATION

Kinetic Measurements Using Conventional Electrodes

In experiments other than those at constant potential the strong non-linearities (in part caused by the complex connectivity of the phenomena) become apparent[32].   The rate of crystal growth is non-linear in potential in the first place, equation 2. Secondly, even in the simple cases of two-dimensional growth, the area of a single centre follows a power law in time and hence implicitly in the overpotential. Progressive nucleation introduces an exponential inverse power law and the total area is obtained by yet a further exponentiation to allow for the overlap of centres. It will be apparent from the previous section that a reasonably complete investigation will require experiments under more than one set of conditions (such as the combination of measurements at a single potential with those employing a succession of two potentials so as to separate A and k). There is therefore a considerable incentive to develop ramp and perturbation methods which can decouple the various parameters notwithstanding the complexities of the interactions.

In the general case of the two dimensional growth of circular centres the extended area is now given by a convolution integral

$$S_{ext}(\eta,t) = \frac{\pi N_o M^2}{\rho^2} \int_o^t A[\eta(t-z)]\left\{\int_o^z k(\eta(y)dy\right\}^2 dz \qquad (20)$$

and expressions have been derived for the current in linear sweep voltammetry for two-dimensional growth both with instantaneous and progressive nucleation. The response to linear (as well as non-linear) perturbations superimposed on growth at constant potential has also been analysed: in the case of progressive nucleation coupled to two-dimensional growth there is an "inductive" component[32].

Measurements on Electrodes of Very Small Dimensions:

Stochastic Effects

In view of the high sensitivity of electrical measurements it is possible to reduce the size of the substrate electrodes to such an extent (typically $10^{-6}- 10^{-5}$ cm^2) that only a small number of nuclei can be formed on the surface during the time scale of an experiment[33]. Nucleation can be regarded as a Poisson process the probability of a birth in the time interval du at time u being

$$P(1 \text{ nucleus in time } du) = \lambda \exp{-\lambda u} du \qquad (21)$$

$$\text{where } \lambda = Aa \qquad (22)$$

and a is the area of the electrode. If we assume hemispherical growth say, equation (3), then we immediately obtain an estimate of the probability g(i) of observing a current i due to a single centre

$$g(i) = \frac{\lambda}{2Ci^{\frac{1}{2}}} \exp{-\lambda}\left(t - \frac{i^{\frac{1}{2}}}{C}\right) \qquad (23)$$

$$\text{where } C^2 = \frac{2\pi n F M^2 k^2}{\rho^2} \qquad (24)$$

The moments of the current can therefore be predicted. The stochastic mean agrees with the deterministic result, equation (6) with A replaced by λ. The higher moments of an ensemble of transients on the other hand provide new kinetic information since the rate constants appear in a different combination to that in the mean. For example, in the case discussed, the variance

$$<I^2>-<I>^2 = \frac{C^4 t}{\lambda^3 [\exp{\lambda t}-1]} [(\lambda^4 t^4 - 4\lambda^3 t^3 + 12\lambda^2 t^2 - 24\lambda t + 24)\exp{\lambda t} - 24]$$

$$\cong C^4 \lambda t^5 \text{ if } \lambda t \gg 1 \qquad (25)$$

$$\text{Thus} \quad \frac{\text{standard deviation}}{\text{mean current}} = \frac{\sigma_I}{\bar{I}} = \left(\frac{1}{\lambda t}\right)^{\frac{1}{2}} \qquad (26)$$

and the product of the rate constants is split into the component parts. Fig. 9 illustrates the applicability of (26). If the size of the substrate electrode is reduced still further (typically to 10^{-10} cm^2) then only a single nucleus can be formed and the plots according to (26) can take a more complicated form, fig 8[24]. The increasing standard deviation with increasing time can be modelled as the "death" of a single growing crystallite which is followed by the "birth" of a further nucleus. The role of death processes in electrocrystallisation reactions is at present not understood but it is certainly important since most deposits are micro crystalline. Stochastic measurements clearly afford a means of investigating these phenomena. The role of overlap of growth centres can also be taken into account, for example, by regarding S_{ext} as a stochastic variable[34,35]. For the growth of right circular cones (see discussion leading to equation (16)) we obtain[24]

$$\frac{\sigma_I}{\bar{I}} = \frac{\pi M^2 k^2}{\rho^2} \left(\frac{At^5}{5a}\right)^{\frac{1}{2}} \frac{\exp{-S_{ext}}}{1-\exp{-S_{ext}}} \qquad (27)$$

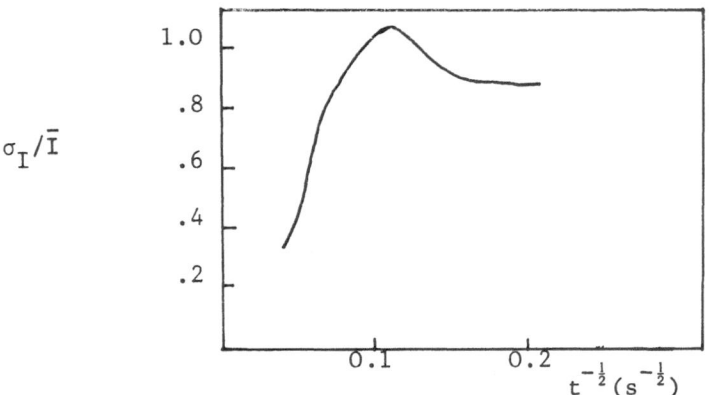

Fig 9a. Statistical analysis of the transients for the deposition of α-PbO$_2$ onto a 10μm diameter Pt electrode.

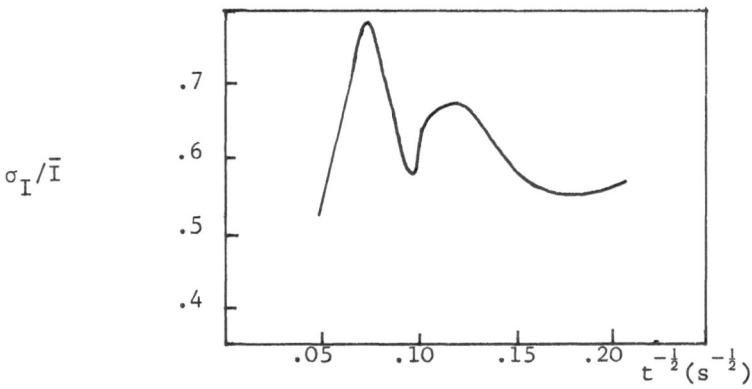

Fig 9b. Statistical analysis of the transients for the deposition of α-PbO$_2$ onto a 2μm diameter Pt electrode.

and the effects of "death" due to overlap can be seen to be oppos-ite to those observed in the experiments, fig 9b.

Stochastic effects may alternatively be observed by analysing temporal experiments for example by determining the autocovariance function

$$\Phi_I(\tau) = \langle I(0)I(\tau) \rangle \tag{28}$$

or the power spectral density

$$G_I(f) = 2\mathrm{Re} \int_{-\infty}^{\infty} \Phi_I(\tau)\exp{-j\omega\tau}d\tau \tag{29}$$

of the random fluctuations in the steady state.   Nucleation, crystal growth (and "death") are predicted to generate white noise at
very low frequencies, the position and shape of the falloff at somewhat higher frequencies being determined by overlap and the nature
of other "death" processes.   The amplitude of the white noise again
permits the division of the composite rate constants.   A second
region of white noise above this frequency band is determined by the
kinetics of lattice formation (the frequency band due to this process is inevitably higher in view of the shorter relaxation time of
these processes compared to the sequence nucleation-growth-overlap
(death)).   Theoretical formulations of the noise spectra can be
obtained for any chosen sequence by solving the appropriate master
equations and these can be compared to the experimental data.   For
example, for the reaction scheme

$$A \underset{k_2}{\overset{k_1}{\rightleftharpoons}} B_{ads} \longrightarrow C$$

leading to the formation of a single adsorbed intermediate B(compare the scheme for the formation of Ag O) the probability of observing x vacant sites is determined from

$$\frac{dP_{(x)}}{dt} = P_{(x+1)}k_1 n_A (x+1) + P_{(x-1)}(k_2+k_3)(x_o-x+1)$$

$$- P_{(x)}k_1 n_A x \quad - \quad P_{(x)}(k_2+k_3)(x_o-x) \tag{30}$$

where x_o is the total number of sites.   Conversion of the set of
equations (30) to the master equation and solution of the problem
shows that for example at low coverage

$$\left(\frac{G_I(f)}{\bar{I}^2}\right)_{f=o} = \frac{4}{k_1 n_A x_o} \tag{31}$$

$$\text{whereas} \quad \bar{I} = \frac{k_3 k_1 n_A x_o}{k_2} \tag{32}$$

Thus the overall rate of lattice formation can also be divided into
its component parts.   The noise at low frequencies is determined
by the fast pre-equilibrium just as the noise power for phase
growth is determined by the first nucleation step.   Fig. 10 illustrates such a measurement[34].   However, lattice formation can, in
fact, be a considerably more complicated process and in some instances may have a non-Markovian character[33].

The Structure of Electrodeposits

The structure of electrodeposits has hitherto been examined
by diffraction methods after removal from the solution.   The aim

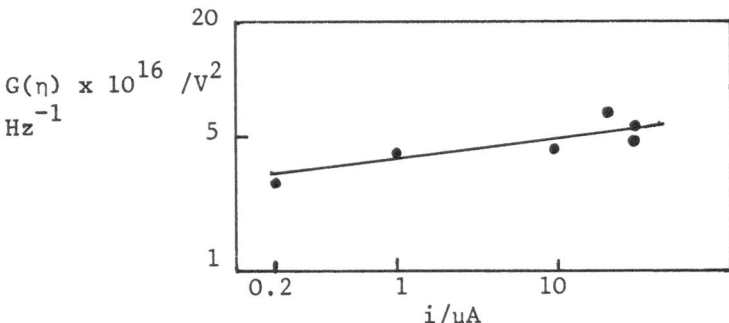

Fig 10. Voltage noise generated in the steady state of deposition of α-PbO$_2$ as a function of the applied current.

of most of the investigations by X-ray and high voltage electron diffraction as well as by X-ray topography etc., of thick deposits is the correlation of the deposit composition, orientation and defect structure with solution and deposition conditions. This extensive area of research will not be reviewed here.

The structure of thin deposits such as those which have been discussed here is conveniently examined by selected area high voltage electron diffraction[6,7,22]. The deposits may be removed from the substrate and mounted on grids using standard replicating techniques. Fig. 11 shows a diffraction pattern of a film of cadmium

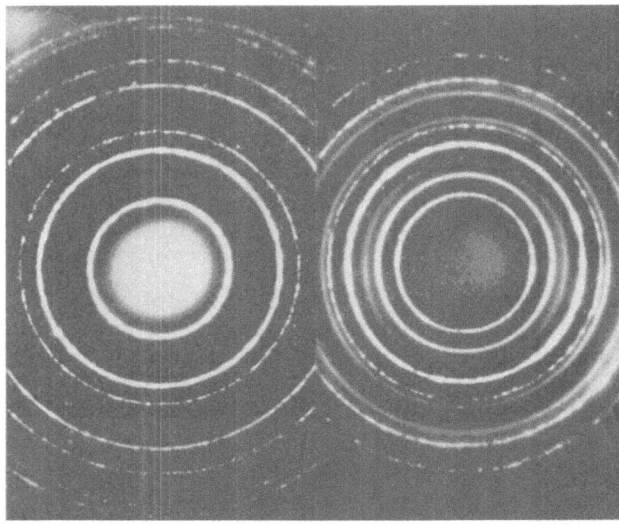

Fig. 11. Electron diffraction pattern of cadmium hydroxide film on 1% cadmium amalgam in 1 N sodium hydroxide. Beam parallel to <00, 1> axes of crystallites.

hydroxide formed by the deposition of three successive layers[36].
The pattern shows (h,k,0) diffractions only in view of the orientation and thinness of the deposit; the missing diffractions appear
on layer lines on tilting the deposit.   Recent developments in the
use of *ex-situ* techniques include the application of ESCA and LEED
to electrodes[37,38,39].   In some cases the electrochemical measurements are conducted in special chambers attached to the ultra high
vacuum system so that the transfer of the electrodes can be effected without exposure to air.   One of the first applications of
ESCA to electrochemistry has been concerned with the detailed
structure of the oxide film formed on platinum anodes[37,38] and how
this structure varies with potential and the nature of the electrolyte.   The major contribution of the method is in the positive
chemical identification of the platinum oxidation state[38].   The initially formed oxide, which grows to a thickness of 0.8 nm, was
shown to have a Pt^{2+} oxidation state and an XPS binding energy which
does not correspond to PtO, $Pt(OH)_2$ or PtO_2.   At higher potentials
the oxide is $PtO_2 \cdot XH_2O$ and at higher concentrations of the acid
electrolyte, it was observed that anions of the acid were incorporated into the structure.   The ESCA technique has also been applied
to the study of the nature of submonolayer amounts of silver and
copper deposited in the underpotential region (see section 3) on
platinum electrodes[39].   The deposits gave spectra which were clearly metal-like rather than ionic although the XPS binding energies
were shifted significantly from those of the bulk metals.   There
was a good correlation, however, with the spectra obtained for very
thin layers of the metal formed by evaporation.

The difficulty with all such *ex-situ* techniques is that the
deposits may change in structure on removal from the solution.   The
only method so far developed which permits the *in-situ* identification
of deposits on electrodes is Raman spectroscopy[40,41].

PHASE TRANSFORMATIONS IN TWO-DIMENSIONAL SYSTEMS

The formation of a number of two-dimensional solid phases via
two-dimensional nucleation and growth has been illustrated in section 2.   In many instances, phase formation in the first layer
takes place after adsorption of an appreciable quantity of the ions
to be incorporated in the lattice and at a positive free energy
(underpotential) with respect to the formation of the bulk phase;
the latter effect is a measure of the interaction energy between
the two-dimensional phase and the substrate.   Phase transformations
have also been observed in simple systems in which only a single
layer is formed and there is no relationship with a three-dimensional deposit, e.g. in the adsorption at a mercury surface of
simple molecules such as quinoline[42] and pyridine[43].   The dissolution of monolayers can also follow this mechanism, fig 2, and indeed the mechanism of transformations via nucleation and growth
extends to the field of bioelectrochemistry.   Thus fig 12 shows
the fluctuations in the conductance of a lipid bilayer due to the in-

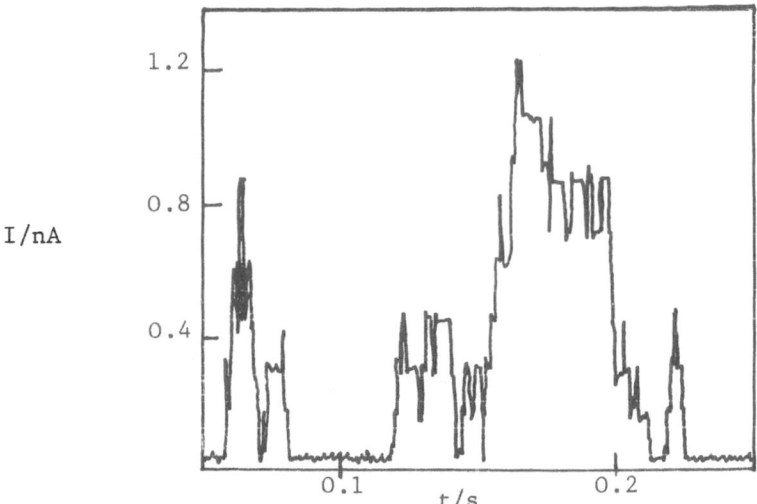

Fig. 12. Fluctuations in the current flowing through a cholesterol/ glycerol monooleate bilayer in contact with 2 x 10^{-8}M alamethicin. Voltage across the layer 0.175V. Electrolyte 0.1M KCl.

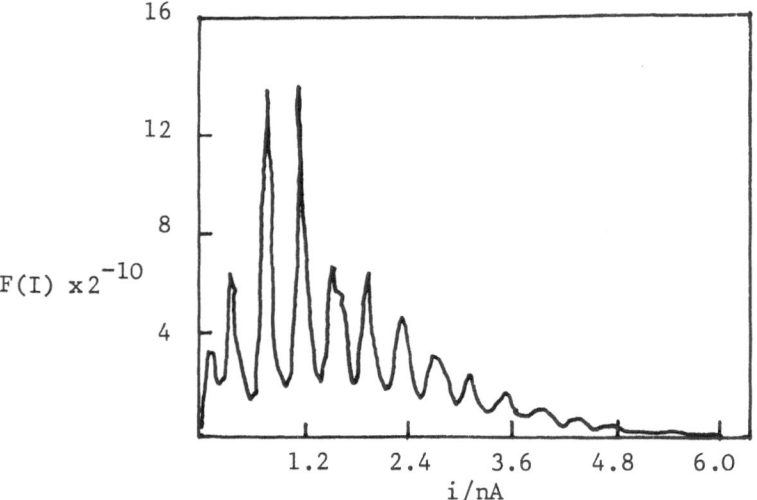

Fig. 13. Frequency distribution of the current flowing through a cholesterol/glycerol monooleate bilayer in contact with 2 x 10^{-8}M alamethicin. Voltage across membrane 0.158V. The zero current level was suppressed and the current sampled at 3μs intervals. Data were accumulated in 100 channels and displayed with inter- polation of 1000 points.

sertion (and removal) of successive molecules of alamethicin[44,45,46].
In this case the addition of each molecule leads to a measurable
stepwise increase in the current and the probability density of the
fluctuations, fig 13, shows that the heights of the major steps are
linearly related to the number of molecules inserted. The phenomenon can be modelled as the growth of a circular hole and the form
of the probability density curves can be predicted by the solution
of the appropriate master equations and choice of hole and edge
energies[46]. It appears likely that details of the molecular phenomena taking place at edges in other two-dimensional phase formation
processes will also become accessible by suitable stochastic measurements (see [25]).

The most extreme examples of the differentiation of the formation of a first layer from the deposition of the bulk metal are seen
in the underpotential and overpotential deposition of metals[47,48,49]
which have been extensively investigated recently. The difference
in the free energies of formation of the two-dimensional and bulk
phases may approach $40kJ \, mol^{-1}$. In these cases metals are deposited
on solid substrates (in contrast to the perfectly random liquid substrates in the examples in fig 2) and the surface phases owe their
existence to complex interactions with the substrate electrode so
that none of the systems are strictly two-dimensional. The major
components of this interaction which will affect the free energy of
the surface phase and also the type of phase transformations in the
layer are: the strength of the bonding between the substrate atoms
and those of the adlayer; the way in which this bonding varies with
the structural correlation between the substrate surface and the adlayer; the conflict between the development in the adlayer of energetically favourable configurations as determined by the bonding
within the layer and the development of configurations imposed by
the structure of the substrate. An understanding of the properties
of two-dimensional layers interacting with a substrate surface is
now a particularly interesting topic in surface physics[50]. Electrochemical techniques are uniquely placed to study dynamics of phase
formation and phase transformation processes in such systems and this
information complements the equilibrium and structural data being obtained by conventional methods and by LEED studies of the surface
layers formed by adsorption from the gas phase onto solid surfaces.

It has been shown[49] that a major contribution to the free energy
difference between the bulk phase and the underpotential phase is
provided by the ionic component in the bond energy between the dissimilar metal atoms arising from the difference between their elec-
tronegativities. In earlier studies, it was assumed that underpotential deposition was simply the formation of an adsorbed layer
of partially charged atoms[51,52] although the epitaxial features of
the process were recognized[47]. More recently, it has been shown
that a variety of phase formation and phase transformation processes
take place and that these are determined by the structure of the

surface of the substrate and by the relative sizes of the substrate
atoms and the depositing atoms[48,53,54,55,56,57].   A considerable
amount of data is now available for the deposition of thallium or
of lead onto single crystal surfaces of silver or of copper, and
those examples will be used to illustrate the types of behaviour
observed.

The spectrum of processes in the underpotential region is shown
clearly by linear sweep voltammetry.   Fig 14. illustrates how this
technique shows up the differences for the deposition of lead onto
three single crystal silver electrodes with exposed surfaces having
the orientations (111), (100) and (110)[54].   The various processes
are seen as current peaks the number, the positions and the shapes
of which vary with substrate orientation.   Measurement of the
electrical charge enclosed by each peak, and therefore the amount
of lead deposited in each separate step, enables deductions to be
made about the structure of the lead layer.   This leads to the conclusion that the first peak (marked A1) observed on the (100) and
(110) surfaces is associated with the formation of a full epitaxial
layer of lead atoms in which all of the favourable adsorption sites
on the silver surface are filled within the limits imposed by the
relative sizes of the lead and silver atoms.   The second peak
(marked A2) is due to the deposition of an additional amount of lead
to form a layer coulometrically equivalent to a close-packed layer
which would not fit epitaxially to the substrate.   In the case of
the (111) surface which does not possess such favourable sites for
adsorption, deposition takes place in a single step to form the
close-packed layer.   A similar pattern was observed[48] for the deposition of thallium onto silver.   In this case, however, a second
underpotential layer is formed at more negative potentials.   The
second deposition process, which occurs on top of the first close-
packed layer of thallium, is still markedly dependent upon substrate
orientation.   These results indicate that the close-packed layer is
partially distorted and moulded by the substrate structure, a process that would be most likely on the (110) surface.   This view is
strengthened by the observation that only on the (110) surface is
there any deposition of a second layer in the case of lead, fig 14c.

A change of substrate from silver to copper results in interesting variations in the deposition pattern[56,58].   The basic features
of the linear sweep voltammograms for lead deposition onto copper,
fig. 15, are similar to those observed on silver, fig. 14.   However,
there are differences in the structures of the layers as indicated
by the coulometry.   On the (111) surface the complete underpotential
layer is formed in a single step but it corresponds to the most
closely packed epitaxial layer, fig 15a and not a close-packed, non-
registered layer.   Deposition on the (100) surface is by a two-step
process leading to a close-packed epitaxial layer, fig 15b, which is
similar to the (100) structure for bulk lead.   The two step process
on the (100) surface corresponds to formation of a full epitaxial

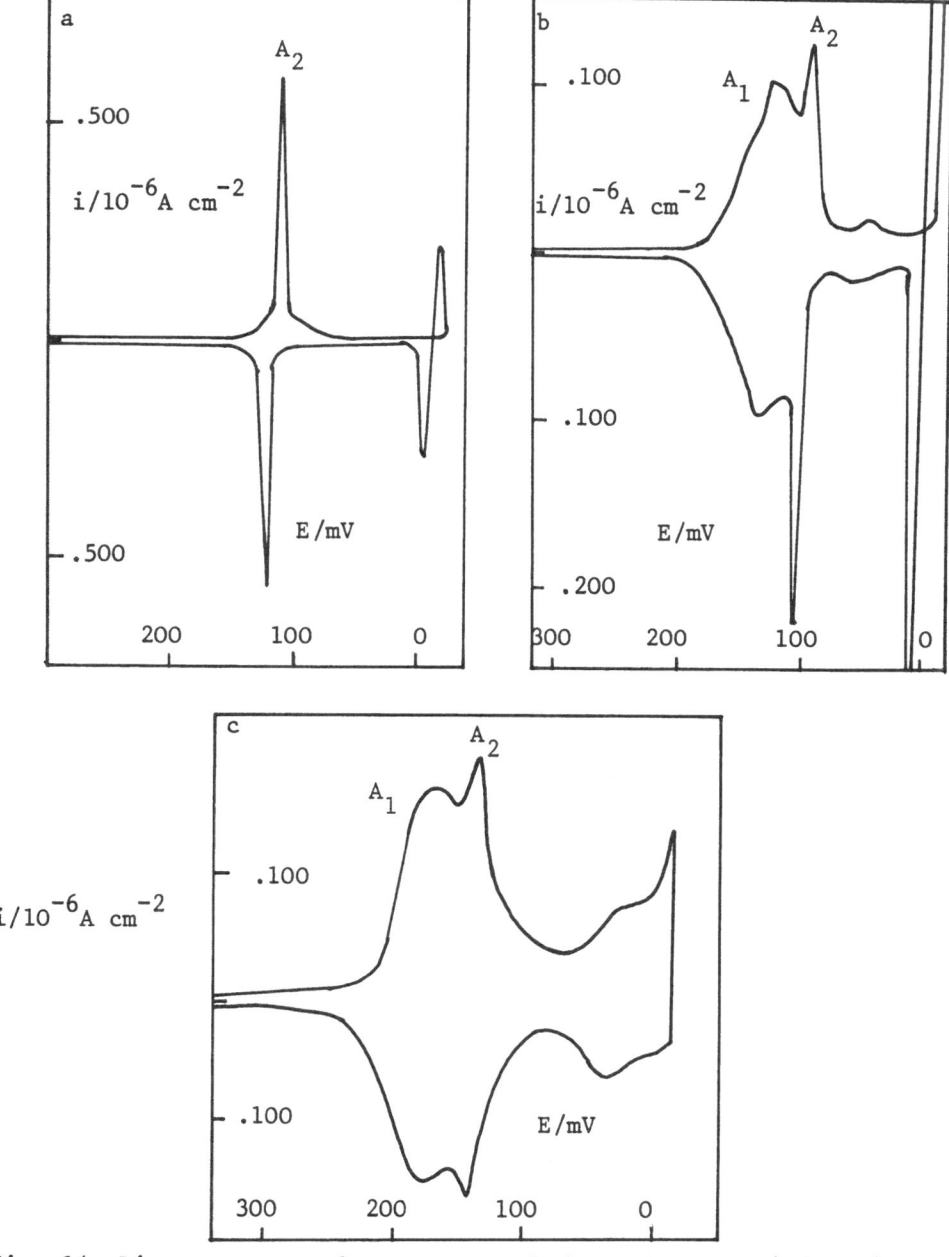

Fig. 14. Linear sweep voltammograms of the underpotential region for the deposition of lead onto silver single crystals of orientation. (a), (111);(b), (100); (c), (110). Solution 5 x 10^{-3}M lead acetate/0.5M sodium acetate/0.1M acetic acid. Sweep speed 30mV s^{-1}

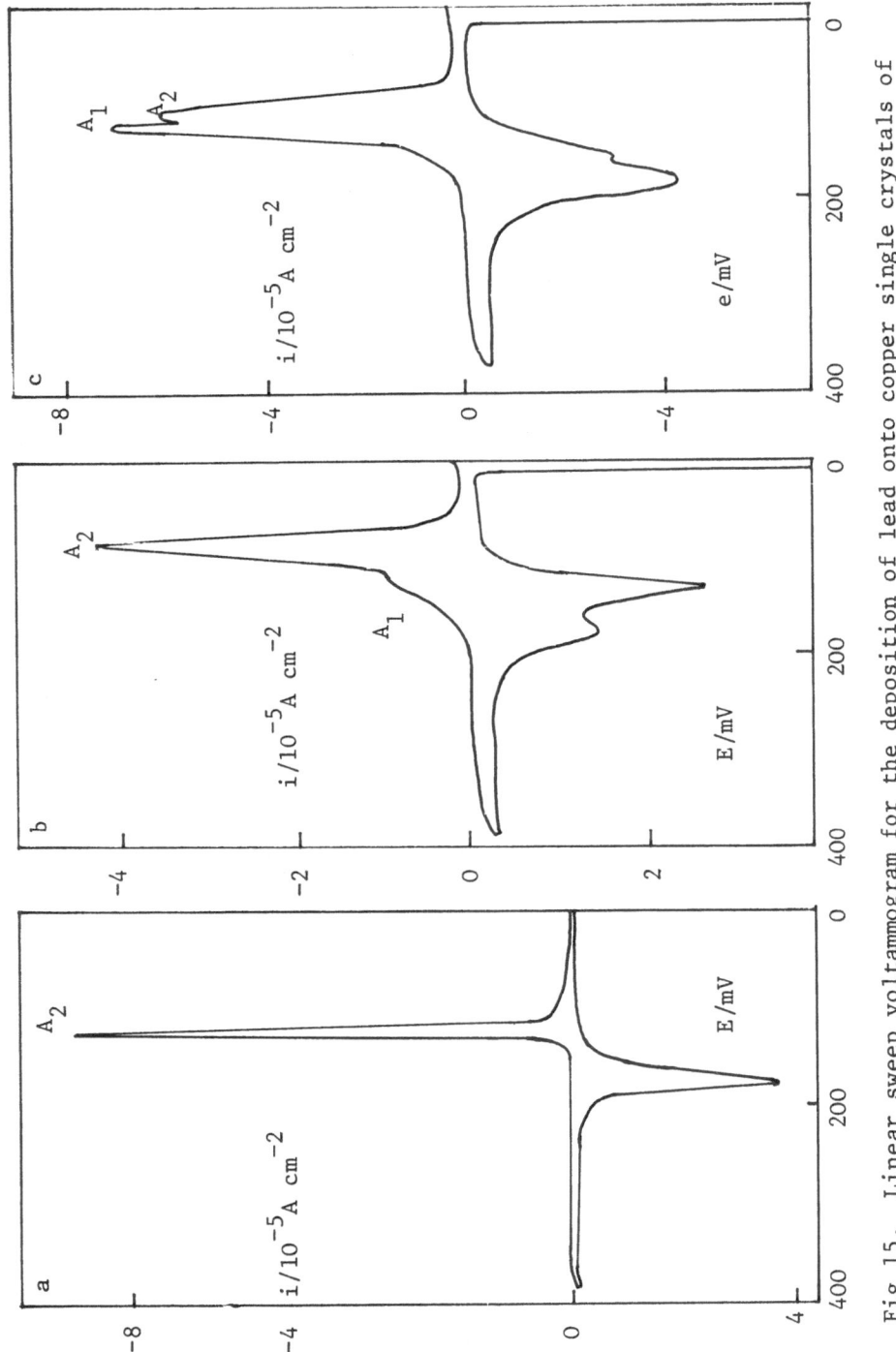

Fig 15. Linear sweep voltammogram for the deposition of lead onto copper single crystals of orientation (a), (111), (b), (100); (c), (110); electrolyte $10^{-2}M$ $Pb(OAc)_2 + 0.5M$ $NaClO_4 + 10^{-3}$ M $HClO_4$. Sweep speed $3mV$ s^{-1}.

layer, fig 15c, followed by transformation to a non-registered,
close-packed layer, i.e. a similar deposition pattern to that observed on silver. Thallium deposition onto copper follows a
similar pattern but with the additional formation of a second underpotential layer as found on silver.

The differences in the behaviour observed on silver and on
copper can be accounted for by the higher electronegativity of
copper, which will lead to stronger bonding to the lead or thallium,
and by the significantly smaller atomic radius (0.128nm compared
with 0.142nm.). As a result, the atoms in an epitaxial layer are
closer together on copper and the structure of a full epitaxial
layer is close to that of a crystal plane of the bulk metal.

The structures observed on the voltammograms for underpotential
metal deposition and the way in which these change with substrate
orientation show that there is a variety of phase transformation
processes which depend on the detailed registration between the two
dimensional layer and the substrate. Some of the current peaks can
be attributed to simple adsorption processes but others are very
sharp and they indicate phase formation[53,56]. This view is confirmed by the nature of the current/time transients obtained by
applying potential steps to form the deposit at a series of constant
potentials spanning the peak systems. As expected, some of these
transients take the form expected for an adsorption process, e.g.
those for potential steps into the peaks labelled A_1 in fig. 14 and
15, and others have the characteristic shapes for phase formation
by the nucleation and growth of two-dimensional centres e.g. those
for potential steps into the peaks labelled A_2 on the voltammograms.
Equations (11) and (12) given in section 1 describe the growth
transients for phase formation by this mechanism. Fig. 16 gives
examples of transients obtained for lead deposition on a (111) copper
surface[56,58] at potentials in the sharp voltammetric peak (fig 15(a))
and fig. 17 is a test of the fit to equation 12. In all cases for
deposition of lead or of thallium onto silver or copper, the initial
deposition process appears to be adsorption and it is seen as a peak
(A_1) on the voltammograms (on the (111) surfaces the adsorption peak
is not always developed and the adsorption process is seen as a
broad foot to the very sharp A_2 peak). Further deposition at
slightly more negative potentials occurs as a nucleative phase formation process and this produces the sharp A_2 peaks on the voltammograms. These phase formation processes are of particular interest.
Only on the (111) surfaces are the A_2 peaks sufficiently narrow to
correspond to simple first order phase transformations[53,56], after
making some allowance for peak broadening due to slow kinetics and
slight surface heterogeneity. On the (100) and (110) surfaces the
A_2 peaks indicate that layer formation proceeds by a nucleative
mechanism. It has been suggested, therefore, that these represent
continuous, higher-order phase transformation processes[53,55,56] instead of the abrupt, first-order changes observed on the (111) sur-

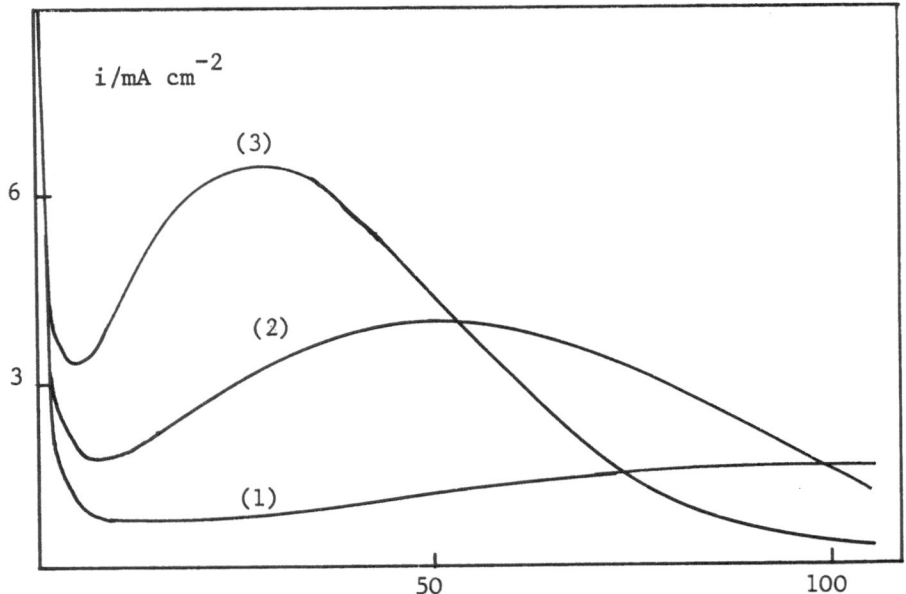

Fig. 16. Current/time transients at Cu (111) electrode in 10^{-2}M
Pb(OAc) + 0.5M NaClO$_4$ + 10^{-3}M HClO$_4$. Potential step from
+390mV to: (1) + 130 mV, (2) + 110 mV, (3) +100 mV.

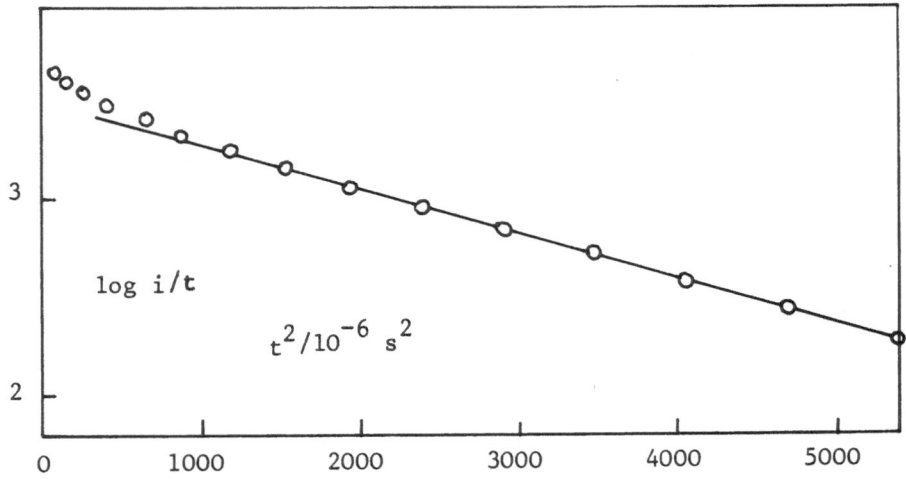

Fig. 17. Plot of log i/t versus t^2 for transient (3) in fig. 16.

face.   Other workers[59] have disputed that phase formation takes
place.   They have attributed the sharp peaks to adsorption processes following a Frumkin adsorption isotherm with a large value
for the attractive interaction parameter.   It is clear, however,
that an adsorption model cannot account for growth transients fitting to the shapes predicted by the nucleative mechanisms represent-
ted by equations (11) and (12).

Related studies of adsorption onto well defined solid surfaces
from the gas phase have also revealed a variety of interesting
phase formation processes,[50] and the special nature of phase transformations in two-dimensional systems has been pointed out.   It has
been suggested[50] that registered films which have long range order
imposed upon them by the substrate appear to show abrupt transitions
whereas non-registered films show continuous transitions.   The data
for underpotential metal deposition does not fit completely into
this classification;   first-order transitions are observed in the
formation both of non-registered and registered layers and higher-
order transitions also appear to be associated with registered
layers in some cases and some non-registered layers in others.   There
is some correlation with the structure of the substrate surface;
first-order transitions are always found for the (111) surface on
which the adsorption energy is relatively low whereas the higher
order transitions occur on the (100) and (110) surfaces which have
particularly high adsorption energies.   When the structure of the
final layer is approximately close-packed, the correlation distance
is likely to be larger for the layer on the (111) surface because
there will be less distortion due to the moulding effects of strong
interaction with the substrate.   However, these interesting questions concerning the detailed structure will only be answered when
proper structural investigations of such layers become possible.   On
the other hand, it is clear that certain aspects of phase-transitions
in two-dimensional systems can be conveniently probed using these
electrochemical systems and, in particular, the detailed investigation of the dynamics of higher-order transitions should be specially
rewarding.

4. ELECTROCRYSTALLISATION AND CATALYSIS

In view of the range of topics discussed at this meeting we
conclude this paper by referring to a number of electrochemical
studies of catalysis which are based on the concepts outlined in the
previous section and, in particular, on the *in situ* deposition of
the catalyst.

The evolution of hydrogen on mercury is very slow while it is
very fast on platinum metals.   The deposition of a monolayer of
ruthenium on a mercury surface[60] in acid solutions is therefore accompanied by a catalysed hydrogen evolution reaction.   It is found
that this is a transient phenomenon, fig. 18, so that hydrogen is

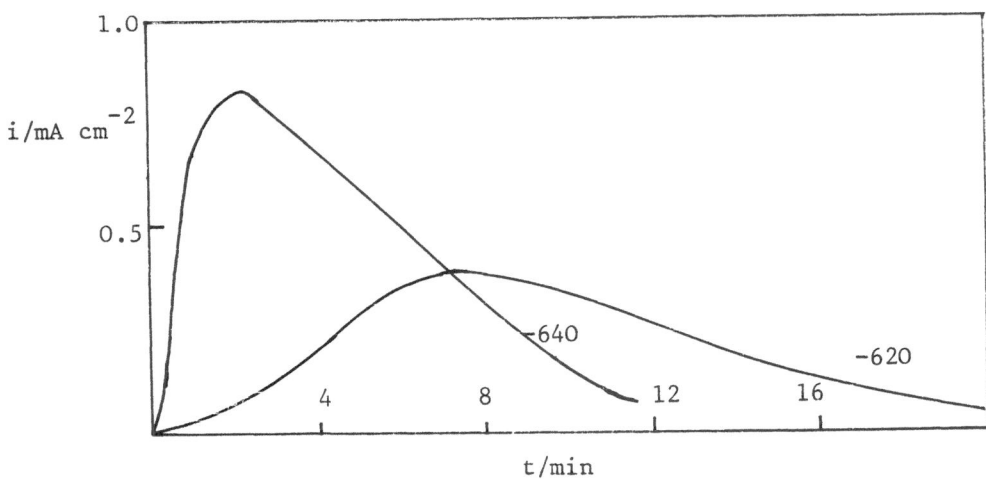

Fig. 18. Current-time transients for the electrodeposition on
mercury of ruthenium and catalytic hydrogen evolution in a solution $10^{-3}M$ RuCl $+10^{-2}M$ HCl$+1M$ KCl. Potentials are given in mV
with respect to the saturated calomel electrode.

only evolved at the edges of two-dimensional growth centres, the
flat top surfaces being inactive. At low negative potentials the
process follows the equation

$$i = \frac{2\pi F M h N_o k_1 (3k_1 + 2k_2) t}{\rho} \exp - \frac{\pi N_o M^2 k_1^2 t^2}{\rho^2} \qquad (33)$$

where k_1 and k_2 are the molar rates per unit area of the ruthenium
deposition and hydrogen evolution reactions. The rate determining
step for the latter is the recombination of two hydrogen atoms at
the edges of the growth centres. On the other hand, the hydrogen
evolution reaction on small three-dimensional crystallites electro-
deposited on vitreous carbon is very slow[61]. Hydrogen evolution
is only observed when these centres have reached a sufficient size
and the rate determining step is surface diffusion to sites where
the adsorbed species can combine. With further increase of the size
of the catalyst centre the rate determining step becomes the discharge of the hydrogen ions, presumably in view of the increase in
the number of sites where combination of atoms can take place. It
is evident that in the case of the growth of the two-dimensional
centres, the diffusion of atoms to the edges is sufficiently fast so
that the combination of two atoms becomes rate determining. Fig.19
shows data for the electrochemical oxidation of a monolayer of chemisorbed carbon monoxide on platinum[62]. In this case the oxidation
takes place at the edges of two dimensional growth centres of platinum oxide which grow in the layer of chemisorbed carbon monoxide,

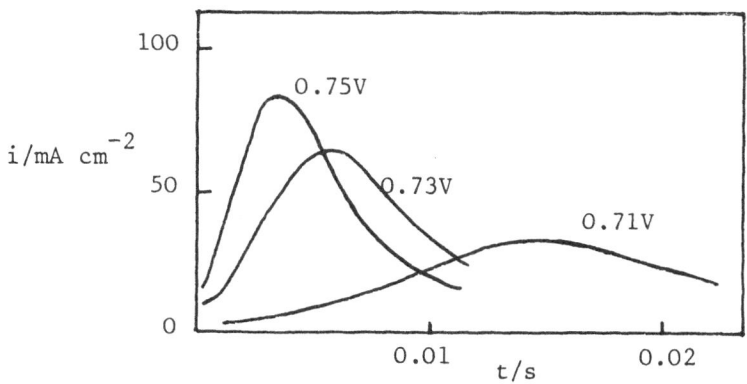

Fig. 19. Current-time transients in response to single potential steps. Pt wire working electrode. Solution phase 1.0M HClO$_4$ saturated with CO (potential stepped from +0.73V and +0.75V versus SCE.

a mechanism first suggested by Gilman[63]. The transients follow the equation

$$i = \frac{2\pi FMhN_o k_1 (zk_1 + 2k_2)}{\rho} \{t - \frac{1}{A}(1 - \exp - At)\}$$

$$x \exp \frac{-\pi N_o M^2 k_1^2}{\rho} [t^2 - \frac{2t}{A} + \frac{2}{A^2}(1 - \exp - At)] \qquad (34)$$

where k_1 and k_2 are the molar rates per unit surface of oxide formation and of oxidation and z is the number of Faradays per mole of oxide formed. The rate determining step has been shown to be the transfer of a single electron[62].

GLOSSARY OF SYMBOLS

a	area of an electrode (cm^2)
A	nucleation rate constant (nuclei cm^{-2}s^{-1})
A'	nucleation rate constant (s^{-1})
c	concentration (moles cm^{-3})
C	a constant
E	electrode potential measured with respect to an arbitrarily chosen reference electrode (V).
F	Faradays constant (96,550 coulombs equiv^{-1})
f	frequency (Hz)

$g_{(i)}$ probability of observing a current i

G power spectral density

h height of a two-dimensional growth centre (cm)

i current density (amps cm^{-2})

I current due to an assembly of growth centres (amps)

j $(-1)^{\frac{1}{2}}$

k rate constant of crystal growth (moles $cm^{-2}s^{-1}$)

k_0 rate constant of crystal growth at the potential E = 0 (units depend on order of reaction)

k' rate constant of crystal growth perpendicular to a substrate (moles $cm^{-2}s^{-1}$).

k_1, k_2, k_3 rate constants of heterogeneous reactions

K frequency factor (nuclei $cm^{-2}s^{-1}$)

M molecular weight (g $mole^{-1}$)

n concentration (number of species cm^{-3})

N number of nuclei (nuclei cm^{-2})

N_0 number of nucleation sites (nuclei cm^{-2})

$P_{(x)}$ probability of observing state x

R gas constant (Joules $mole^{-1}$ $degree^{-1}$)

S surface area of a growing electrode (cm^2)

S_{ext} surface area per unit area of substrate neglecting the effects of overlap

t time (s)

T temperature (degrees)

u a time (s)

x number of vacant sites (number cm^{-2})

x_0 maximum number of sites (number cm^{-2})

y a time (s)

z a time (s)

α fraction of electrical energy driving a cathodic reaction ($0<\alpha<1$)

κ a constant (cm^6 $ergs^{-3}$ V^2)

λ nucleation rate constant (nuclei s^{-1})

ρ density (g cm^{-3})

σ surface energy (ergs cm^{-2}) or standard deviation

η overpotential (V)

τ a time (s)

ν_i order of reaction with respect to species i

ϕ autocovariance function

ω angular frequency (radians s^{-1})

REFERENCES

1)   M. Volmer, Kinetik der Phasenbildung Steinkopff, Dresden und Leipzig, 1939.

2)   M. Fleischmann and H.R. Thirsk, Advances in Electrochemistry and Electrochemical Engineering Vol III ed. P. Delahay, Interscience Publishers Inc., New York 1963.

3)   M. Fleischmann and M. Liler, Trans Faraday Soc., 54 1370, 1958.

4)   G.W.D. Briggs, M. Fleischmann and H.R. Thirsk, Fourth In-
     ternational Symposium on Batteries, Pergamon Press, Oxford,
     1964.

5)   M. Avrami, J. Chem. Phys., 7, 1103, 1939; 8, 212, 1940; 9
     177, 1941.

6)   A. Bewick, M. Fleischmann and H.R. Thirsk, Trans. Faraday
     Soc., 58, 2200, 1962.

7)   R.D. Armstrong, M. Fleischmann and H.R. Thirsk, J. Electro-
     anal. Chem. 11, 208, 1966.

8)   M. Fleischmann and H.R. Thirsk, Electrochimica Acta, 9, 757,
     1964.

9)   E. Budevski, W. Bostanov, T. Vitanov, Z. Stoinov, A. Kotzewa
     and R. Kaishev, Phys. Stat. Sol. 13, 577, 1966. Electrochim.
     Acta 11, 1967; T. Vitanov, A. Popov and E. Budevski, J.
     Electrochem. Soc., 121, 207, 1974.

10) V. Bostanov, R. Roussinova and E. Budevski, Chem. Ing. Techn.
     45, 179, 1973.

11) V. Bostanov, G. Staikov and D.K.Roe, J. Electrochem. Soc.,
     122, 1301, 1975.

12) M. Abyaneh and M. Fleischmann to be published.

13) M. Fleischmann and H.R. Thirsk, Electrochimica Acta, 1, 146,
     1959.

14) S. Toshev, A. Milchev, K. Popova and I. Markov, L.R. Acad.
     Bulg. Sci., 22, 1413 (1969).

15) A. Milchev, S. Stoyanov and R. Kaishev, Thin Solid Films,
     22, 255, (1974).

16) A. Milchev and S. Stoyanov, J. Electroanal. Chem. 72, 33
     (1976).

17) D. Walton, J. Chem. Phys. 37, 2182 (1962).

18) I. Dugdale, M. Fleischmann and W.F.K. Wynne-Jones, Electro-
     chimica Acta, 5, 229, 1961.

19) M. Fleischmann, D.J. Lax and H.R. Thirsk, Trans. Faraday Soc.,
     64, 3137, 1968.

20) W.K. Burton, N. Cabrera and F.C. Frank, Phil. Trans. A243,
     299, 1951.

21) W. Lorenz, Z. Phys. Chem. 17, 136, 1959; Z. Naturf., 94, 716,
     1954.

22) M. Fleischmann and H.R. Thirsk, Electrochemica Acta, 2, 22,
     1960.

23) W. Allgaier and K.E. Heusler, Z. Phys. Chem. N.F. 98, 161,
     1975.

24) M. Fleischmann, J.A. Harrison and H.R. Thirsk, Trans. Faraday
     Soc., 61, 2742, 1965.

25) R. Kaishev and E. Budevski, Contemp. Phys. 8, 489, 1967.

26) T. Vitanov, A. Popov and E. Budevski, J. Electrochem. Soc.,
     121, 207 1974.

27) H. Gersicher, Z. Elektrochem., 62, 256, 1958.

28) J. O'M Bockris and W. Mehl, Canad. J. Sci., 37, 190, 1959.

29) H. Gerischer and R.P. Tischer, Z. Elektrochem. 61, 1159, 1957.

30) M. Fleischmann and J.A. Harrison, Electrochimica Acta, 11,
     749, 1966.

31)   A. Bewick et al to be published.
32)   S.K. Rangarajan, Faraday Symposium of the Chemical Society
       No.12. 1977, to be published.
33)   P. Bindra, M. Fleischmann, J.W. Oldfield and D. Singleton,
       Faraday Discussions of the Chemical Society, 56, 180, 1973.
34)   M. Fleischmann, J.W. Oldfield and D. Singleton, to be
       published.
35)   S.K. Rangarajan, J. Electroanal. Chem., 46, 119, 1973.
36)   M. Fleischmann, K.S. Rajagopalan and H.R. Thirsk, Trans.
       Faraday Soc., 59, 741, 1963.
37)   G.C. Allen, P.M. Tucker, A. Capon and R. Parsons, J. Electro-
       anal. Chem., 50, 335, 1974.
38)   T. Dickinson, A.F. Povey and P.M.A. Sherwood, J.C.S. Faraday
       I, 71, 298, 1975; J.S. Hammond and N. Winograd, J. Electro-
       anal. Chem, in press.
39)   J.S. Hammond and N. Winograd, J. Electroanal. Chem. in press.
40)   M. Fleischmann, P.J. Hendra and A.J. McQuillan, J. Chem. Soc.
       Chem. Comm. 80, 1973.
41)   R. Cooney, M. Fleischmann and P.J. Hendra, to be published.
42)   L. Gierst.   Paper presented at Spring Informal Meeting,
       Society for Electrochemistry, Birmingham, 1973.
43)   R.D. Armstrong, J. Electroanal. Chem., 20, 168, 1969.
44)   P. Mueller and D.O. Rudin, Nature, 217, 713, 1968.
45)   L.G.M. Gordon and D.A. Haydon, Nature, 225, 451, 1970.
46)   M. Fleischmann, M. Labram and A. McMullen, to be published.
47)   F. Hilbert, C. Meyer and W.J. Lorenz, J. Electroanal. Chem.,
       47, 167, 1973.
48)   A. Bewick and B. Thomas, J. Electroanal. Chem., 65, 911, 1975.
49)   D.M. Kolb, M. Przasnski and H. Gerischer, J. Electroanal.
       Chem., 54, 25, 1974.
50)   See for example, J.G. Dash, Films on Solid Surfaces, Academic
       Press, New York, 1975.
51)   E. Schmidt and H. Gygax, J. Electroanal. Chem., 12, 300, 1966.
52)   E. Schmidt, M. Christen and P. Beyeler, J. Electroanal. Chem.,
       42, 275, 1973.
53)   A. Bewick and B. Thomas, J. Electroanal. Chem., 70, 239, 1976
54)   A. Bewick and B. Thomas, J. Electroanal. Chem., in press.
55)   A. Bewick and B. Thomas, J. Electroanal. Chem., 84, 127, 1977.
56)   A. Bewick, J. Jovićević and B. Thomas, Faraday Symp. Chem.
57)   No.12, in press.
57)   D. Dickertmann, F.D. Koppitz and J.W. Schultze, Electrochim.
       Acta, 21, 967, 1976.
58)   A. Bewick and J. Jovićević, in preparation.
59)   H.D. Herrmann, N. Wuthrich, W.J. Lorenz and E. Schmidt, J.
       Electroanal. Chem., 68, 273 and 289, 1976. W.J. Lorenz,
       E. Schmidt, G. Staikov and H. Bart, Faraday Symp. Chem.
       Soc., No.12, in press.

MONOLAYER ASSEMBLIES

Dietmar Möbius

Max-Planck-Institut für biophysikalische Chemie
Abteilung molekularer Systemaufbau
D 34 Göttingen-Nikolausberg, Germany

ABSTRACT

Techniques of organizing molecules at interfaces
and of assembling and manipulating monolayers are described. Complex systems of cooperating molecules have
been constructed and it is shown that the planned functions of the artificial units depend strongly on the
spatial and energetic coordination of the constituent
molecules.

INTRODUCTION

The construction of functional units of molecular
dimensions is a fascinating task. In biological systems
extremely complex arrays of cooperating molecules exist,
e. g. in the photosynthetic membrane, where the conversion
of light energy in chemically stored energy is achieved
by a large team of molecules with a peculiar energetic
and spatial coordination. Artificial functional units
can be constructed by organizing appropriately designed
molecules by the action of intermolecular forces. The
properties of the obtained systems are determined by the
specific cooperation of the interlocked molecules.

Widely used structural subunits in artificial functional systems are monomolecular layers, which can easily be formed at interfaces. These monolayers can consist
of only one kind of molecules or be composed of different

molecules which may organize themselves into molecular
functional units. By combining different monolayers in a
stepwise procedure, monolayer assemblies of planned structure and properties can be built.

PREPARATION OF MOLECULAR FUNCTIONAL UNITS

Three methods are available to produce designed monolayer assemblies. The self-assembly technique using
appropriately designed molecules which organize themselves
at an interface; the coating technique in which monolayers
formed at the air-water interface or liquid-liquid interface are transferred to the solid substrate; the manipulating technique where prefabricated assemblies are con-
tacted to obtain complex systems.

The Self-Assembly Technique

Molecules with a hydrophobic part and a polar group
like long chain fatty acids can be adsorbed from a solution in hexadecane to a glass or metal surface in contact
with the solution (see Fig. 1). In the case of fatty acids
the surface is hydrophobic as well as oleophobic [1].

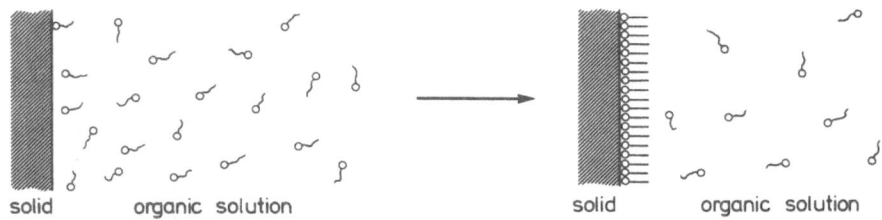

Fig. 1: Self-assembling of a monolayer at a solid surface
        from an organic solution. The circles represent
        hydrophilic groups or functional groups that can
        react chemically with functional groups on the
        solid surface; the coiled and stretched lines
        represent the long chain hydrocarbon parts of
        the monolayer forming molecules.

This means that a drop of clean water runs off the coated glass surface, when the surface is somewhat tilted with respect to the horizontal position. The same phenomenon is observed with a drop of pure hexadecane instead of water. Bigelow, Pickett and Zisman [1] interpreted these results by assuming that the molecules assemble themselves to a monolayer with the hydrophilic groups in contact with the glass surface and the long hydrocarbon portion oriented vertically to the surface.

Evidence for monolayer formation at the interface was obtained from a measurement of the area which can be coated on repeated dips of a platinum sheet (with cleaning of the sheet between the dips) with a well-determined amount of n-octadecylamine dissolved in dicyclohexyl. The area of the amine on the platinum was 30 $Å^2$/ molecule [1], which is larger than the cross-sectional area of the molecule (20 $Å^2$). Bartell and Ruch found that an appreciable amount of the solvent hexadecane may adhere to the adsorbed monolayers of octadecylamine when pulling the platinum plate out of the solution [2]. The layers according to ellipsometric measurements were much thicker than a monolayer but reached this monolayer thickness after an hour by solvent evaporation.

Monolayers have been formed at the solid-liquid or solid-vapor interface with many long chain hydrocarbon compounds containing a hydrophilic group like aliphatic alcohols, acids and amines of different chain length [3].

Recently Sagiv has modified this technique by using a surface chemical reaction to fix the adsorbed molecules on substrates like glass, aluminum with a thin surface layer of aluminum oxide and films of polymers, e. g. polyvinylalcohol [4]. The chemically fixed monolayers remained on the solid surface on rinsing with water, whereas monolayers of long chain fatty acid like arachidic or stearic acid produced in this way on glass floated off.

New convincing evidence for the formation of densely packed monolayers on adsorption or surface chemical reaction was obtained from studies of the electrical properties of self-assembled fatty acid monolayers [5].

For the measurement of the electrical properties the self-assembled monolayers were sandwiched between metal electrodes. Monolayers of various fatty acids and of perfluorated fatty acids were produced on the base electrode (aluminum or nickel evaporated on glass) by

dipping the plate with the electrode in a solution of
the fatty acid in hexadecane. The plate is pulled out
and transferred into a high vacuum chamber where the
top electrode is deposited. Such sandwich structures
show the capacitance values anticipated for one insulating monolayer of the various fatty acids used. More
sensitive to defects like holes or cracks due to incomplete monolayer formation than the a.c. capacitance
measurements are the results of d.c. conductance measurements. The monolayers produced by the self-assembly
technique exhibit electron tunneling. The observed d.c.
conductance is smaller than found in the case of fatty
acid monolayers produced at the air-water interface and
transferred to the base electrode in the usual monolayer
technique [6].

Further, Sagiv has assembled dye monolayers and
mixed monolayers of cyanine dyes with long chain substituents and fatty acids at the solid-liquid interface
using appropriate solutions of the components in n-hexa-
decane and glass or polymer films as solids [4]. From
the measurement of absorption and fluorescence spectra
and from studies of energy transfer between separated
donor and acceptor layers it was concluded that densely
packed monolayers can be formed in this very convenient
way.

At the hexadecane-water interface appropriate molecules like long chain fatty acids or cyanine dyes provided with long chain substituents assemble to densely
packed monolayers which can be transferred to a hydrophilic solid surface by traversal of the solid from the
aqueous phase through the hexadecane-water interface [4].

Many substances like cationic dyes with long chain
substituents can be organized to insoluble monolayers
at the air-water interface by dropping a solution of the
material in a volatile solvent on the water surface [7].
The solvent, generally benzene, chloroform or methylen-
chloride, evaporates, and the solute molecules remaining
at the surface are packed by applying an appropriate surface pressure. This surface pressure is taken from the
measurement of surface pressure - area isotherms. The
widely used fatty acids, spread on water with 3×10^{-4} M
cadmium chloride at pH = 6.6 exhibit the behaviour of a
solid densely packed monolayer with a cross sectional
area of about 20 $\overset{o}{A}{}^2$ per molecule in the case of arachidic
acid (see Fig. 2, curve 1 for 10 arachidate molecules).
For assembling complex systems we have prepared mixed

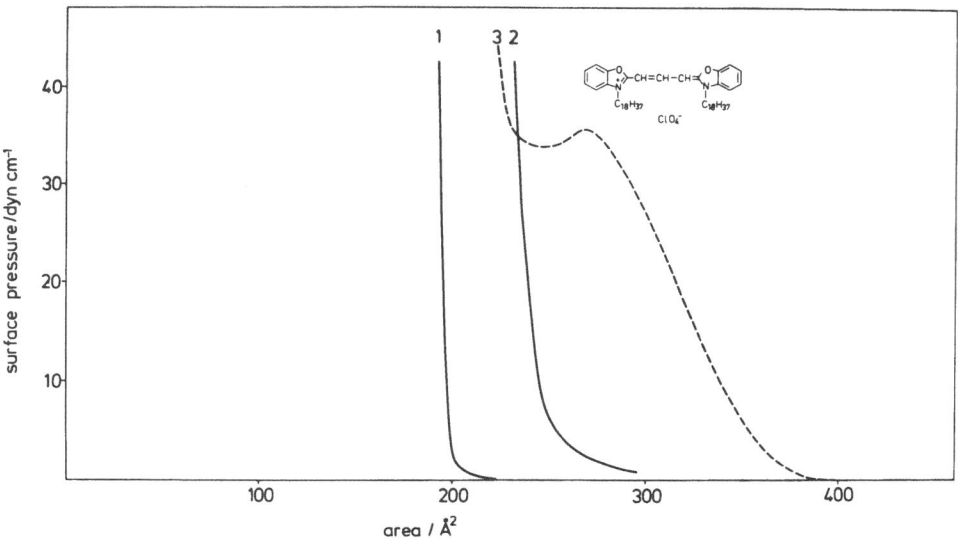

Fig. 2: Surface pressure - area isotherms of an arachidic
        acid monolayer (curve 1, area corresponding to
        10 molecules of $C_{19}H_{39}COOH$) and a mixed mono-
        layer of arachidic acid and the oxacarbocyanine
        dye in the molar ratio acid: dye =10:1 (curve 2).
        Curve 3 represents the isotherm calculated as a
         superposition of curve 1 and the isotherm of
        the pure oxacarbocyanine monolayer. The non-addi-
        tive behaviour of the molecular area indicates
        molecular mixing. Substrate: Bidistilled water
        with 3 x 10^{-4} M $CdCl_2$ and 5 x 10^{-5} M $NaHCO_3$, 22° C.

monolayers of cyanine dyes and arachidic acid [8]. As a
typical example the isotherm of a mixed monolayer of arachidic acid and N,N'-dioctadecyl-oxacarbocyanine perchlo-
rate, molar ratio acid : dye = 10:1, is shown in Fig. 2,
curve 2. From this we obtain about 40 $Å^2$ for the area of
the dye molecule in this mixture at surface pressures between 10 and 40 dyn/cm . This corresponds to the cross
sectional area of the hydrocarbon substituents oriented
perpendicular to the air-water interface. By addition of
the area of one dye molecule (taken from the isotherm of
the pure monolayer) to the area corresponding to 10 arachidate molecules the calculated isotherm is obtained
(Fig. 2, curve 3). The difference between the constructed

and the measured isotherms reveals the organization of the
dye molecules in the fatty acid matrix, such that no contribution of the chromophore to the total area is observed.
In the mixed monolayer the dye molecules do not form a
monolayer separate from the fatty acid. The formation of
a separate dye phase in mixed monolayers of dye and arachidate has been observed in the case of a merocyanine
dye, and in this case the isotherm of the mixed monolayer
is identical with the isotherm calculated as a superposition of the fractional areas of fatty acid and dye
from their separately measured isotherms [9].

Not all cyanine dyes with long chain substituents
behave as demonstrated in Fig. 2. Monolayers of arachidic acid and N,N'-dioctadecyl-indocarbocyanine of molar
ratio acid : dye = 10:1, have an isotherm (see Fig. 3,
curve 2) which differs considerably from the isotherm
shown in Fig. 2, curve 2. The relatively large areas
measured at low and medium surface pressures indicate
a contribution of the dye chromophores to the total area.
By addition of the area of one dye molecule in the pure
dye monolayer to the area corresponding to 10 arachidate
molecules (Fig. 3, curve 1) the isotherm represented by
curve 3 in Fig. 3 is obtained. It is evident that the
measured isotherm closely resembles the constructed isotherm and therefore the dye molecules are not homogene-
ously distributed but form patches of separate monolayer.

The reason for the separation of the monolayer into two phases could be the different solubilities of dye
and fatty acid, respectively, in the solvent chloroform.
After spreading, fatty acid and dye precipitate at different rates, and the resulting monolayer is inhomogene-
ous and remains so since the mobilities of the molecules
in the packed monolayer are too small for effective mixing. A sort of molecular lubricant would be appreciated
which is slowly squeezed out under the applied surface
pressure and evaporates but keeps the molecules of the
mixed monolayer mobile when the spreading solvent has
already disappeared. Hexadecane is an appropriate liquid
for this purpose. If this is added to the binary mixture
of dye and fatty acid and spread along with these non-
volatile components the film area at any surface pressure
is initially larger than without hexadecane. The area
decreases with time at constant pressure and reaches a
constant value after some 10 to 20 minutes after applying the surface pressure. We have taken these constant
values of area for a plot of the surface pressure vs.
area in Fig. 3, curve 4 for a mixed monolayer of arachi-

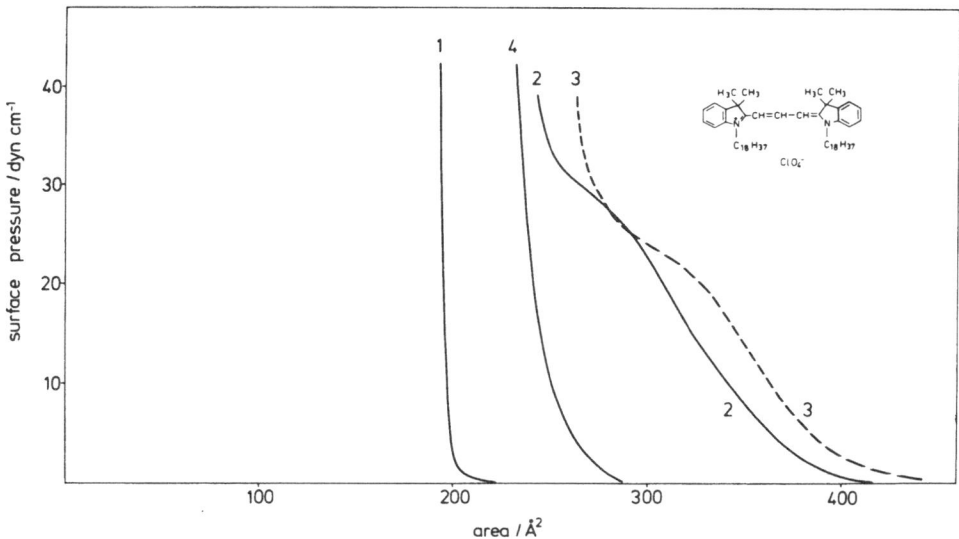

Fig. 3: Surface pressure - area isotherms of an arachidic acid monolayer (curve 1, area corresponding
to 10 molecules of $C_{19} H_{39}$ COOH) and a monolayer
of arachidic acid and the indocarbocyanine dye
in the molar ratio acid : dye = 10:1 (curve 2).
Curve 3 represents the isotherm calculated as
superposition of curve 1 and the isotherm of the
pure indocarbocyanine monolayer. The close re-
semblance of curves 2 and 3 reveals the presence
of separated pure indocarbocyanine phase in the
monolayer (mosaic structure). Molecular mixing
is achieved by addition of hexadecane to the
spreading solution. After application of a surface pressure the hexadecane is squeezed out of
the monolayer and the isotherm 4 is obtained.
Substrate: Bidistilled water with 3×10^{-4} M $CdCl_2$
and 5×10^{-5} M $NaHCO_3$, 22° C.

dic acid, hexadecane and the N,N'-dioctadecylindocarbo-
cyanine in the molar ratio acid : hydrocarbon : dye =
10:10:1. It is clearly seen from this plot that the dye
chromophores no longer contribute to the area of the
mixed monolayer. The isotherm has a similar shape as in
the case of the oxacarbocyanine dye (Fig. 2, curve 2),
and the additional area compared with that of 10 arachidate molecules is also about 40 $Å^2$. These observations

demonstrate the important role of kinetics in the formation of monolayers at the air-water interface and show
how the arising difficulties can be overcome [10].

Different molecules within a multicomponent monolayer can associate and interlock if they meet the spatial and energetic requirements. The resulting unit may
have new interesting properties. Molecular models of
the N,N'-dioctadecylthiacarbocyanine and the azo dye
4'-dimethylamino-4-nitro-3-carboxy-azobenzene attached
to 12-hydroxy-stearic acid via ester formation can be
interlocked as shown schematically in Fig. 4. Thus, it
should be possible to obtain a functional unit of the
cyanine dye and the azo dye with the azo chromophore
perpendicular to the cyanine chromophore when spreading
the solution of both dyes in a surplus of fatty acid on
water [11]. The specific arrangement should cause a non-
additive behaviour of the components in the surface pres-
sure-area isotherm.

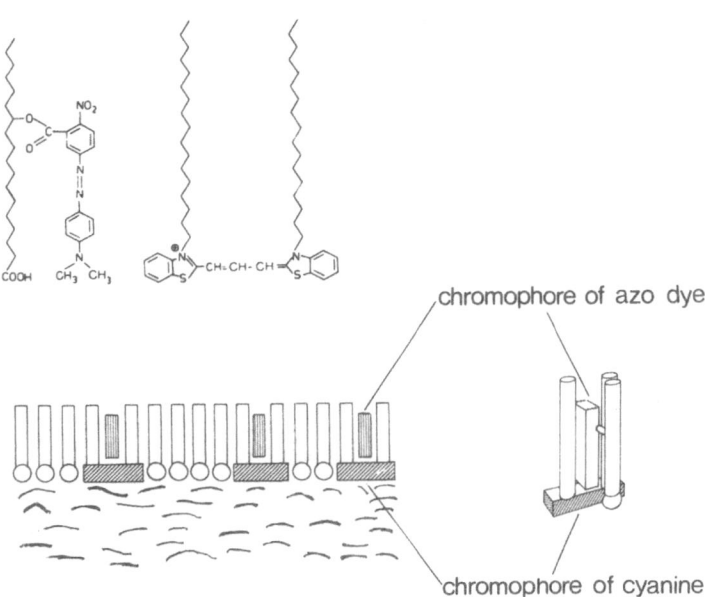

Fig. 4: Schematic representation of a functional unit
        formed by interlocking a molecule of the thiacarbo-
        cyanine dye and the azo dye with the depicted for-
        mulas. The organization of the components to the
        units is achieved in a mixed monolayer with ara-
        chidic acid at the air-water interface.

The isotherm of the pure arachidic acid is shown
again in Fig. 5, curve 1 (area of 10 arachidic acid molecules). When the azo dye is mixed to arachidic acid,
molar ratio acid : dye = 10:1 (Fig. 5, curve 2) the steep
increase in surface pressure on decrease of the area is
observed at a surplus area of 40 $\overset{o}{A}{}^2$ with respect to the
fatty acid isotherm. This increase of 40 $\overset{o}{A}{}^2$ corresponds
to the sum of the cross sectional areas of the azo chromophore and the hydrocarbon chain. On further decrease
of the area a shoulder appears at a surface pressure

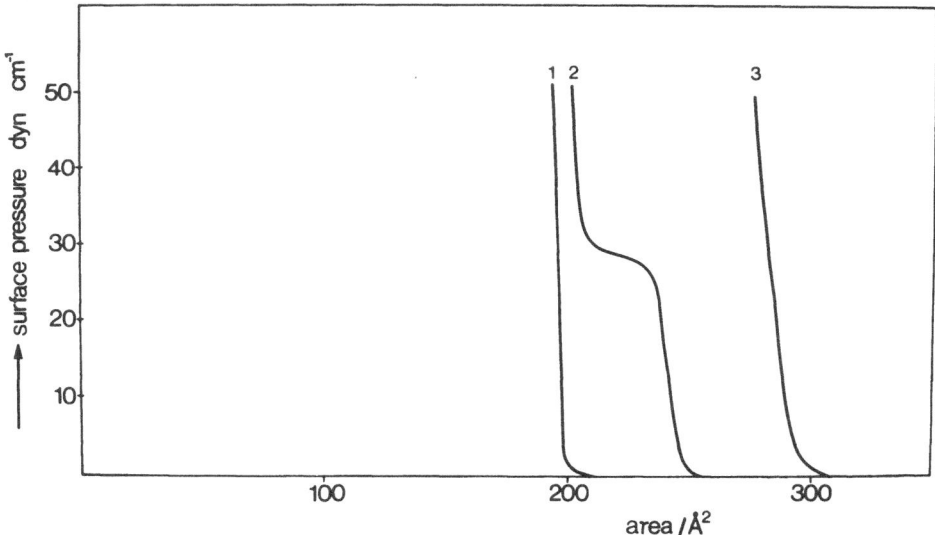

Fig. 5: Surface pressure – area isotherms of an arachidic
         acid monolayer (curve 1, area corresponding to
         10 molecules of arachidic acid) a mixed mono-
         layer of arachidic acid and the azo dye, molar
         ratio acid : azo dye = 10:1 (curve 2) and a mixed
         monolayer of arachidic acid, the azo dye and the
         cyanine dye, molar ratio acid : azo cyanine =
         10:1:1 (curve 3). Formulas of the dyes are given
         in Fig. 4. The inflection of curve 2 between 25
         and 30 dyn/cm is no longer observed in curve 3.
         This indicates a change in the monolayer proper-
         ties due to the interaction of the azo dye and
         cyanine dye molecules in the 3-component mono-
         layer. Substrate: Bidistilled water with 3×10^{-4} M
         $CdCl_2$ and 5×10^{-5} M $NaHCO_3$.

between 25 and 30 dyn/cm and at about 30 dyn/cm another steep increase in surface pressure with decreasing
area is observed. This part of the isotherm corresponds
to the isotherm of the arachidic acid. Therefore, the
azo dye molecules are squeezed out of the mixed monolayer
at a surface pressure of about 30 dyn/cm. After addition of the cyanine dye (Fig. 5, curve 3 for acid : azo
dye : cyanine dye = 10:1:1) the steep increase of the
surface pressure with decreasing area is observed at a
surplus area of another 40 $\mathring{A}^2$. This is to be expected
since the surplus area corresponds exactly to the cross
sectional area of the two hydrocarbon substituents of
the cyanine dye packed in the film. As we further decrease the surface area the steep increase in surface
pressure continues without an indication of a shoulder.
This means that the azo dye is not squeezed out of film
but is retained in the monolayer through its interaction
with the cyanine dye.

The Coating Technique

The transfer of monolayers from the air-water interface to a solid substrate is achieved by immersing the
solid into the water through the packed monolayer or by
withdrawal of an immersed solid substrate from the water
through the packed monolayer [12]. In the first case a
monolayer is transferred if the solid surface is hydrophobic. The hydrophobic methyl end groups of the monolayer hydrocarbon chains attach themselves to the hydrophobic surface. In the second case (withdrawal) the solid surface is hydrophilic and the polar groups of the
monolayer attach themselves to the hydrophilic surface.
In fact, in this case three different interactions influence the fixation of the monolayer to the solid. The
interaction between the solid surface and water, between
the monolayer polar groups and water and between the
monolayer polar groups and the solid surface. The attach-
ment of the monolayer can therefore be influenced by
changing the composition of the monolayer which changes
the interaction of the monolayer polar groups with the
water subphase and with the solid surface, respectively
[13].

Complex layer systems for spectroscopic investigations and optical studies have been built by consecutive
dips through different monolayers prepared at the air-
water interface [8]. In such systems the energy transfer
from an excited donor dye molecule to an appropriate

acceptor allows to assemble a simple functional unit. The donor is excited with light that is exclusively absorbed by the donor D. The excitation energy is transferred to the acceptor A over distances of the order of d = 50 Å. Consequently, the fluorescence emission of D is prohibited and the emission of fluorescence of A sensitized. The emission of A on excitation of D is due to the cooperation of D and A in a functional unit with molecular organization.

Both, the quenching of the donor fluorescence and the sensitization of the acceptor fluorescence depend strongly on the thickness d of the spacer layer between the planes of the chromophores of S and A [14]. This thickness can be varied by deposition of spacer layers of fatty acid molecules between the monolayers of D and A. The distance dependence of the donor fluorescence intensity I is given by

$$(I/I_\infty) = [1 - (d_0/d)^4]^{-1} \qquad (1)$$

where I_∞ is the fluorescence intensity of D when A is missing and d_0 the critical distance at which the probability of deactivation of excited D via energy transfer is 1/2. The distance dependence of the sensitized fluorescence of A is given by

$$(I'/I'_0) = [1 - (d/d_0)^4]^{-1} \qquad (2)$$

where I'_0 is the intensity of sensitized fluorescence of A in case of complete energy transfer. The critical distance d_0 can be calculated from spectroscopic data, the refractive index of the medium and the relative orientation of the transition moments of D and A [14].

An example of an energy transfer system is shown in Fig. 6 with a mixed monolayer of arachidic acid and N,N'-dioctadecyloxacyanine, molar ratio acid : dye = 10:1, as donor and a mixed monolayer of arachidic acid and N,N'-dioctadecylthiacyanine, molar ratio acid : dye = 10:1, as acceptor (formulas given in Fig. 6). The experimental data are represented by bars (fluorescence of D) and circles (fluorescence of A) and the lines (broken line in case of sensitized fluorescence of A) are drawn according to the relationships given above with the adjusted value d_0 = 60 Å [8]. From energy transfer studies information on pathways of deactivation of excited molecules can readily be obtained [15, 16, 17].

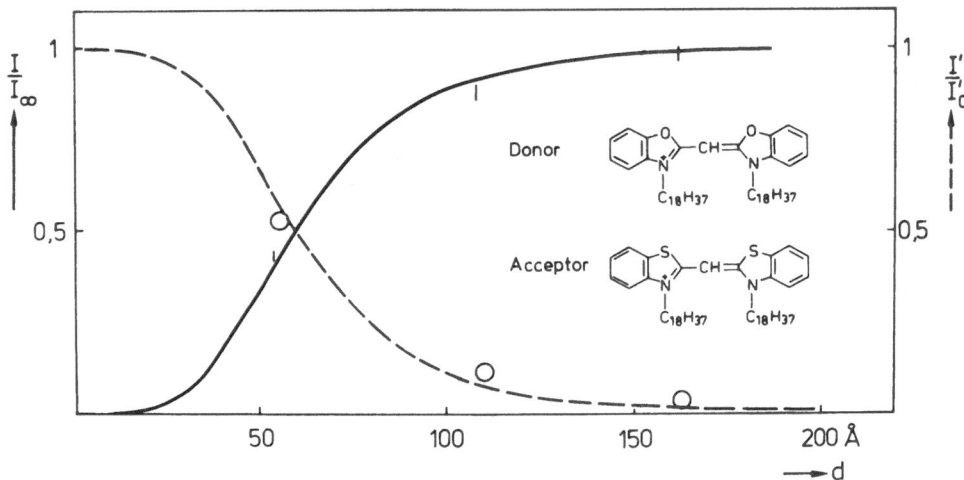

Fig. 6: Energy transfer from excited molecules of the
        donor oxacyanine to a monolayer of the acceptor
        thiacyanine kept at distance d. Relative fluor-
        escence intensity I/I_∞ of the donor (bars) and
        relative intensity of the sensitized fluores-
        cence, I'/I'_0 of the acceptor. The lines are
        drawn according to Eq. 1 (full line) and Eq. 2
        (broken line) with the same adjusted parameter
        $d_0 = 60$ Å.

The Manipulating Technique

        The coating technique is restricted to substrates
which can be brought in contact with water and to whose
surface the spread monolayer can be attached. Many solid
substrates can be coated with only one monolayer which
floats off the solid on immersion in the water during
the subsequent coating step. Such limitations of the
coating technique can be overcome in the following way:
Subunits for larger systems are assembled on glass plates
under well controlled conditions using the self-assembly
and the coating techniques. Then, the subunits are separated from the glass plates with molecular precision and
brought in contact with each other or with the new solid
substrate. Thus, the manipulating technique comprises
methods of cleavage of multilayer systems at well defined
interfaces and the achievement of molecular contact [18].

The separation of a multilayer system from a glass plate requires a flexible film as intermediate support for the separated system. This intermediate support should be removable after transfer of the separated system to the new substrate. A thin flexible cover of polyvinyl-alcohol (PVA) can be prepared on top of the hydrophilic groups of a deposited fatty acid monolayer. When lifting this PVA-film from the glass plate the adhering layer system is removed from the glass plate with exception of the initial layer (see Fig. 7). The separated layer system can be laid down on the hydrophobic surface of another monolayer or multilayer system and the molecular contact at the hydrophobic interface is achieved in this simple way.

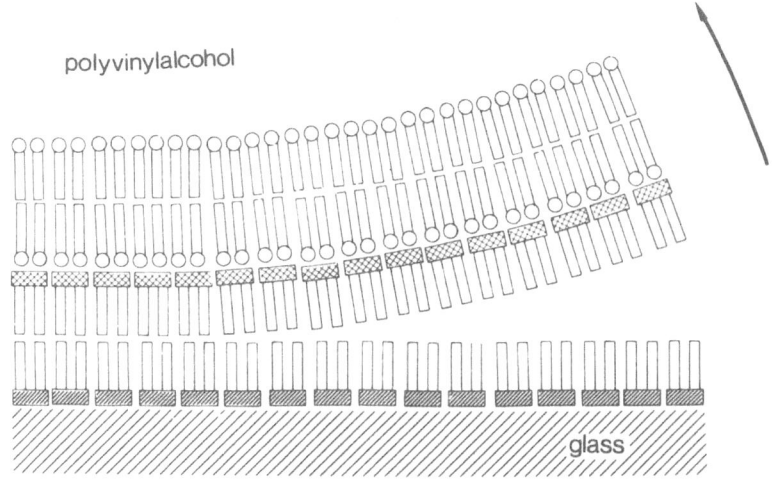

Fig. 7: Separation of a multilayer system from a glass plate by lifting a thin film of polyvinylalco-hol formed on top of the last monolayer. The system with exception of the monolayer deposited initially on the glass surface is separated. The molecular precision of cleavage of the multilayer system at this interface can be controlled by investigation of the energy transfer between mixed arachidic acid and cyanine dye monolayers positioned appropriately at both sides of the cleavage plane.

These manipulations can be controlled by energy
transfer studies in systems with separation and contacting, respectively, at an interface between an excited
donor monolayer and an acceptor monolayer [18]. Fig. 8
shows donor fluorescence intensity traces measured across
a glass plate covered partially with a layer system according to Fig. 7 (section of plate with donor + acceptor)
and partially with the corresponding layer system with
an arachidate monolayer instead of the mixed arachidate/
acceptor monolayer.

The same pair of donor and acceptor dyes was used
as shown in the energy transfer system of Fig. 6. The
left trace shows the quenching of the donor fluorescence
to nearly 1/2 of the intensity of the unquenched donor
monolayer. After separation of the acceptor monolayer the
donor fluorescence intensity is constant across the whole
plate (second trace from left). Another glass plate coated
with a sensitizer monolayer exhibits also a reasonably
homogeneous donor fluorescence (third trace from left).
After contacting the separated monolayer a quenching of
the donor fluorescence to the same extend as in the original layer system is observed in the section containing
the acceptor dye (right trace). Since the energy transfer
at the distance between sensitizer and acceptor chromophores in the system used here is very sensitive to changes in distance these measurements clearly demonstrate
the precision of cleavage and contact formation.

Methods have been developed for separation at the
hydrophilic interface [18] and have been controlled with
energy and electron transfer studies [19]. The manipulating technique was used for transferring prefabricated
multilayer systems to vacuum deposited layers of silver
bromide in a quantitative investigation of the spectral
sensitization of the photographic process in silver bromide [20]. The transfer of the information represented
by the ordering of the chromophores of a dye monolayer
at an anisotropic surface to another dye monolayer was
achieved by using the manipulating technique [21]. Thus,
the manipulating technique has become a valuable tool
in assembling complex systems from subunits built by
the self-assembly and the coating techniques.

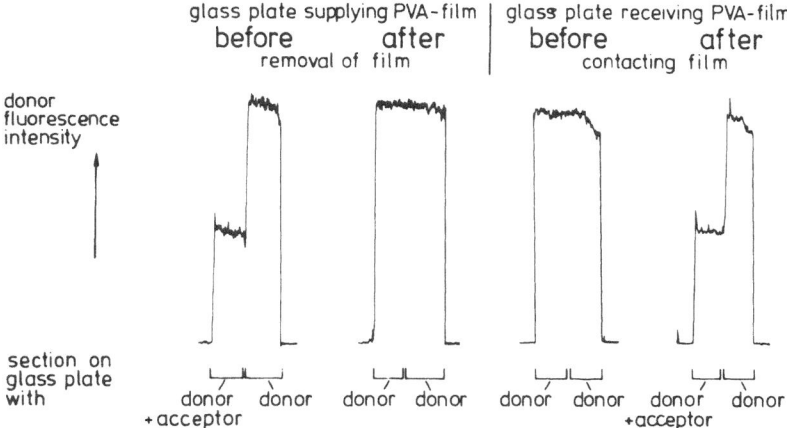

Fig. 8: Manipulation of a multilayer system with incor-
        porated monolayer of the energy donor oxacyanine
        and the energy acceptor thiacyanine (formulas
        see Fig. 6). The acceptor layer is separated
        from one plate with the PVA-film and transferred
        to a second plate. Donor fluorescence intensity
        traces across the glass plates with different
        sections as indicated demonstrate complete sepa-
        ration of the acceptor layer with the PVA-film
        and achieval of molecular contact of the trans-
        ferred acceptor layer with the donor layer on
        the second glass plate.

MODULATION OF DYE FLUORESCENCE BY COMBINATION OF ENERGY TRANSFER AND ELECTRON TRANSFER PROCESSES

Quenching of fluorescence due to electron transfer
from the excited electron donor to an electron acceptor
is observed in monolayer assemblies if donor and acceptor
are in the same monolayer or at the hydrophilic interface between adjacent monolayers (see Fig. 9 above). When
donor and acceptor are separated by the long hydrophobic
portions of the monolayers or by two fatty acid interlayers no fluorescence quenching is observed (Fig. 9,
below) in the case of N,N'-dioctadecyl-oxacyanine as
donor and N,N'-dioctadecyl-4,4'-bipyridinium as acceptor
(formulas given in Fig. 9) [22].

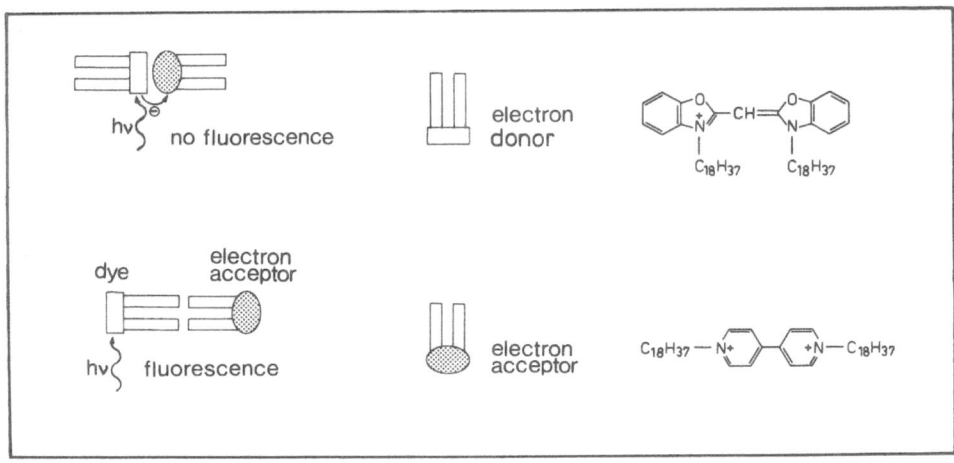

Fig. 9: Light induced electron transfer. The fluorescence of the electron donor is quenched if the
excited donor and the electron acceptor are at
the same interface (above). The donor fluorescence is unquenched in the case of insulating
interlayers (long hydrocarbon chains) between
the excited donor and the electron acceptor
(below).

The electron acceptor, the bipyridinium ion, forms
the viologen radical by electron transfer from the excited
cyanine. This type of radical has absorption bands in
the range of 330 to 400 nm and 500 to 700 nm. The amount
of radical present in the steady state during continuous
illumination is limited by the back transfer of the electron from the reduced acceptor to the oxidized cyanine
dye molecule. Therefore, the absorption of the radical
in a monolayer assembly was not directly detectable. The
viologen radical layer, however, might be used as energy
acceptor in a system with another dye monolayer having
an emission around 600 nm.

This consideration leads to the conception of a
functional unit of different components combining electron transfer and energy transfer processes. The structure of the assembled system is shown in Fig. 10 along
with the formulas of the components. The electron transfer subunit is the system of cyanine dye and viologen

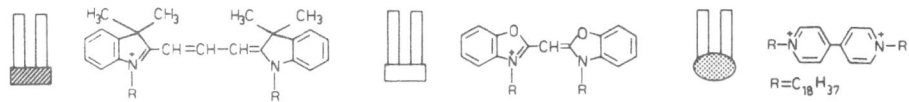

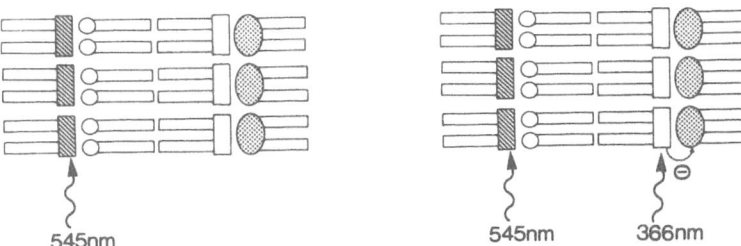

Fig. 10: Complex functional unit combining electron and
         energy transfer processes. The indocarbocyanine
         dye is excited with green light (545 nm) and
         emits a red fluorescence (590 nm). The electron
         transfer system represented in Fig. 9 is sepa-
         rated from the indocarbocyanine by two insula-
         ting layers. On excitation of the electron donor
         with UV radiation (366 nm) electron transfer
         takes place and the resulting viologen radical
         (reduced electron acceptor) acts as energy
         acceptor for the spaced away indocarbocyanine:
         The red fluorescence should be partially quenched
         in the case of simultaneous illumination of the
         system with green light and UV radiation.

given in Fig. 9. The chromophores of the energy transfer
sensitizer monolayer, a mixed monolayer of the N,N'-dioc-
tadecyl-indocarbocyanine, molar ratio of acid : dye =
10:1, are separated from this subunit by two layers of
long chain hydrocarbon substituents. This prevents any
electron transfer from the excited indocarbocyanine chromophores to the viologen.

    In the situation shown in Fig. 10, the indocarbocy-
anine is excited with green light and emits a red fluorescence (wavelength of maximum 590 nm). On simultaneous

illumination of the system with green light and UV radiation (at 366 nm) the viologen radical is formed which
acts as energy acceptor for the excited indocarbocyanine
and consequently quenches the red fluorescence. Therefore,
the red fluorescence of the system on excitation with
green light should be modulated by turning the UV radiation on and off.

    This phenomenon is actually observed as shown in
Fig. 11. The fluorescence intensity at 590 nm on illumination with green light is recorded with time, and the
UV radiation is periodically turned on and off. The modulation of the red fluorescence is clearly seen from
the plot. The fluorescence intensity drops when the UV
radiation is turned on and recovers to the initial value
after turning the UV radiation off.

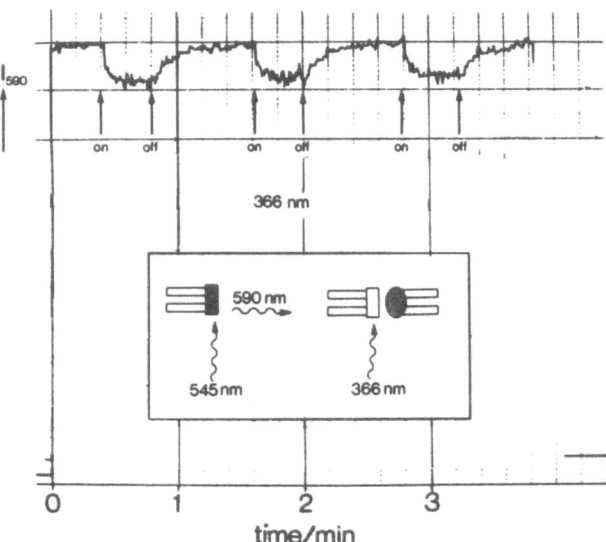

Fig. 11: Modulation of dye fluorescence by combining
         electron and energy transfer processes. Inten-
         sity of red fluorescence (I_{590}) on excitation
         with green light against time of the layer sy-
         stem according to Fig. 10. On simultaneous il-
         lumination with UV radiation the red fluores-
         cence is reduced by ca. 10 % and recovers when
         the UV radiation is turned off.

This behaviour of the assembly is determined by the
adequate molecular organization of appropriate molecules.
When the electron donor and the electron acceptor are
separated by arachidate interlayers, no modulation of
the red fluorescence is observed. When the electron transfer subunit is spaced away farther from the energy donor
by two additional arachidate interlayers no modulation
of the red fluorescence is observed. This demonstrates
the importance of the organization in achieving new
functions by planned interaction of several different
molecular components.

LONG RANGE ENERGY TRANSPORT
IN EXTENDED COOPERATIVE SYSTEMS

The molecules of the cyanine dye N,N'-dioctadecyl-
oxacyanine can form insoluble monolayers at the air-water interface. When the long chain hydrocarbon octade-
cane is added to the dye solution in chloroform the dye
molecules and the hydrocarbon organize themselves to
extended two-dimensional units [23]. The peculiar organization is indicated by the high and narrow absorption
band shifted with respect to the monomeric absorption
band by 18 nm to larger wavelengths (see Fig. 12). This
feature, along with the corresponding narrow resonance
emission (Fig. 12, broken line) is characteristic for
large aggregates of dye molecules, the J-aggregates [24].

Excited aggregates of this type can be treated as
a large array of coupled resonant oscillators. A different molecule with resonance at smaller frequency would
act as an energy trap provided it is incorporated into
the array of cooperating molecules. This is possible
in the case of the cyanine dye monolayers characterized
by the spectra in Fig. 12 by addition of the N,N'-dioc-
tadecylthiacyanine dye to the solution of the oxacyanine
dye and hexadecane in chloroform (molar ratio of oxacyanine (host) : thiacyanine (guest) > 10^3).

In fact, quenching of the fluorescence of the host
by the guest incorporated into the host monolayer is
observed up to a ratio of host : guest of 5×10^4. The
ratio of intensities of the J-aggregate fluorescence of
the host in presence of the guest (I) and without guest
(I_o) is plotted in Fig. 13 (circles) against the average
distance $\bar{R}$ of guest molecules as calculated from the
ratio host : guest and the molecular areas of host and
guest molecules in the monolayer.

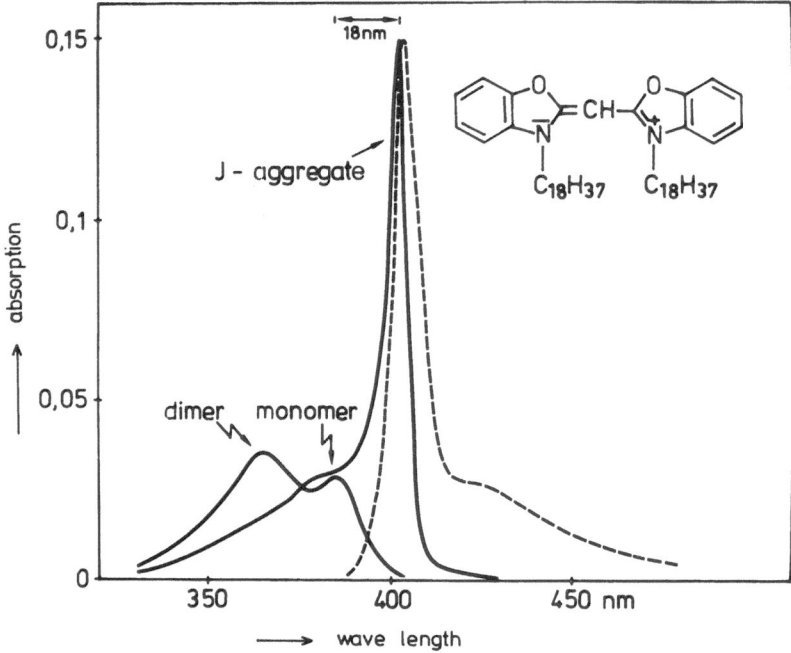

Fig. 12: Absorption (full line) and fluorescence (broken
         line) spectra of a transferred monolayer of oxa-
         cyanine and hexadecane (molar ratio dye : hydro-
         carbon = 1:1). The narrow absorption band shifted
         to longer waves with respect to the monomer band
         and the narrow resonance fluorescence band are
         characteristic for organization of the chromo-
         phores in J-aggregates.

        Since the guest has an absorption band in the range
of the J-aggregate emission band Förster type energy
transfer is expected to occur which requires no incorporation of the acceptor molecules in the host aggregates.
In order to separate this type of energy transfer from
the energy trapping due to incorporation of the guest in
the host aggregate the guest was packed in the monolayer
adjacent to the aggregate monolayer. Then, the distance
between the layer of J-aggregates and the layer of acceptor chromophores is negligible (d = 3 Å) and the efficiency of energy transfer depends on the average distance
of the acceptor chromophores in the acceptor layer. The
J-aggregate fluorescence is indeed quenched by the acceptor chromophores in the adjacent monolayer and the values

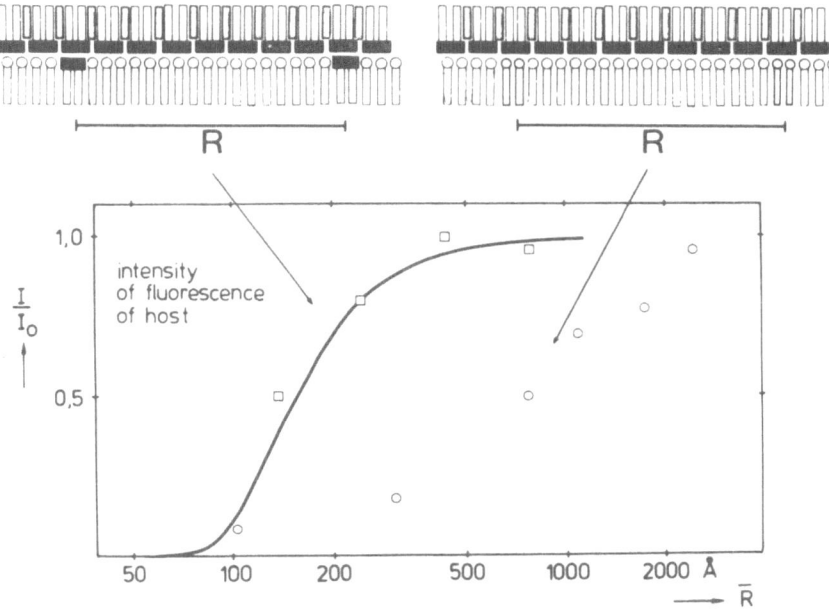

Fig. 13: Fluorescence intensity of the host J-aggregate
(oxacyanine) plotted against the average distance
between guest molecules (thiacyanine). Squares:
guest molecules in the adjacent monolayer, energy transfer from host to guest according to a
Förster mechanism (full line represents the
theoretical dependence). Circles: Guest incorporated into the host J-aggregate. The increased
range of fluorescence quenching compared with
Förster type energy transfer indicates the peculiar interaction of the incorporated guest
with the excited J-aggregate.

of I/I_o obtained are plotted in Fig. 13 against the average distance $\bar{R}$ of acceptor chromophores (squares). The
full line represents the distance dependence which is
expected according to Förster type energy transfer.

The prominent result of the experiments is the large
range of fluorescence quenching in the case of guest molecules incorporated in the host J-aggregates. It must

be concluded from this observation that the J-aggregates
consist of some 5000 to 10 000 host molecules if we assume
that one guest molecule in such a large array is sufficient to quench the luminescence completely.

Besides energy trapping by the guest molecules in
the J-aggregate there is the possibility of electron
transfer from the excited J-aggregate to the guest molecule which would also cause a quenching of the J-aggregate fluorescence. The two deactivation processes can
easily be discriminated by measuring the amount of sensitized fluorescence of the guest. In the case of energy
transfer the number of light quanta emitted by the guest
should be approximately equal to number of light quanta
not emitted by the host aggregate due to the action of
the guest. In the case of electron transfer the guest
is reduced to the dye radical which transfers the electron back into the ground state of the J-aggregate without emission of fluorescence.

In the case of the ratio host : guest of 10^4 the
fluorescence quenching of the host was measured, and
from the fluorescence spectra of the host aggregates
and the (host + guest) aggregates the ratio ϕ of the
number of quanta emitted by the guest and the number of
quanta not emitted by the host due to quenching was determined. The ratio $\phi = 1.1$ was obtained [25] in accordance with an energy transfer mechanism of J-aggregate
fluorescence quenching by the guest incorporated in the
host aggregate.

PHOTOCONDUCTION BY COOPERATION OF SENSITIZING AND CONDUCTING MOLECULAR COMPONENTS

The system of a cyanine dye and an azo dye organized
in a monolayer at the air-water interface as represented
in Fig. 4 is characterized by the different orientations
of the cyanine and azo chromophores. The cyanine chromophore is oriented in the layer plane whereas the azo
chromophore is oriented perpendicular to this plane. The
spatial correlation of these components of the system
was deduced from the surface pressure - area isotherm
and absorption spectrum of the mixed monolayer. It seems
promising to investigate the possibility of vectorial
charge separation in assemblies where such a mixed monolayer is sandwiched between fatty acid monolayers and

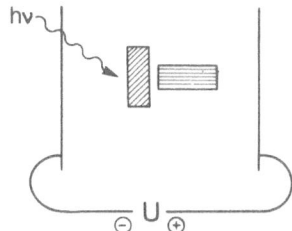

Fig. 14: Schematic representation of the combined system
of the cyanine dye and the chain like π-electron
system (azo dye) between metal electrodes (formulas given in Fig. 4). In the study of photo-
conduction an electrical field is applied and
the system is illuminated with light of wavelength absorbed by the cyanine dye.

metal electrodes (Fig. 14). The cyanine dye is excited
by light and the electrons move from the excited state,
of the cyanine chromophor through the π-electron system
of the azo dye which should act as a conducting element.

A photocurrent was indeed observed in such systems
[11], and the action spectrum of the photocurrent coincides with the absorption spectrum of the cyanine dye.
The photocurrent is proportional to the light intensity
and its logarithm increases linearly with the applied
voltage (see Fig. 15). Conduction takes place according
to different mechanisms depending on the temperature.
In the low temperature mode the logarithm of the photocurrent decreases linearly with (1/temperature)$^{1/2}$, while
in the high temperature mode it decreases linearly with
(1/temperature).

If the conducting π-electron system is absent, the
photocurrent is about an order of magnitude smaller but
again proportional to the light intensity. Its logarithm
increases linearly with the square root of the applied
voltage.

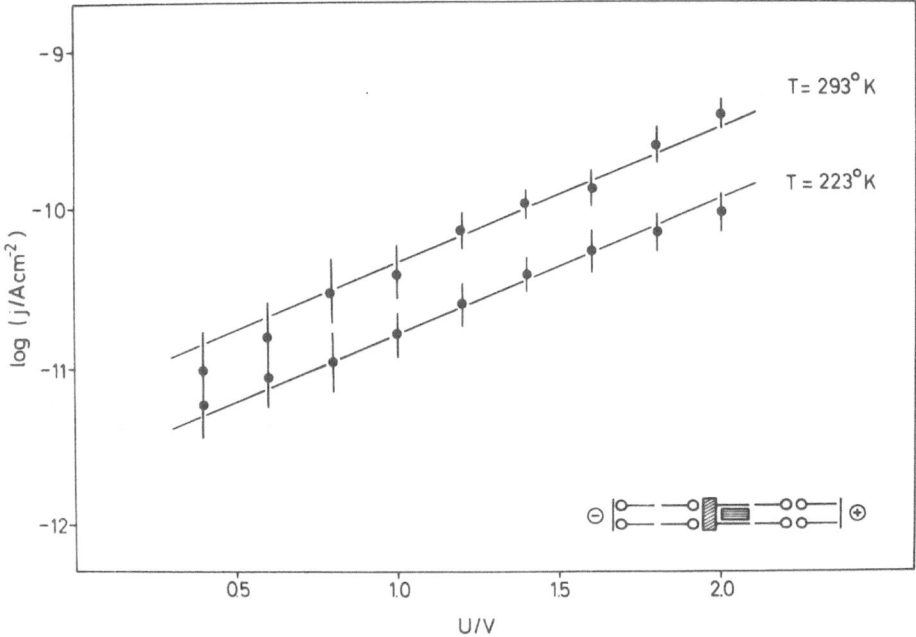

Fig. 15: Logarithmic photocurrent density j versus applied
voltage U for the system arachidate: cyanine
dye : azo dye in a molar ratio of 10:1:1 at
293° K and 223° K.

        The results in the complex assembly of cyanine dye
and azo dye can be interpreted by assuming that the electron is transferred from the excited cyanine dye (I*)
to the π-electron system (II) by tunneling or by thermal
activation over a barrier of E_{act} = 0.26 eV (see Fig. 16).
The cyanine dye is regenerated by accepting an electron
which is effectively supplied by the negatively biased
electrode. The cyanine dye, therefore, does not influence the subsequent processes. The additional electron
in the reduced azo chromophore tunnels to an accepting
site in the next interface (S_1) in a fast process. The
existence of such interface states has been deduced
from investigations of the electron conducting properties of multilayer systems with and without incorporated dye molecules [26, 27, 28]. From the state S_1 the
electron tunnels through the next fatty acid interlayer
to the interface state S_2, and then hops to the positively biased electrode.

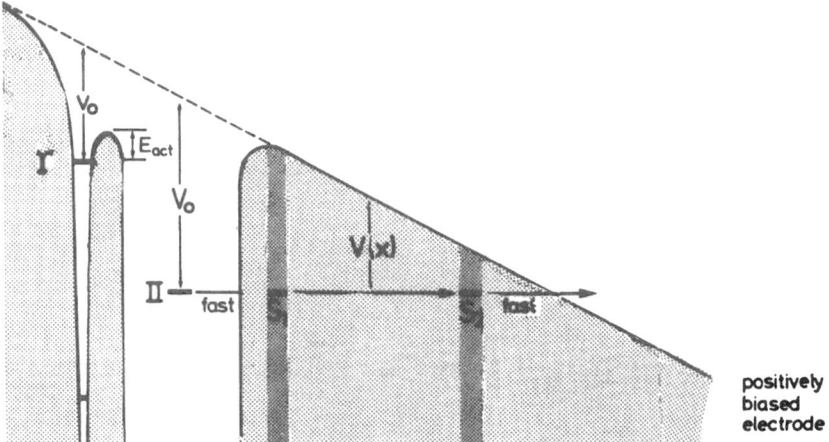

Fig. 16: Monolayer assembly with the cyanine dye and the
         azo dye (formulas given in Fig. 4) as functional
         unit. Electron conduction model based on the
         existence of two-dimensionally arranged traps
         at the interface between neighbouring monolayers.
         The electron in the excited state I* of the cy-
         anine dye jumps over the barrier or tunnels
         through the barrier to the azo dye, from where
         it proceeds to an interface state S_1, and then
         tunnels to some state S_2 at the next interface.
         Subsequently it reaches the positively biased
         electrode.

        From this model the voltage dependence and the dif-
ferent temperature dependences of the photocurrent can
easily be deduced. The model has been checked by specifically changing the thickness of the tunneling barrier
between the states S_1 and S_2 [11]. This can easily be
done by deposition of palmitic acid or a mixture of my-
ristic acid and cholesterol instead of arachidic acid at
this position of the monolayer assembly. The increase in
photocurrent anticipated from the model (see Fig. 16) is
indeed observed only if this monolayer of smaller thickness is situated between the conducting π-electron system and the positively biased electrode.

The cooperation of the cyanine chromophore and the azo chromophore depends strongly on the energetic position of both components of the unit, i. e. the quantities v_o and V_o in the model according to Fig. 16. Therefore, not only spatial requirements must be fulfilled by the components but also energetic conditions in order to achieve the planned function.

CONCLUSION

Appropriately designed molecules can be arranged by a variety of techniques to form cooperative functional units. This was demonstrated by different systems of increasing complexity:

(1) A simple system of two different dyes cooperating in a limited spatial range by energy transfer;

(2) a monolayer of cyanine dye molecules which arrange themselves to large arrays of coupled oscillators (J-aggregates);

(3) a complex system combining light induced electron transfer and energy transfer processes;

(4) a cooperative unit of sensitizing and conducting molecular components capable of achieving vectorial electron transport.

The functions of these systems seem not to be directly related with practical application or adequate of model systems for biological processes. However, the fact that molecules can be designed and successfully assembled to complex cooperative units may stimulate the construction of systems with more interesting properties.

REFERENCES

[1] W. C. Bigelow, D. L. Pickett and W. A. Zisman,
     J. Coll. Sci. 1, 513 (1946).
[2] L. S. Bartell and R. J. Ruch, J. Phys. Chem. 60,
     1231 (1956).
[3] O. Levine and W. A. Zisman, J. Phys. Chem. 61, 1068
     (1957).
[4] J. Sagiv, publication in preparation.
[5] E. E. Polymeropoulos and J. Sagiv, publication in
     preparation.

[6] B. Mann and H. Kuhn, J. Appl. Phys. $\underline{42}$, 4398 (1971); E. E. Polymeropoulos, J. Appl. Phys. $\overline{\underline{48}}$, 2404 (1977).
[7] A. Pockels, Nature $\underline{46}$, 418 (1892).
[8] H. Kuhn, D. Möbius $\overline{and}$ H. Bücher, in Physical Methods of Chemistry, edited by A. Weissberger and W. B. Rossiter, Vol. I, Part IIIB, Wiley, New York 1972.
[9] H. Bücher, O. v. Elsner, D. Möbius, P. Tillmann and J. Wiegand, Z. physik. Chem. N. F. $\underline{65}$, 152 (1969).
[10] D. Möbius, publication in preparation.
[11] E. E. Polymeropoulos, D. Möbius and H. Kuhn, J. Chem. Phys. in print.
[12] I. Langmuir, Trans. Faraday Soc. $\underline{15}$, 62 (1920).
[13] J. Petrov, H. Kuhn and D. Möbius, in print.
[14] H. Kuhn, J. Chem. Phys. $\underline{53}$, 101 (1970).
[15] O. Inacker and H. Kuhn, Chem. Phys. Letters $\underline{27}$,317,471 (1974); $\underline{28}$, 15, (1974).
[16] D. Möbius and G. Dreizler, Photochem. Photobiol. $\underline{17}$, 225 (1973).
[17] D. Möbius and G. Debuch, Chem. Phys. Letters $\underline{28}$, 17 (1974).
[18] O. Inacker, H. Kuhn, D. Möbius and G. Debuch, Z. physikal. Chem. N.F. $\underline{101}$, 337 (1976).
[19] D. Möbius, Chemie in unserer Zeit $\underline{9}$, 173 (1975).
[20] D. Möbius, Photogr. Sci. Eng. $\underline{18}$, 413 (1974).
[21] D. Möbius and G. Debuch, Ber. Bunsenges. $\underline{80}$, 1180 (1976).
[22] D. Möbius and H. Kuhn, publication in preparation.
[23] H. Bücher and H. Kuhn, Chem. Phys. Letters $\underline{6}$, 183 (1970).
[24] G. Scheibe, Z. Angew. Chem. $\underline{49}$, 563 (1936); E. E. Jelley, Nature $\underline{138}$, 1009 (1936).
[25] D. Möbius and H. Kuhn, publication in preparation.
[26] M. Sugi, K. Nembach and D. Möbius, Thin Solid Films $\underline{27}$, 205 (1975).
[27] M. Sugi, K. Nembach, D. Möbius and H. Kuhn, Solid State Comm. $\underline{15}$, 1867 (1974).
[28] M. H. Nathoo and A. K. Jonscher, J. Phys. C. $\underline{4}$,L301 (1971).

FAST RADIATION-INDUCED PROCESSES IN MICELLAR ASSEMBLIES

M. Grätzel

Institut de chimie physique, Ecole Polytechnique Fédérale

Lausanne, Switzerland

Laser photolysis and pulse radiolysis techniques are employed to unravel specific effects exerted by micellar aggregates on the rate of fast redox processes. These comprise the reduction of dimensionality in reaction space in radical decay kinetics and the kinetic control of photo redox reactions. In the latter type of process micelles, in particular functional surfactant assemblies, may be used to achieve light-induced charge separation and conversion of light into chemical energy.

INTRODUCTION

In the preceding paper attention was drawn to some unique photochemical phenomena observed when photoactive components are embedded into monolayer assemblies. This report addresses itself to another type of system containing organized molecular structures, i.e. micellar or microemulsion aggregates in aqueous solution. Micelles are agglomerates which are formed spontaneously by surfactant molecules above a certain critical concentration (CMC). Detergents such as sodium lauryl sulfate (NaLS) or cetyl trimethyl ammonium

bromide (CTAB) are the most common classes of micelle-forming agents.
In dilute aqueous solution these "synthetic" micelles have a spherical shape with a radius of ca. 10-28 Å. Figure 1. The hydrocarbon
tails of the surfactants form the core of the sphere, with the polar
head groups facing the surrounding aqueous phase. By addition of a
co-surfactant, i.e. alcohol and a hydrocarbon to a surfactant solution, the radius of the micellar aggregate may be increased more
than ten times. These systems which are thermodynamically stable
and optically clear are called microemulsions or micellar emulsions.
Figure 2.

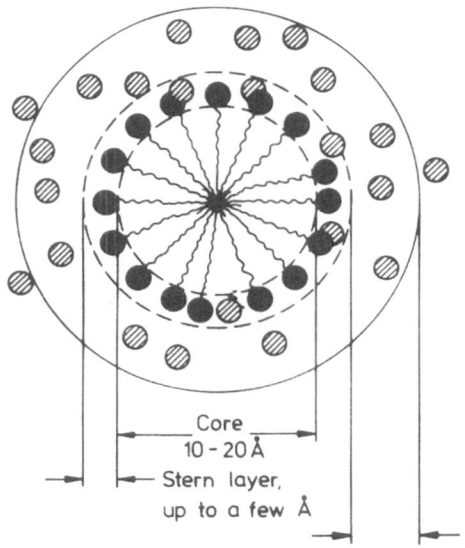

Core
10 - 20 Å

Stern layer,
up to a few Å

Gouy-Chapman layer,
up to several hundred Å

s u r f a c t a n t s :

cationic: alkyltrimethylammonium halides

$$CH_3(-CH_2)_n - \overset{\overset{\displaystyle CH_3}{|}}{\underset{\underset{\displaystyle CH_3}{|}}{N^+}} - CH_3, \; X^-, \qquad \begin{array}{l} X^- = Br^-, \; Cl^- \\ n = 12, 16 \end{array}$$

anionic: sodium lauryl sulfate

$$CH_3(-CH_2)_{11} - O - SO_3^-, \; Na^+$$

Figure 1. Schematic representation of a surfactant micelle with
its ionic atmosphere.

MICROEMULSION SYSTEMS

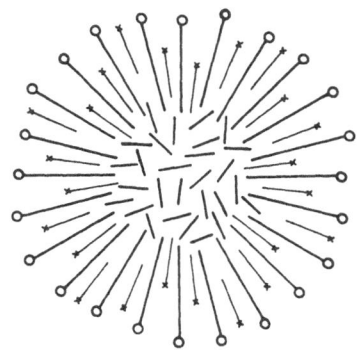

○———— **SURFACTANT**

POTAS. OLEATE $(CH_3-(CH_2)_7 CH=CH-(CH_2)_7 COO\ K^+)$

✱——— **COSURFACTANT**

CYCLOHEXANOL

—— **HYDROCARBON**

Figure 2. Schematic representation of a microemulsion assembly consisting of surfactant and hydrocarbon components.

One important feature of an aqueous micellar solution is its
intrinsic heterogeneous character. First, two domains of different
polarity exist: hydrophobic regions of minute dimensions constituted
by the interior of the aggregate are in contact with the polar aqueous
bulk phase. Moreover, when charged surfactant aggregates are considered, these two regions differ also in electrostatic potential.
For example, the interior of an ionic micelle has a potential which
typically differs by 200 to 300 mV from that of the bulk aqueous
phase.[1] The potential drop occurs in the electrical double layer
present at the micelle-water interface. This situation is depicted
in Figure 3.

Another interesting aspect which bears particularly on the topic
of this symposium is the existence of large charged interfaces in the
micellar systems. The surface area per micelle amounts to several
thousand A^2 and that of a microemulsion aggregate can be one hundred
times larger. Thus, in these surfactant solutions well defined
amphipatic interfaces are available which frequently display unique
catalytic effects.[2]

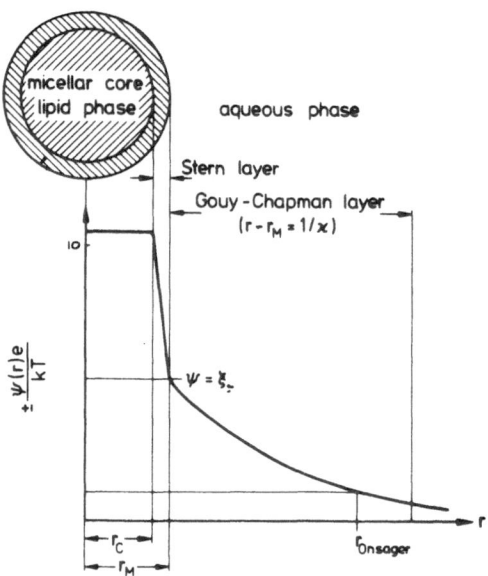

Figure 3. Qualitative representation of the potential distance function in the vicinity of a charged surfactant aggregate.

EXPERIMENTAL

Materials

Sodium lauryl sulfate (Merck, "for tenside investigations") was purified by extraction with diethylether and subsequent repeated recrystallization from water. Hexadecyltrimethylammonium bromide was recrystallized from alcohol-ether mixtures. All dyes employed in the photochemical studies were purified by recrystallization and column chromatography. The preparation of the functional surfactant Copper lauryl sulfate (CuLS) has been described elsewhere.[3]

Apparatus

Laser photolysis experiments were carried out using a Q-switched Korad K1QP ruby laser. The 347.1 nm pulse had a duration of 15 ns and maximum energy of 100 mjoules as measured by a bolometer. Pulse radiolysis experiments were performed with a 18 MEV Linac. Transient species were detected by fast kinetic spectroscopy. A detailed description about experimental details has been given elsewhere.[4]

RESULTS AND DISCUSSION

Reduction of Dimensionality in Radical Decay Kinetics

The first part of my talk is used to illustrate a case of specific surface catalysis exerted by micellar aggregates. The reaction considered is the bimolecular dismutation of bromide anion radicals:

$$Br_2^- + Br_2^- \xrightarrow{k_1} Br_3^- + Br^- \qquad (1)$$

Br_2^- may be produced by pulse radiolysis in an aqueous solution of NaBr saturated with N_2O.[5] In our experiments the widths of the radiation pulse was 5 to 20 ns corresponding to an initial Br_2^- concentration of $0.3 - 3 \times 10^{-5}$ M. Br_2^- possesses an absorption maximum at 360 nm while that of Br_3^- is located at a wavelength of 270 nm. Thus, the dissappearance of reactants and formation of products in equ. (1) can be followed readily by monitoring the absorption of the solution at these two wavelengths.

Results are displayed in Figure 4. In the absence of micelles (left column), the transient absorption at 360 nm decays in a smooth fashion and plots of the reciprocal absorption versus time yield linear relations from which a second-order rate constant $k_2 = 2.4 \times 10^9$ $M^{-1}s^{-1}$ is derived. This value is in close agreement

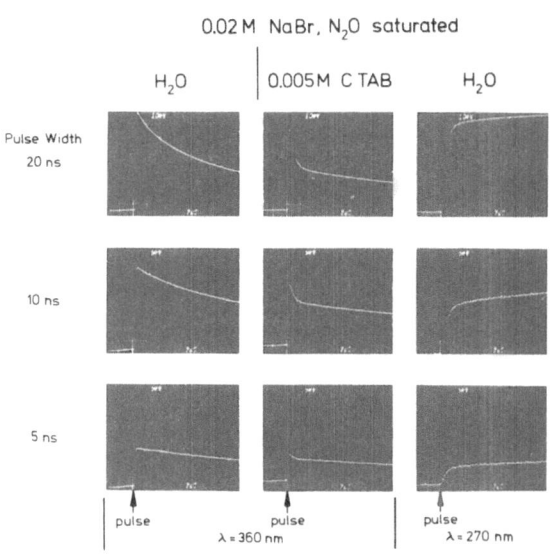

0.02 M NaBr, N₂O saturated

H₂O | 0.005M C TAB | H₂O

Pulse Width 20 ns

10 ns

5 ns

pulse | pulse λ = 360 nm | pulse λ = 270 nm

Figure 4. Oscilloscope traces of the kinetics of Br_2^- radicals (left and middle columns) and Br_3^- ions (right column). Left column: aqueous 0.02 M NaBr. Middle column: aqueous 0.02 M NaBr and 5×10^{-3} M CTAB saturated with N_2O. Right column: aqueous 0.02 M NaBr and 5×10^{-3} M CTAB saturated with N_2O. Upper row: Abscissa, 2 μs per large division. Middle row: Abscissa, 2 μs per large division. Lower row: Abscissa, 2 μs per large division.

with literature values and indicates the diffusion-controlled nature
of the dismutation process. In the presence of cationic CTAB micelles
the kinetic behaviour of the Br_2^- species is characterized by a two-
step decay. The fast process, indicated by the rapid decrease in
the 360 nm absorption, is completed within 2 ns and is followed by
a second much slower decay occurring over a period of more than 100
ns. Concomitant with these events, one observes a pronounced increase in the absorption at 270 nm (right column), which consists
also of two well-separated components. The fast portion of the
270 nm decay and 360 nm growth was found to obey first order kinetics.
Conversely, the slower reaction of Br_2^- was found to obey second
order kinetics similar to the results obtained in the absence of
CTAB micelles. The rate constants for the two processes derived
from this analysis are k_1(fast) = 2.1 x $10^6 s^{-1}$ and $2k_1$(slow) =
1.6 x $10^9 M^{-1} s^{-1}$. As the pulse width decreases, and therefore the
initial Br_2^- concentration, the fast decaying fraction of the 360 nm
absorption also diminishes. The same observation is true for the
fast and slow components of Br_3^- formation.

The disproportionation of Br_2^- in CTAB micellar solution is en-
visaged to proceed via three different pathways as illustrated
schematically in Figure 5.

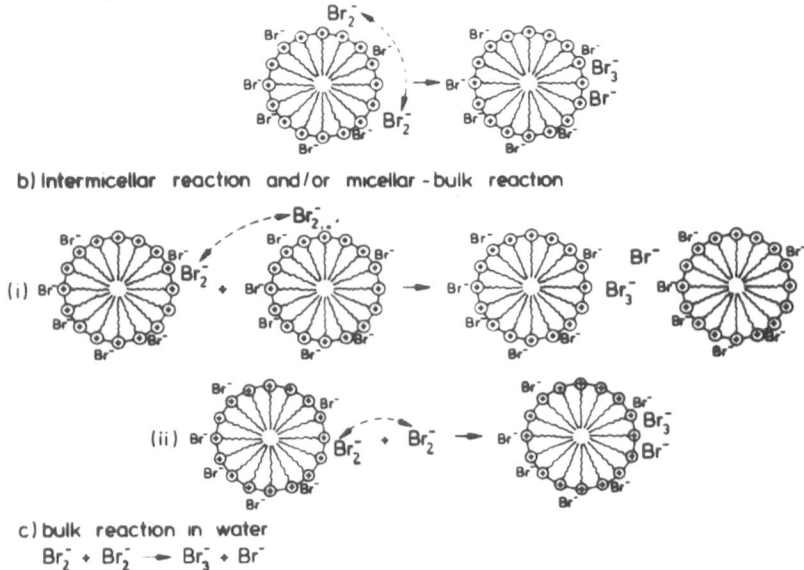

Figure 5. Schematic illustration of possible pathways for the dismutation reaction of Br_2^- ion radicals in micellar CTAB solution.

Due to hydrophobic and electrostatic interactions, most of the Br_2^- produced will be trapped immediately after the radiation pulse on the micellar surface. From the total number of micelles having Br_2^- radicals absorbed on the surface, a certain fraction will have single associations and the remainder will have multiple associations with Br_2^-. The distribution is actually governed by Poisson statistics. The experimental conditions were selected in such a way that among the micelles with multiple Br_2^- associations, the majority will have only two Br_2^- bound to them. In these aggregates, intramicellar reaction involving two-dimensional surface diffusion is feasible. Since this reaction occurs between pairs of Br_2^- radicals on single host aggregates, it should obey first order kinetics. Apparently, the initial and rapid Br_2^- decay originates from this two-dimensional reaction while the slower portion reflects the three-dimensional diffusion of the reactants. The relative rates of the two processes can be interpreted theoretically using predictions derived from a theory of random walk on two- and three-dimensional lattices with traps.[5] The ratio of specific rates for the surface and bulk reaction obtained from the Montroll equation is in surprisingly good agreement with the experimental results.

Light-Induced Electron Transfer Reactions

Considerable effort has recently been directed towards finding chemical systems that are capable of quantum storage of light energy. A very attractive process to consider is the light-induced electron transfer reaction:

$$A + D \; \underset{\Delta}{\overset{h\nu}{\rightleftarrows}} \; A^- + D^+ \qquad (2)$$

In this system light can be regarded as an electron pump which drives the reaction against a positive gradient of free energy change. The initial conversion efficiency, i.e. the ratio of free energy of reaction over light energy input can be made impressively high ($>$ 80%) if suitable donor acceptor pairs are selected. However, a common feature of systems that are endoergic in the forward direction is their ability to back react at a diffusion-controlled rate in homogeneous solution. This leads to rapid thermal dissipation of chemical energy. One method whereby inhibition of the undesirable back transfer of electrons may be achieved is to employ heterogeneous solutions such as micellar or membrane systems. Here, ultra-thin electrostatic barriers present as charged phase boundaries make the kinetic control of the electron transfer events feasible. In the following typical examples the micellar effects will be illustrated.

We shall consider first a situation where the two species A, D
are both incorporated in the particular aggregate. Light quanta are
used to excite one of the probes and promote electron transfer from D
to A, Figure 6. A radical ion pair A⁻ ... D⁺ is thereby produced within
the aggregate. For the elucidation of the further fate of these radical ions, it is instructive to examine their electrostatic interaction with the environment. In a negatively charged aggregate A⁻
is clearly destabilized with respect to the aqueous bulk solution.
Therefore, once it reaches the interface, it will be ejected from
the surface into the water. Conversely, the cation radical is electrostatically more stable in the micellar than in the aqueous phase and
will thus remain associated with the surfactant aggregate. Once A⁻
and D⁺ are separated, their diffusional re-encounter **will be obstructed**
by the electrostatic ultra-thin barrier of the micellar double layer.

The conjecture which evolves from these considerations is that
ionic micelles may be used as mediators to achieve light-induced
charge separation. The efficacy of such a process will crucially

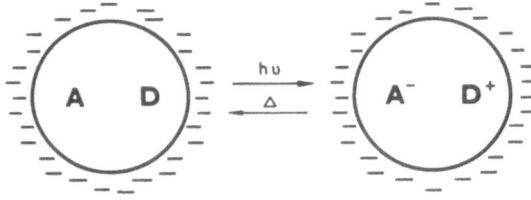

ELECTROSTATIC INTERACTION ENERGY

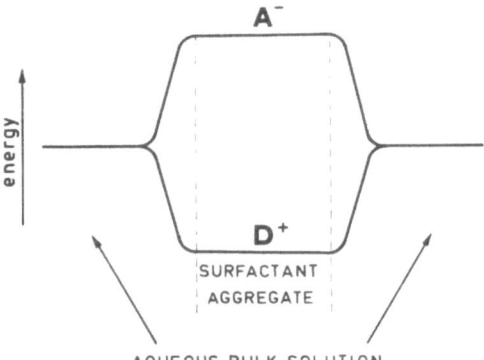

Figure 6. Light-induced charge transfer reaction within an anionic
surfactant aggregate. Electrostatic interaction of anion and cation
radical with the local environment.

depend on the relative rates of A⁻ ejection and intramicellar back
transfer of electrons from A⁻ to D⁺. The latter reaction is thermodynamically favourable and can in principle occur very rapidly. The
rate of ejection of an anion from a negatively charged micelle is on
the other hand expected to depend critically on the degree of hydrophobic interaction of A⁻ with the aggregated surfactant molecules.

This situation was examined experimentally by carrying out flash
photolysis experiments with aqueous solutions of NaLS micelles containing chlorophyll-a (chl-a) as a photoactive electron donor and duro-
quinone (DQ) as an acceptor.[6] Experimental data obtained with red
light excitation are presented in Figure 7. The upper two traces represent absorption versus time curves from micellar solutions containing chl-a alone. They reflect formation and decay of chl-a triplet
states. The time course of both the 685 and 465 nm absorption is
drastically effected when DQ is co-solubilized with chl-a in the
micelles. In particular, a long-term bleaching of chl-a absorption
is noted which can be attributed to chl-a⁺ cation radical formation

CHLOROPHYLL IN ANIONIC MICELLES

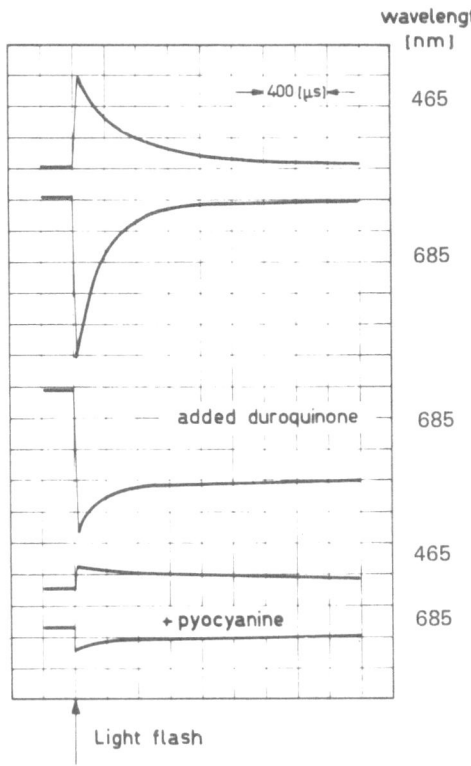

Figure 7. Flash photolysis
results obtained from solutions of 3×10^{-5} M chl-a in
0.1 M sodium lauryl sulfate.
Behaviour of the absorption
versus time function under
red light excitation.
From the top: trace 1,2:
no additive; trace 3,4:
added 3×10^{-3} M DQ; trace 5:
added 3×10^{-3} M DQ + 8×10^{-5}
M pyocyanine.

via:

$$chl-a^+ + DQ \longrightarrow chl-a^+ + DQ^- \qquad (3)$$

Apparently, a large fraction of $chl-a^+$ escapes from geminite recombination with DQ^- inside the aggregate. This must be due to the efficient ejection of DQ^- from the micellar into the aqueous phase as was suggested above in our model considerations. Once in the aqueous phase DQ^- cannot return into its native micelle since it is electrostatically rejected from the micellar surface. Hence, it will undergo disproportionation into durohydroquinone and DQ. **The cation radical** $chl-a^+$ is slowly reconverted into chlorophyll-a as indicated by the bleaching results.

The effect of pyrocyanine (PC^+) observed supports the suggested mechanism. PC^+ is present in the aqueous phase and is reduced by DQ^-. The neutral PC produced can enter the micelle and reduce in turn $chl-a^+$. Thus, the fraction of chl-a initially removed from the sphere of observation is replenished rapidly via this sequence of redox reactions. Hence, no long-term bleaching of chl-a is expected to occur as indeed is observed.

The above mentioned light-induced charge separation effect may be applied in a variety of fields. For example, it has been used in the ionic polymerisation of vinyl carbazol (VC) in microemulsion systems.

We next consider a situation where the photoactive donor D is solubilized in the apolar interior of the aggregate while the acceptor is located in the surrounding aqueous phase. The two cases will be distinguished in which the water itself acts as an electron acceptor and where an electron acceptor metal ion is added to the solution. Since the former reaction leads to the formation of hydrated electrons, it is designated as a photoionisation process. Figure 8 illustrates schematically these two different pathways. In both cases the light-induced charge transfer reaction will **involve movement of** electrons across a charged interface separating the micellar lipid from the aqueous bulk phase.

The photoionisation reaction:

$$D^+ + H_2O \longrightarrow D^+ + e^-_{aq} \qquad (4)$$

has attracted attention as a possible pathway for photochemical production of hydrogen from water since the hydrated electrons produced are known to undergo the diffusion-controlled reaction:

$$2 \, e^-_{aq} \longrightarrow H_2 + 2OH^- \qquad (5)$$

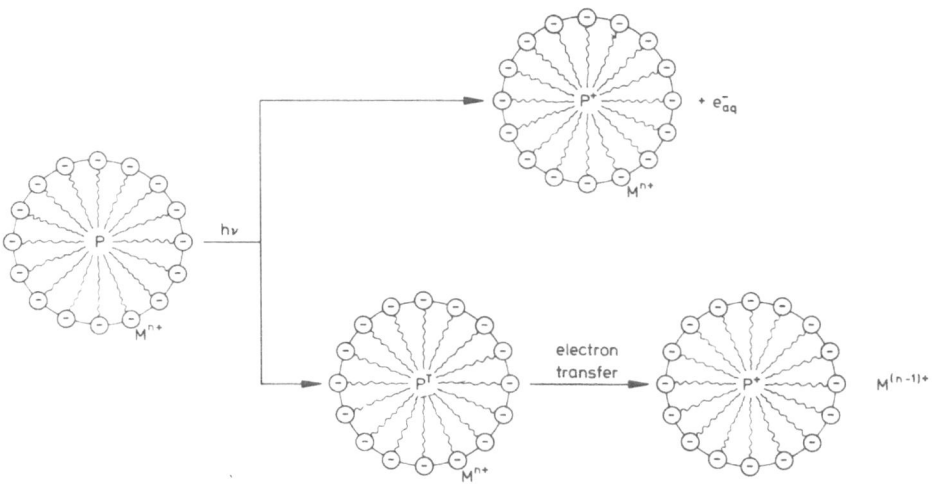

Figure 8. Schematic representation of photoionization and electron transfer processes in solutions of surfactant micelles containing a solubilized photoactive probe D. The electron acceptor is M^{n+} located in the Stern layer of the micelle and the electron is transferred through the Stern layer from the triplet (D^T).

In order to be able to test such a system, two major obstacles have first to be overcome:

(a) the majority of organic sensitizers with ionization potentials low enough to permit electron ejection by visible or near-u.v. light are sparsely soluble in water;

(b) in homogeneous systems rapid recombination of e^-aq/D^+ ion pairs leads to annihilation of electrons before hydrogen can be formed.

If anionic micellar solutions are used as a solvent, a hydrophobic donor will be incorporated into the interior of the aggregates. Subsequent to photoionization, the donor cation remains associated with the micelle while hydrated electrons reside in the aqueous phase. As the reentry of e^-aq into the parent micelle is prevented by the negative surface potential, the desired charge separation is achieved.

This effect is illustrated in Figure 9 which shows laser photolysis results obtained with N,N,N',N' - tetramethyl benzïdine in NaLS micellar solution.

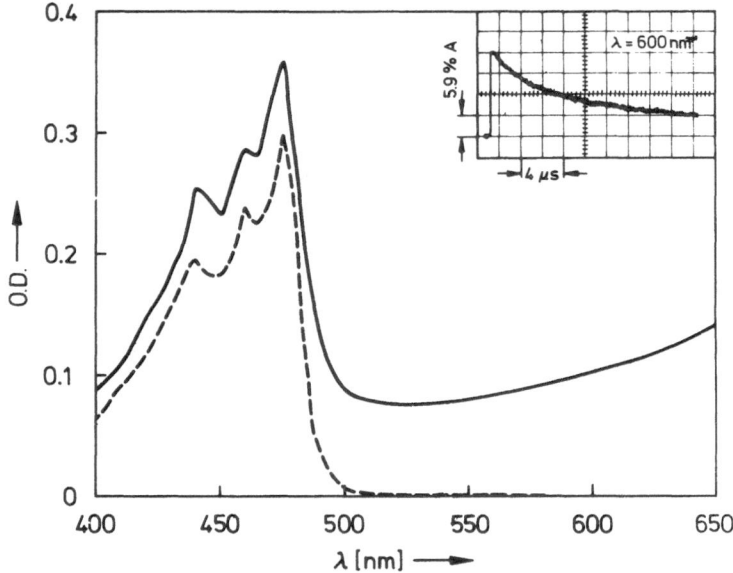

Figure 9. Spectra of transients obtained in the laser photolysis of 5×10^{-5} M TMB in aqueous 0.1 M sodium lauryl sulfate. Solid line: end of pulse; Dashed line: 10 ms after pulse. Insert: oscilloscope trace showing kinetic behaviour of solvated electrons.

The end of pulse spectrum contains two contributions arising from hydrated electrons and TMB cation radicals. These species are produced with a quantum yield of almost unity [8] during the laser flash according to the equation:

$$TMB \longrightarrow TMB^+ + e^-_{aq} \qquad (6)$$

The oscillogram inserted in Figure 9 shows the time dependency of the absorption at 600 nm, where the hydrated electron contributes more than 90% to the optical density change. The photoejection of electrons into water and their subsequent hydration is a very rapid process which is reflected by the immediate increase of optical density occurring simultaneously and within the laser pulse. Thereafter, the absorption signal decays via second order kinetics with a first halflife time of $6 \mu s$. In contrast to the kinetic behaviour of e^-_{aq}, the TMB^+ absorption at 475 nm is stable over a period of many hours. As has been pointed out before, recombination of e^-_{aq} with the parent cation inside the micelle, i.e. the reverse of photoreaction (6), is prevented by the negative micellar surface potential. The most likely reaction of e^-_{aq} in the absence of other scavengers is therefore the conversion into hydrogen via reaction (5).

Photoionisation of TMB in micellar systems can also be achieved with sunlight. However, as the intensity of solar radiation is relatively weak, only a small stationary concentration of e^-aq will be produced. As a consequence, the rate of the bimolecular reaction (5) will be retarded drastically and scavenging of e^-aq by impurities will become prominent. This renders difficult any attempt to utilize a photoionisation reaction for the production of hydrogen from sunlight. Moreover, the conversion efficiency of such a system would be unattractively small due to the high energy loss in reaction (5). On the other hand, alternative ways of exploiting the strongly reducing power of hydrated electrons are conceivable. For example, e^-aq may be employed to reduce carbon dioxide to oxalate.

Monophotonic [9,10] and biphotonic [11] photoionisation processes provide good examples for successful prevention of undesirable back reactions by micellar systems. Pronounced micellar effects are also noticeable in the photoinduced reduction of a metal ion by an organic donor:

$$D + M^{n+} \xrightarrow{h\nu} D^+ + M^{(n-1)+} \qquad (7)$$

In the following we shall investigate the kinetics of this reaction for D = Phenothiazine and $M^{n+} = Cu^{2+}$ under three conditions:

a) the reactants are dissolved in a homogeneous solution, ie. a mixture of water and alcohol (3% V/V);

b) Phenothiazine (PTH) is dissolved in NaLS micelles while Cu^{2+} is present in the aqueous phase;

c) PTH is introduced in functionalized surfactant aggregates where Cu^{2+} acts as a micellar counter ion.

The reactive excited state of Phenothiazine is the lowest triplet state (PTH^T). Hence, equation (7) can be formulated more specifically as

$$PTH^T + Cu^{2+} \longrightarrow PTH^+ + Cu^+ \qquad (8)$$

The kinetics of this electron transfer reaction will be analyzed by monitoring the optical absorption at 460 nm and 520 nm, i.e. the wave lengths of the characteristic absorptions of PTH^T and Phenothiazine cation radicals respectively.

In Figure 10 are displayed oscilloscope traces obtained from the laser photolysis of phenothiazine in ethanol water in the presence of 2×10^{-3} M Cu SO_4. The optical density at 460 and 520 nm increases during the laser pulse reflecting the formation of triplets. The signal at 460 nm decreases thereafter while that at 520 nm increases until a plateau is attained. The kinetics of the 520 nm growth match

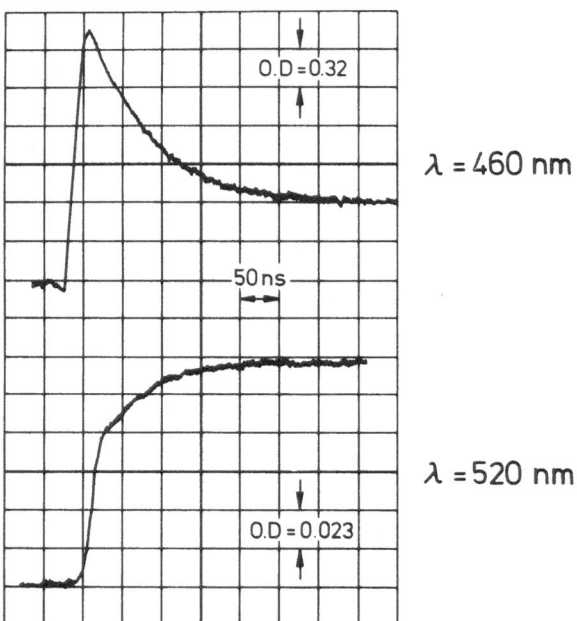

O.D = 0.32

λ = 460 nm

50 ns

λ = 520 nm

O.D = 0.023

Figure 10. Oscilloscope traces showing PTH^T decay (upper trace) and the PTH$^+$ formation (lower trace) during the laser photolysis of 5×10^{-5} M PTH in ethanol/water containing 2×10^{-3} M Cu SO$_4$.

those of the 460 nm decay indicating the formation of PTH$^+$ during the reaction of Cu^{2+} with PTH^T. The rate of the **electron transfer reaction** was found to be essentially diffusion-controlled. Figure 11 shows transitory spectra obained immediately after the laser pulse and after completion of reaction (8). In agreement with **the mechanism postulated** above, these spectra correspond to the PTH^T and PTH$^+$ absorptions respectively.

The peculiarity of the situation encountered in case b) lies in the fact that Cu^{2+} ions are strongly absorbed on the surface of NaLS micelles. Hence, the electron transfer reaction (8) does not obey simple second order kinetics. Since it constitutes a summation of intramicellar events occurring between donor-acceptor pairs, it is expected to obey first order kinetics, the rate depending on the degree of coverage of the micelle by the Cu^{2+} ions.

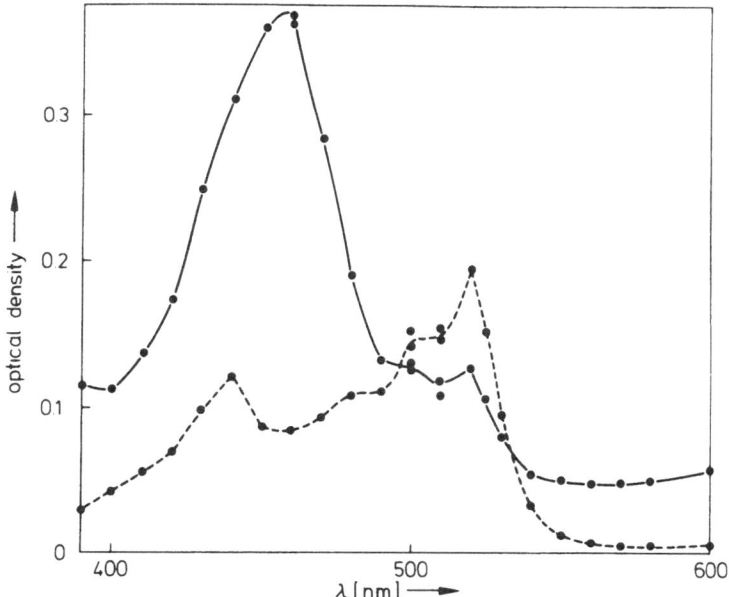

Figure 11. Transitory spectra obtained from the laser photolysis of
5×10^{-5} M PTH in ethanol/water containing 2×10^{-3} M CuSO$_4$.
Solid line: spectra immediately after the pulse. Dashed line:
spectra 300 nsec after the pulse.

The optical events depicted in Figure 12 indicate the very rapid
nature of the electron transfer quenching (8) in the micellar system.
In fact, a considerable fraction of the PTH^T reacts so fast with
Cu^{2+} that it disappears already during the laser pulse. For the remaining part, the half lifetime of the reaction is only 26 ns. Thus,
a drastic catalytic effect of anionic micelles on the rate of the
electron transfer process (8) is noted.

Most interesting and promising from the viewpoint of light
energy conversion is case c) where the photoactive donor species is
incorporated into a functional surfactant aggregate[12] Figure 13 describes this situation in more detail. The prominent feature of such
a system is that the micellar counter ion, i.e. Cu^{2+}, can itself act
as an electron acceptor. Since the local concentration in the spherical double layer region is very high, drastic catalytic effects may
be anticipated.

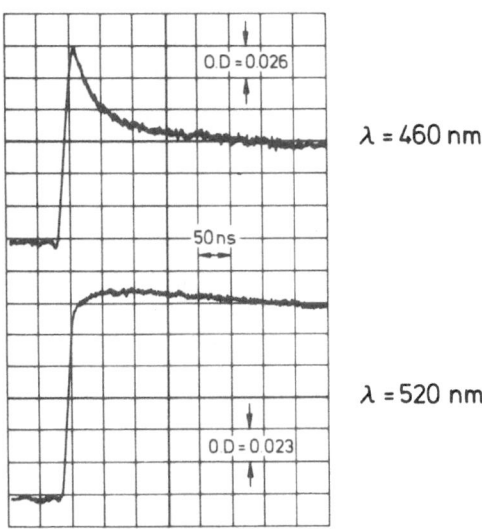

Figure 12. Oscilloscope traces showing PTH^T decay (upper trace) and the PTH^+ formation (lower trace) during the laser photolysis of 5×10^{-5} M PTH in 0.1 M NaLS containing 2×10^{-3} M Cu SO_4, deoxygenated.

PHOTOINDUCED ELECTRON TRANSFER IN FUNCTIONAL SURFACTANT SYSTEMS

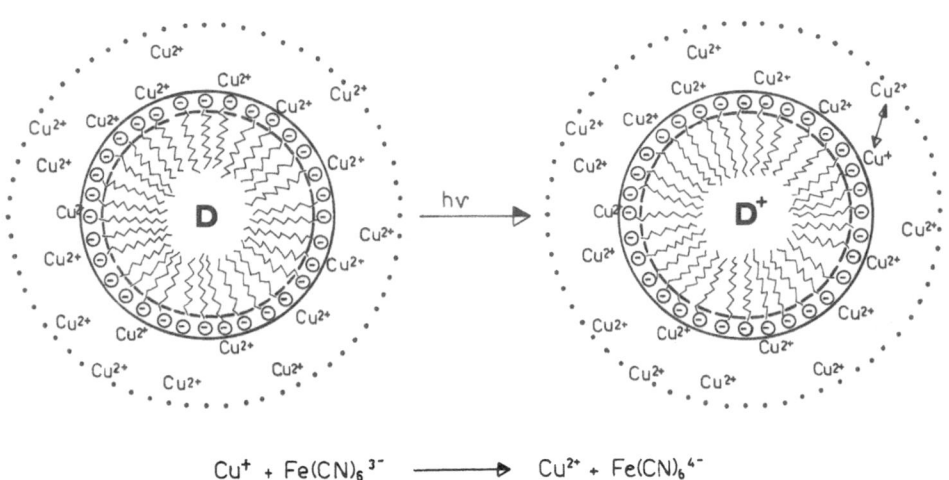

$$Cu^+ + Fe(CN)_6^{3-} \longrightarrow Cu^{2+} + Fe(CN)_6^{4-}$$

Figure 13.

The experimental data obtained with the PTH/Cu^{2+} system fully confirm these expectations as becomes apparent from inspection of Figure 14. A striking difference exists between these data and the ones obtained with solutions a) and b) inasmuch as the transitory spectrum present after the laser pulse is identical with that of PTH^+. This implies that the rate of reaction (8) is so rapid that it is already completed during the laser pulse. It is suggested that efficient electron tunneling from the donor inside the micelle to the acceptor states present in the double layer is responsible for this drastic rate enhancement.

The significance of such a functional organization becomes evident also when the subsequent reaction of Cu^+ and PTH^+ is considered. The favourable situation in case c) is that Cu^+, as soon as it is formed on the micellar surface, will exchange with a Cu^{2+} ion present in the Gouy-Chapman layer. This is caused by the fact that electrostatic attraction of Cu^{2+} towards the surface is twice as much as for the single charged cation. Through this exchange

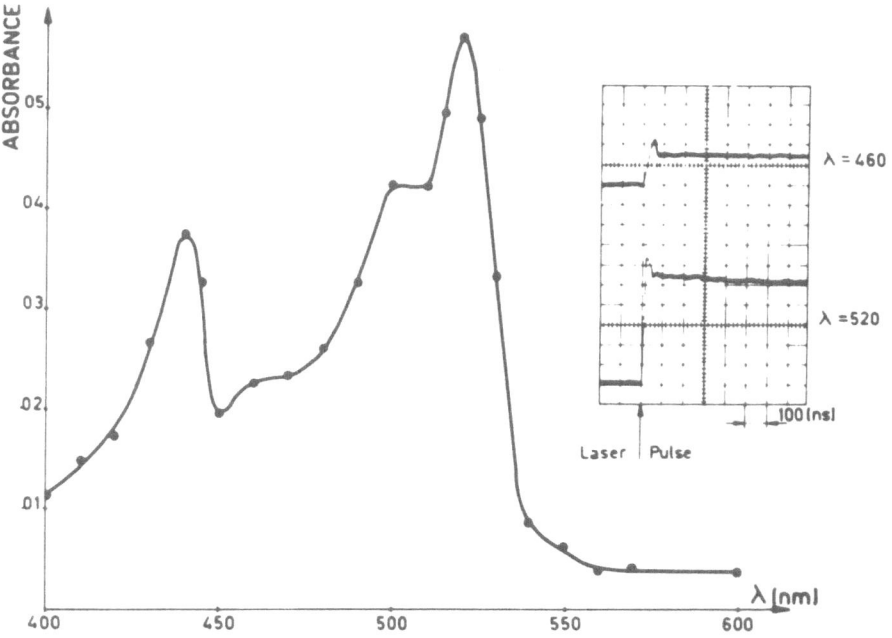

Figure 14. Laser photolysis data from PTH (5×10^{-5} M) solution in CuLS micelles (2×10^{-2} M). End of pulse spectrum and transient absorption vs. time traces showing immediate formation of PTH^+.

mechanism Cu escapes from its native micelle into the bulk medium before back reaction can occur (c.f. Figure 13). There it may be used for a redox reaction with a negative ion or zwitter ion. For example, the following reaction was found to take place:

$$Cu^+ + Fe(CN)_6^{3-} \longrightarrow Fe(CN)_6^{4-} + Cu^{2+} \qquad (9)$$

The advantage of this second redox process is that the reduced positive ion is converted into a reduced negative ion. The back reaction of the latter with PTH^+, although thermodynamically possible, is prevented by the micellar surface potential. Hence, this system is successful in storing the light energy originally converted into chemical energy during reaction (8).

LITERATURE REFERENCES

1.   D. Stigter and K.J. Mysels, J.Phys. Chem., 59, 45 (1955);
     M.F. Emerson and A. Holker, J.Phys. Chem., 69, 3718 (1965).

2.   C. Tanford, "The Hydrophobic Effect; Formation of Micelles
     and Biological Membranes", Wiley-Interscience, New York,
     N.Y., 1973;
     E.J. Fendler and J.H. Fendler, Adv. Phys. Org. Chem., 66,
     1472 (1970);
     E.H. Cordes and C. Gitler, Progr. Bioorg. Chem., 2, 1 (1973).

3.   Y. Moroi, T. Oyama and R. Matuura, J. Coll. Interface Sci.,
     60, 103 (1977).

4.   G. Beck, J. Kiwi, D. Lindenau and W. Schnabel, Eur. Polym. J.,
     10, 1969 (1974).

5.   A. Frank, M. Grätzel and J. Kozak, J. Amer. Chem. Soc., 98,
     3317 (1976).

6.   C.H. Wolff and M. Grätzel, Chem. Phys. Lett., in press

7.   J. Kiwi and M. Grätzel, to be published.

8.   S.A. Alkaitis and M. Grätzel, J. Amer. Chem. Soc., 98, 3549
     (1976).

9.   S.A. Alkaitis, G. Beck and M. Grätzel, J. Amer. Chem. Soc.,
     97, 5723 (1975).

10. S.A. Alkaitis, M. Grätzel and A. Henglein, Ber. Bunsenges.
     Phys. Chem., 79, 541 (1975).

11. M. Grätzel and J.K. Thomas, J.Phys. Chem., 78, 2248 (1974).

12. Y. Moroi and M. Grätzel, to be published.

POLYMERIZATION OF DIACETYLENES IN MULTILAYERS

B. Tieke and G. Wegner

Institut für Makromolekulare Chemie der
Universität, Stefan-Meier-Str. 31
D-7800 Freiburg (West Germany)

SOLID STATE CHEMISTRY OF DIACETYLENES

The synthesis of extended and oriented very thin layers of polymers by solid-state polymerization of multilayers built up from monomolecular films of suitable monomers by means of the Langmuir-Blodgett (LB) technique has been felt to be a particular challenge for a long time. Attempts to make use of radiation-induced free radical polymerization of long-chain vinyl derivatives have been successful only in part (1-5).

The topochemical polymerization of diacetylenes, however, serves the purpose much better since it leads with retention of molecular packing in a strictly lattice-controlled reaction to highly regular polymers which do not bear the problem of inherent disorder by lack of tacticity. According to Fig. 1 starting from suitably substituted diacetylene polymers with chains containing conjugated double and triple bonds are formed (6,7). The reaction proceeds thermally or on exposure to UV-light or high energy radiation via carbenes as active intermediates (8). The fact, that single crystals of polydiacetylenes studied so far have proven themselves to belong to the class of one-dimensional semiconductors (9,10) aroused additional interest in the construction of extended multilayers of these polymers.

Table 1: Spreading and polymerization behaviour of long-chain diacetylenes with ability to form monomolecular films on water or on 10^{-3}m $CdCl_2$.

	Compound $R-C\equiv C-C\equiv C-R'$		m.p. [°C]	multi-layer formation	Photopolymerization ($\lambda < 300$ nm)	
	R	R'			in crystal	in mul-tilayer[1]
1	n-C_9H_{19}-	-$(CH_2)_8$-COOH	48		++	
2	n-$C_{10}H_{21}$-	-$(CH_2)_8$-COOH	57	yes	++	+
3	n-$C_{12}H_{25}$-	-$(CH_2)_8$-COOH	61	yes	++	+
4	n-$C_{14}H_{29}$-	-$(CH_2)_8$-COOH	46/59 [2]	yes	++	+ [3]
5	n-$C_{12}H_{25}$-	-$(CH_2)_3$-COOH	57	no	++,+	-
6	n-$C_{16}H_{33}$-	-$(CH_2)_2$-COOH	93	yes	++	-
7	n-$C_{16}H_{33}$-	-COOH	75	yes	+++	+
8	n-$C_{12}H_{25}$-	-COOH	58	no	+++	-
9	n-$C_{10}H_{21}$-	-$(CH_2)_9$-OH	49	no	+	-

1) dark blue in less than 10 sec.: +++; dark blue in 10-30 sec.: ++; dark red after exposure >2 min.: +; no reaction; 2) phase transition at 46°C; 3) 5 has two modifications with different reactivity.

At this point it is worth mentioning that polymerization proceeds in a homogeneous manner in most cases such that isolated polymer chains grow independent from each other in the beginning of the reaction. A kind of solid solution of polymer chains extended in the yet unreacted monomer lattice is thus formed and polymerization proceeds without ever leaving the status of a single crystal.

The whole process of polymerization was best investigated using hexadiinediol-bis-(p-toluene sulfonate) as an example. For the sake of clarity the pertinent results of a crystal structure analysis of this monomer and the respective polymer are summarized in Fig. 2(11).

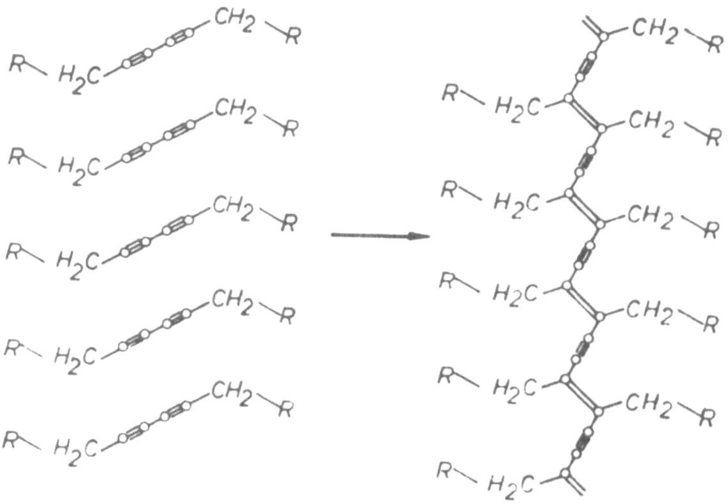

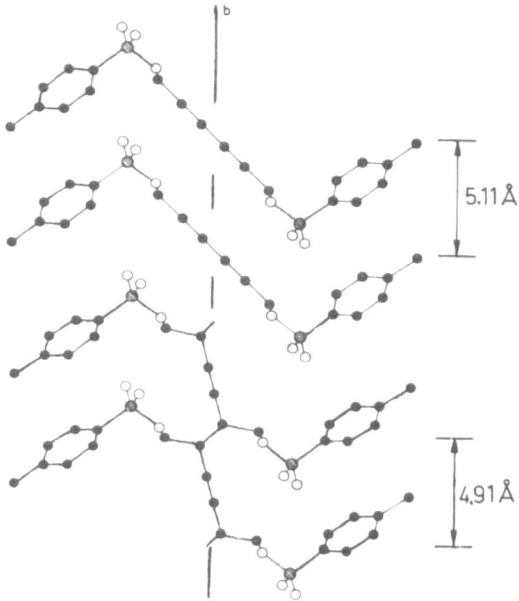

Fig. 1. Scheme of the topochemical polymerization of diacetylenes (6). R:$-CH_2-O-SO_2-C_6H_4-CH_3$ (compare Fig.2).

Fig. 2. Projection of adjacent monomer molecules (above)
and two corresponding repeat units of the polymer onto
the plane of the polymer backbone. Atomic positions derived ● C, ○ O, ◉ S.

In recent publications we have demonstrated that suitable diacetylenes can be spread at the air-water interphase and multilayers can be build up by the LB-technique. These are subsequently polymerized by exposure to UV-light (12-14).

The purpose of the present paper is to report in details on spreading behaviour, multilayer formation and polymerization phenomena of various long chain diacetylene monocarbonic acids and on the properties of the respective polymer multilayers.

MONOMOLECULAR FILMS OF SUBSTITUTED DIACETYLENES

Using a commercial film balance (MGW-Lauda) compounds 1 to 9 were spread on a subphase consisting either of water of pH 5.85 or of 10^{-3}m $CdCl_2$. Structure and some relevant properties of compounds 1-9 are summarized in Table 1. The synthesis is reported elsewhere (12-14).

Typical force-area-curves for compounds 2, 3 and 4 shown in Fig. 3. As expected, the monomolecular films on $CdCl_2$-subphase are much more stable and more densely packed than on pure water. Compound 4 shows a solid-solid-transition at about 15 dyn/cm indicated by a shoulder in the force area curve. The monomolecular films in their condensed state become more stable as the total number of carbon atoms increases and as the temperature of the subphase decreases, e. g. a monolayer of acid 3 on pure water does not bear sufficient surface pressure at ambient temperature in order to be used for build up of ordered multilayers. Contrary, monolayers on a 10^{-3}m $CdCl_2$ subphase are very well suited for our purpose. Generally, a surface pressure of 15 dyn/cm was applied during the build up of multilayers on solid supports.

As a rule of thumb, a total number of 20 C-atoms and a melting point >45[6]C seems to be a necessary requirement to form suitable surface states of diacetylene monocarbonic acids. Thus, acid 8 with total number of 17 C-atoms did not form a condensed surface state on pure water but on $CdCl_2$-subphase only. It is also worth mentioning, that the compounds 1 - 9 polymerize readily in their condensed surface state on exposure to UV-light and monomolecular films of polymers are thus formed (15).

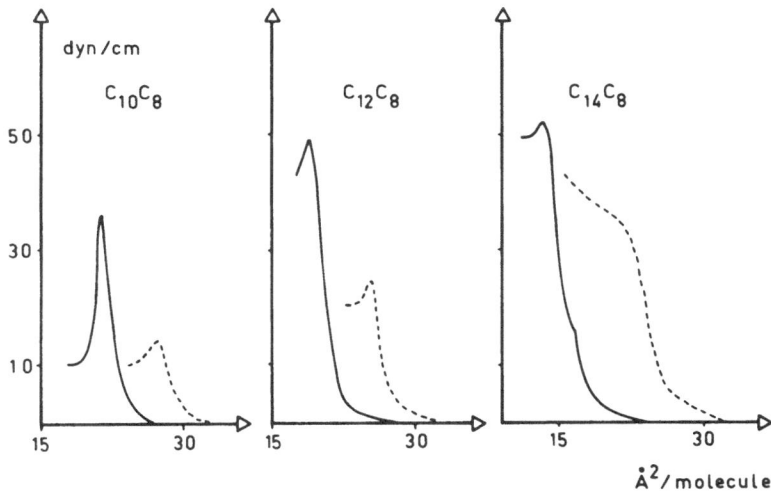

Fig. 3. Force area curves for monomolecular films of 2, 3 and 4 at the air water interphase on water (---) at T=12.5°C, pH 5.85 and on 10^{-3}m $CdCl_2$ (——) at T=24°C, pH 6.1.

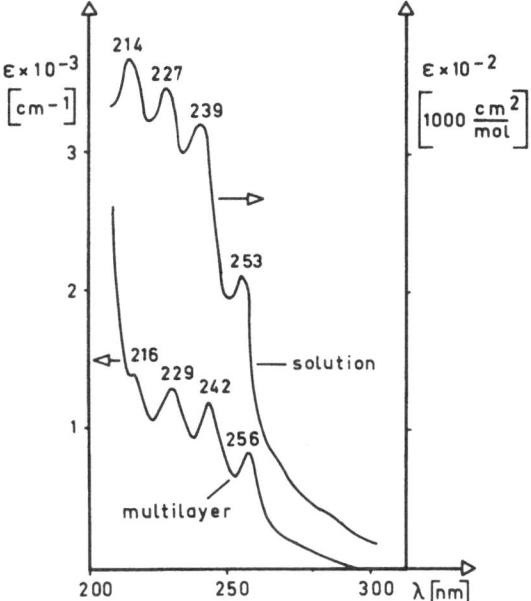

Fig. 4. UV-spectra of monomer 2 in cyclohexane and in form of multilayer (Cd-salt).

CHARACTERIZATION AND POLYMERIZATION OF MULTILAYERS

The monomolecular films that were formed were successively transferred by the LB-technique onto ultra
pure quartz plates until a multilayer of defined thickness was obtained. The plates were cleaned thoroughly
by first soaking in boiling methanol/conc.HCL(1:1 v/v)
for 12 h and then in chromic acid/conc. sulfuric acid
mixture at 120°C for additional 12 h. Finally, they
were rinsed with distilled water followed by short plasma cleaning for 10 min. Prior to the deposition of the
diacetylenes the clean hydrophilic quartz plates were
rendered hydrophopic by deposition of 3 layers of Cd-
arachidate. These layers were prepared by spreading of
arachic acid on a subphase of 10^{-3}m $CdCl_2$ followed by
transfer at a subphase temperature of   24°C. Film
pressure was 25 dyn/cm. A pressure of 15 dyn/cm was used
in all following experiments. Transfer speed was 5mm/min.

Best results with regard to reproducibility of
transfer and stability of the monomer multilayers and
reproducibility of polymerization behaviour was obtained with acids 2 and 3 transfered from $CdCl_2$-sub-
phase. Transfer of 2 is best achieved at a subphase
temperature of 15°C. The multilayers were characterized
by their UV-spectra immediately after the deposition.
Figs. 4 and 5 show spectra of acids 2 and 7 as Cd-salt
multilayers and in dissolved state. The absorption
peaks are slightly shifted to the red in the multilayers by about 3-5 nm as compared to the solutions. X-ray
small angle scattering data of multilayers of 2 shown
in Fig. 6 allow calculation of the characteristic spacing perpendicular to the layers. A value of 5.15 nm
was found. Assuming normal bond lengths and angles a
molecular length of about 3.0 nm is estimated for the
Cd-salt of 2, thus indicating that Y-layers have been
formed where the molecular long axis is inclined to
about 59° with regard to the layer plane. Knowing the
thickness of one double-layer the absolute thickness
of the multilayer sample is calculated from the known
number of dippings applied during the transfer procedure. The absolute value of the extinction coefficient
can then be derived as indicated in Fig. 4. It is recognized that very high values of the extinction coefficient are observed which is one reason for the fast
polymerization on exposure to an appropriate light
source.

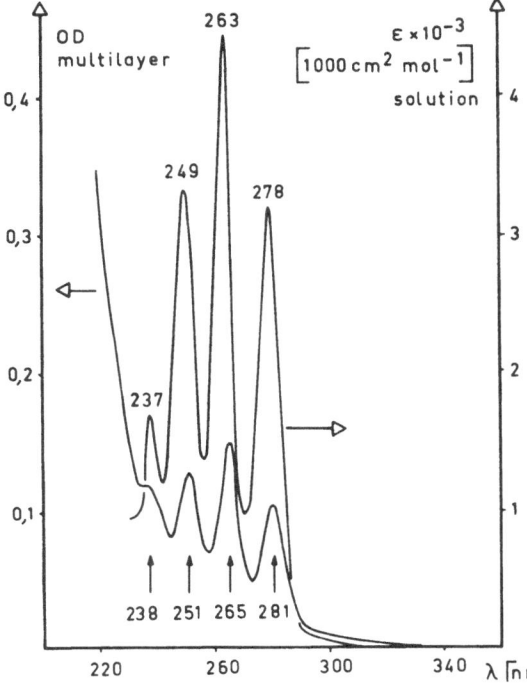

Fig. 5. UV-spectra of monomer 7 in chloroform and as multilayer (80 layers/Cd-salt).

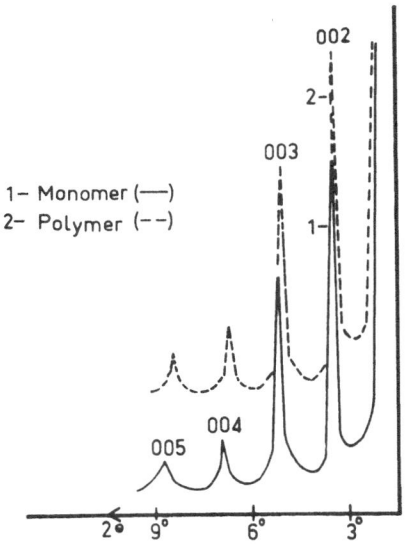

Fig. 6. Long spacing (00ℓ) reflections of multilayers of 2 prior to polymerization (1) and after polymerization (2).

The multilayers were readily polymerized by exposure to a high pressure mercury lamp (Q300,240 W, Hera-
eus, Hanau) at a distance of 30 cm. Within a few minutes
the maximum change in absorption was reached and the
plates looked deep purple. It is worth mentioning that
polymerization seemingly is not affected by the presence
of oxygen as already known for solid state polymerization of similar compounds (7). All samples were therefore irradiated in air and no care was taken to apply a
nitrogen atmosphere. However, if the multilayers were
exposed more than 30 min. colour changes were observed
which indicated a slow photooxidative decomposition of
the polymer multilayers. It is assumed that this decomposition is assisted by the presence of ozone which is
always formed under the applied radiation conditions.

Within the few seconds of irradiation time the multilayers show the characteristic blueish colour also
observed in solid state polymerization of a number of
other diacetylenes (7,16). As an example, Fig. 7 shows
the spectral changes recorded in a spectrometer after
irradiating multilayer assemblies of 28 layers of
acid 3 (Cd-salt) for 30, 60 and 120 sec. A rather broad
absorption with maxima at 638 nm and 585 nm is found
which increases in optical density with increasing polymerization time. Maximum optical density is reached
after about 20 min. irradiation time under the experimental conditions described above.

If a multilayer in its "blue" form is treated by
ethanol, chloroform or similar non-solvents of the polymer immediate and irreversible colour change to
bright red is induced as documented by the second set
of absorption spectra in Fig. 7 exhibiting peaks at
500 nm and 535 nm. A similar irreversible colour
change without destruction of the multilayer assembly
is obtained on heating the photopolymerized sample to
about 90°C.

In order to prove rigorously that the multilayer
structure does survive the polymerization procedure a
set of samples with different number of layers of
acid 3 (Cd-salt) was photopolymerized for 20 min. when
no further increase in optical density at 638 and 585nm
did occur. The samples were then treated by ethanol to
bring about the "red" form and the optical density was
determined at 500 and 535 nm. This is plotted in Fig.8.
vs. the number of layers. The points fall on straight
lines going through the origin as expected for a true
layer structure, each layer adding a constant term to

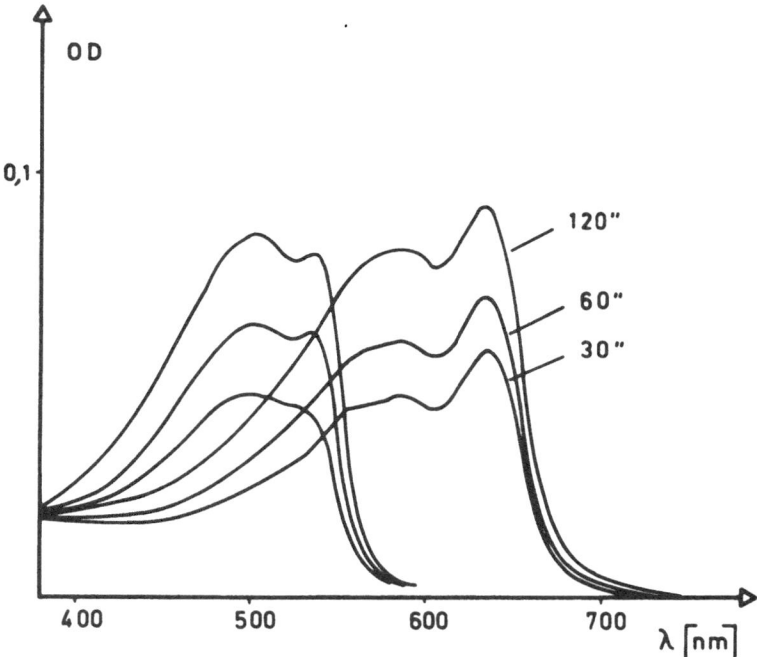

Fig. 7. Spectral changes observed on polymerization of 28 layers of acid 3 (Cd-salt) for various times. The set of curves with maxima at 638 and 585 nm is due to the layers as polymerized. The set of curves with maxima at 500 and 535 nm is obtained after treating the layers as polymerized with ethanol for 5 min.

the total optical density. Assuming quantitative conversion to polymer an optical density of $4.56 \cdot 10^{-3}$
($\lambda = 500$ nm) and of $4.25 \cdot 10^{-3}$ ($\lambda = 535$ nm) per monomolecular layer is calculated from the slope of the lines cor-
responding to extinction values of $1.61 \cdot 10^4 cm^{-1}$ and
$1.50 \cdot 10^4 cm^{-1}$ respectively if one takes into account the
value of 5.66 nm as the thickness of one double layer
as derived from X-ray small angle scattering.

Further proof comes from the comparison of thickness D_I directly measured by interference microscopy(17)
and the one calculated from the number of Y-double layers transfered times the long spacing L determined by
X-ray small angle scattering. A fair agreement is observed as indicated in Fig. 9. On plotting D_I vs. the
calculated thickness for multilayer assemblies with varying number of layers of acid 3 a straight line is obtained which cuts the ordinate at 8.2 ± 0.3 nm. The latter
is exactly the thickness expected (18) of the three layers of Cd-arachidate transfered at first to the quartz
plate in order to render it sufficiently hydrophobic as
mentioned above. It thus can be concluded that true
multilayer structures of polydiacetylenes can be obtained. The "blue" form of the polymer may be ascribed
to the presence of active carbene chain ends in the
sample which are destroyed thermally or by adding a
swelling agent like ethanol or chloroform most probably
by reaction with a neighbouring monomer to form a cyclopropene according to

Although cyclopropene or other possible reaction products of the carbene chain end have not been detected
so far, the presence of carbene chain ends in polyme-
rizing diacetylene crystals has been well established
by ESR-spectroscopy (8,19,20). It is further worth mentioning that relative intensities of the two peaks observed in either the "blue" or the "red" form are very
temperature dependent. With decreasing temperature the

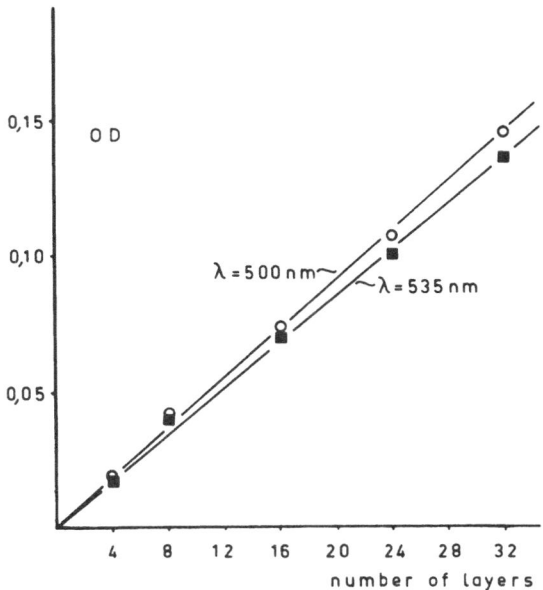

Fig. 8. Optical density vs. number of layers for multi-layer assemblies of acid 3 treated by solvent after photopolymerization of 22 min.

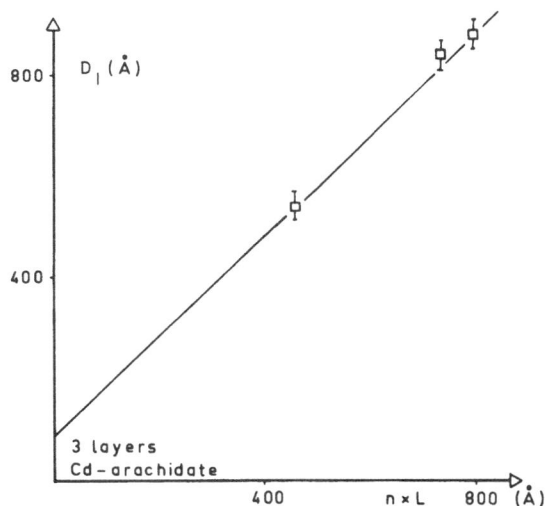

Fig. 9. Thickness (D$_I$) of polymerized multilayer assemblies of acid 3 determined by interference microscopy vs. thickness calculated from number of bilayers n and long spacing L.

long wave length peak decreases considerably in intensity and the short wave length peak increases accordingly. The intensity change is completely reversible.

CRYSTALLINITY AND ORDER IN POLYMER MULTILAYERS

The topochemical polymerization of diacetylenes
is strictly controlled by the packing of the monomers.
The reaction does not proceed in solution melt nor in
liquid crystalline state. It may, therefore, be safely
infered that the packing in the monomer multilayers is
essentially the same as in a perfect lattice with regard to molecular mobility and intermolecular forces;
otherwise the polymerization would not proceed in such
a perfect manner as observed. It is, however, the question to what is the state of order and the size of uniformly scattering elements in the polymerized multilayer assemblies. Some preliminary investigations have
been undertaken in order to clarify this point by
means of electron microscopy. For that purpose the layers were prepared and polymerized on quartz plates as
for spectroscopy. The polymer layers were shadowed by
evaporation of a carbon/Pt film in case micrographs
were taken and were floated onto a clean water surface
containing a trace of HF.For diffraction experiments
the polymer multilayers were. floated without previous
shadowing. The multilayer films were picked up in the
usual manner and then looked at using a Philips 400-
transmission EM. The multilayers were found to be homogeneous in the sense that no holes or traces of phase
boundaries nor crystal edges were detected, except
when the multilayer was accidently destroyed by scrat-
ching or the layers had been irregularly transfered.
The surface exposed to the air during polymerization,
however, was found to be covered with small irregular-
ities looking as if the topmost layer was destroyed
forming small valleys of an average diameter of about
100 nm. Electron diffraction experiments were also
performed and indicated high order within the layers.

Generally, patterns consisting of single spots as
due to scattering of a single crystal like specimen
were observed from selected areas of several μm diameter. We assume that this is roughly the diameter of
a domain of uniform orientation although we have never
observed morphological features due to domain boundaries. Two types of patterns could be observed from
areas present independently from each other in the

same sample. Type a looking as from a paraffinic material with normal three dimensional order and type b looking as from a structure with defined disorder as recognized from the continous streaks and the arcing of only some of the reflections. That the multilayer films must consist of islands of rather small size comes also from investigation by polarizing microscopy. The multilayer shows a rather homogeneous red colouration independent of the plane of polarization of the incident light even at the highest magnification. In other words, dichroic behaviour is not observed as would be expected, if the domains would be larger than a few microns.

At present, we cannot decide whether the individual layers are in register with each other perpendicular to the layer plane although it is fairly clear that rather perfect order must exist inside each monolayer within a given domain. The domains, however, are of small size and of different mutual orientation.

ACKNOWLEDGEMENT

Financial support by the Deutsche Forschungsgemeinschaft and by the Fonds der Chemischen Industrie is greatly acknowledged. Furthermore, we thank Profs. H. Ringsdorf and J. B. Lando as well as Dr. Naegele for helpful discussions.

REFERENCES

1) A. Cemel, T. Fort and J. B. Lando, J. Polymer Sci. A-1, 10, 2061 (1972)
2) M. Puterman, T. Fort and J. B. Lando, J. Colloid Interface Sci. 47, 705 (1974)
3) R. Ackermann, D. Naegele and H. Ringsdorf, Makromolekulare Chem. 175, 699 (1974)
4) V. Enkelmann and J. B. Lando, J. Polymer Sci., Polymer Chem. Ed. 15, 1843 (1977)
5) D. Naegele, J. B. Lando and H. Ringsdorf, Macromolecules, in press
6) G. Wegner, Makromolekulare Chem. 154, 35 (1972)
7) G. Wegner, Pure and Appl. Chem. 49, 443 (1977)
8) H. Eichele, M. Schwoerer, R. Huber and D. Bloor, Chem. Phys. Letters, 42, 342 (1976)
9) B. Reimer, H. Baessler, J. Hesse and G. Weiser, Phys. Status Solidi B, 73, 709 (1976)

10) G. Wegner in "Chemistry and Physics of One-Dimensio-
    nal Metals", H. J. Keller, Ed., Plenum Publ. Corp.,
    New York, 1977, p. 297
11) V. Enkelmann and G. Wegner, Angew. Chem. 89,
    432 (1977)
12) B. Tieke, H.-J. Graf, G. Wegner, B. Naegele,
    H. Ringsdorf, A. Banerjie, D. Day and J. B. Lando,
    Colloid and Polymer Sci., 255, 521 (1977)
13) B. Tieke, G. Wegner, D. Naegele and H. Ringsdorf,
    Angewandte Chem. Int. Ed. Engl. 15, 764 (1976)
14) B. Tieke, G. Lieser and G. Wegner, J. Polymer, Sci.,
    Polymer Chem. Ed., in press
15) B. Tieke, D. Day and B. Naegele, unpublished
    observations
16) K. Takeda and G. Wegner, Makromol.Chem.160,349(1972)
17) S. Tolansky, Surface microtopography, London 1960
18) G. Bücher et al. Molecular Crystals and Liquid
    Crystals 2, 199 (1967)
19) G. C. Stevens and D. Bloor, Chem. Phys. Letters 40,
    37 (1976)
20) W. Hersel, U. C. Wolf and H. Sixl, to be published

ATOMIC AND MOLECULAR SCATTERING FROM SURFACES

ELASTIC SCATTERING

H. Wilsch

Physikalisches Institut der Universität
Erlangen-Nürnberg, Erwin-Rommel-Str. 1,
D 8520 Erlangen, W-Germany

ABSTRACT

The scattering of neutral thermal atoms and molecules (typical data: temperature 300K; kinetic energy 50
meV, De Broglie wave length 1 Å) from solid surfaces has
born out in the last 15 years many interesting and detailed results. With regard to physical (not chemical)
interaction of atoms with surfaces elastic scattering
events are possible among them most strikingly diffraction. An important special diffraction process is the
resonant transition of atoms to bound states at the
solid surface. The analysis of elastic events yields information on the gasatom-crystal surface interaction potential, on the surface topology (i.e. lattice constant
and steps), on the thermal mean square displacement of
surface atoms, on surface phonons and surface impurity.

This paper presents some theoretical background,
introduces to experimental methods and discusses selected
and very recent results as examples for the efforts towards a better understanding of gasatom-solid surface
interaction, which is important in fundamental research
as well as in aerodynamics, corrosion and heterogeneous
catalysis.

1.   INTRODUCTION

     The scattering and diffraction of atomic or molecu-
lar beams from crystal surfaces has taken a development
similar to low energy electron diffraction: the first
succesful experiments have been performed to prove the
De Broglie wave length hypothesis for atomic particles
/1-4/. With the progress of UHV techniques and other surface analytical methods, the molecular beam scattering
has shown a continous increase in number, refinement of
performance and significance of results. The same of
course is true for theoretical models and calculations,
and hence it can be stated that many phenomena in this
field are well understood. Information was gained in
different aspects: the diffraction of gasatoms from
crystal planes /5-9/, the interaction potential of thermal gas atoms with single crystal surfaces /10,11,6,9/,
the attenuation of elastically scattered intensities by
the Debye-Waller factor for increasing crystal temperatures /12-14,7/; translational-rotational energy transfer in the scattering of molecules from ordered surfaces
/15,16/; and some other problems, e.g. concerning band
structure effects for atoms in bound surface states
/17,18/, the thermal expansion coefficient of the topmost
layer of a crystal parallel to the surface /19/ etc.
There can be found a number of rather recent review
papers, stressing one or another aspect of gas-surface
interaction, some of them are /20-24/.
Other techniques and results in the field of molecular
scattering as time of flight measurements for inelastically scattered neutrals and mass analysis of reactively
scattered molecules are emerging rapidly but will not be
discussed in this paper.

2. PHYSICAL BASICS OF GASATOM-CRYSTAL SURFACE INTERACTION
              AND SELECTED EXAMPLES

2.1   The Interaction Potentional

     We will now shortly discuss the fundamental physi-
cal principles underlying the diffraction of atoms and
molecules from solid surfaces. Consider a gas particle
(e.g. H, H_2, He, Ne,...) approaching a crystal surface
along the surface normal and exclude chemical interaction. Then there will be an attractive interaction of
Van der Waals type which is composed of the pairwise
(momentary electric dipole-induced dipole) interactions
between the gas particle and many crystal atoms. Finally

at very small distance to the topmost layer of the crystal there will be repulsion produced by overlap of the
electron density distributions. Quantitatively the interaction potential will depend on the exact position of the
gas atom relativ to the unit cell of the surface lattice
resulting in a periodic change of the potential parallel
to the surface. (Note that the probing gas atom itself
has a diameter in the order of 2.5Å). These facts are
illustrated schematically in fig.1 and more quantitatively in fig.2, where an equipotential surface (1000
cal/mol) for the system He→ LiF (001) is shown (calculated, taken from ref.25) this representing the surface
where a He atom of kinetic energy equal to 44 meV has to
turn around.

With $\underline{r} = (\underline{R} = (x,y),z)$, the gas atom position (z-
direction = surface normal, $\underline{R}$ parallel to surface) and
$G = (m,n)$ the reciprocal surface lattice vectors a convenient mathematical description of the interaction
potential $V(\underline{r})$ is given by

$$V(\underline{r}) = \sum_{\underline{G}} v_{\underline{G}}(z) \cdot e^{i\underline{G}\cdot\underline{R}} \qquad (1)$$

The main term for $\underline{G} = \underline{0}$ only depends on z and may be
represented by a Morse potential

$$v_o(z) = D \left\{ e^{-2\varkappa z} - 2 e^{-\varkappa z} \right\} (2)$$

with depth D and reciprocal range parameter $\varkappa$ (other
model potentials are also possible). The periodicity
of the potential parallel to the surface is contained in
the higher order terms for $\underline{G} \neq \underline{0}$, with the coefficients
$v_{\underline{G}}(z) = \beta_{\underline{G}} \cdot D \cdot e^{-2\varkappa z}$ (3), β gives the relative strength
(for alkalihalides in the order of a few percent). A
realistic example is shown in fig.1c, demonstrating that
these higher order terms represent small corrections to
$v_o(z)$. Regarding only the main potential term (2) (equivalent to a mean potential averaged parallel to the surface) the gas atom has discrete bound states with binding
energies (note that the atom is bound only in z-direction)

$$E_j = - \left(\frac{\sqrt{2mD}}{\varkappa \cdot \hbar} - j - \frac{1}{2} \right)^2 \cdot \frac{\varkappa^2 \cdot \hbar^2}{2m} \qquad (4)$$

where m = mass of gas atom, $j = 0,1,2\ldots$ (restricted by
the condition that the expression within the brackets

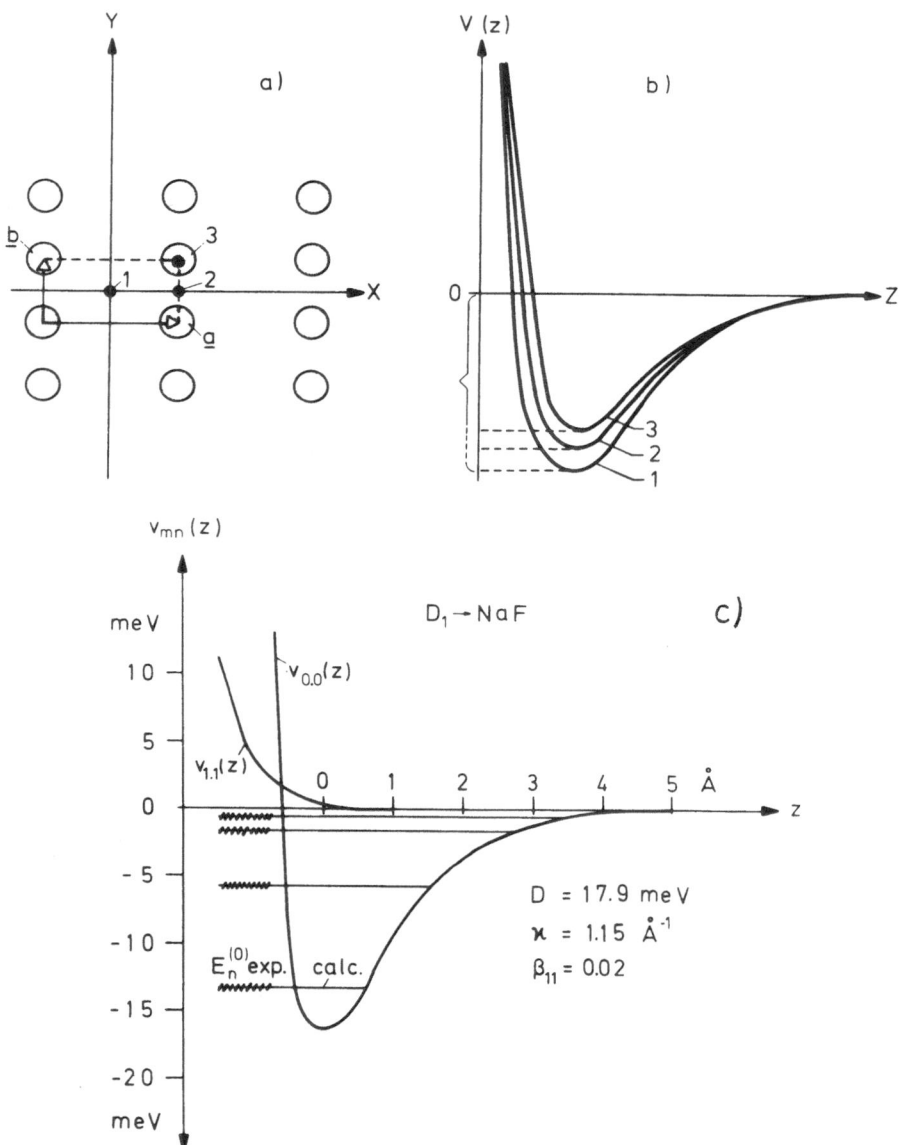

Fig.1   Gas atom-crystal surface interaction potential
        (b) as a function of the coordinate perpendicular
        to the surface for three different positions re-
        lative to the unit cell as indicated in part a.
        c) Morse potential with bound state energies for
        deuterium on NaF(001), showing $v_o(z)$ and the
        higher order term $v_{(1,1)}(z)$.

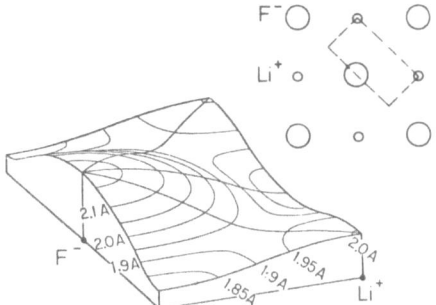

Fig.2   Equipotential surface for He atoms of 44meV on a
        LiF(001) plane.

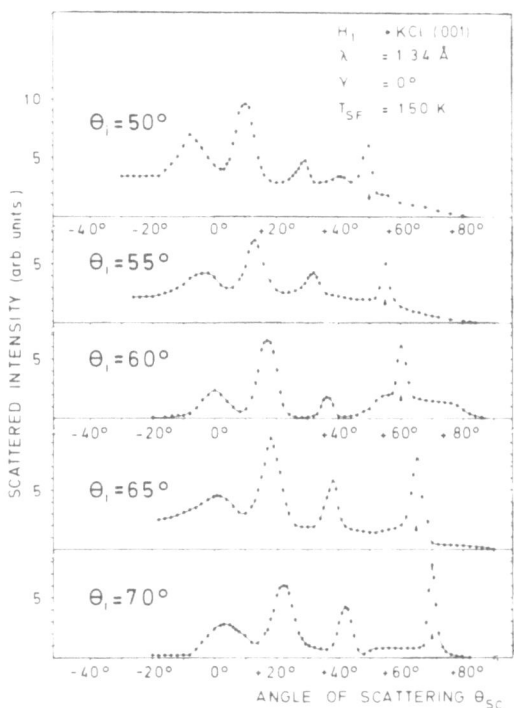

Fig.3   Diffraction peaks in the angular distribution of
        hydrogen atoms (λ_i=1.34Å), scattered from water
        covered KCl(001), T_{SF} = 150K, for several angles
        of incidence θ_i as indicated in the figure; $\langle 110 \rangle$
        crystallographic direction in the scattering
        plane (γ = 0°), peaks from left to right: (-3,0);
        (-2,0);(-1,0);(0,0).

must be non - negative). This is illustrated by the results for atomic deuterium on NaF(001) as shown in fig.
1c. The experimental determination of the binding energies will be discussed in a following section.

2.2 Diffraction, Surface Corrugation, and Debye-Waller
Attenuation

The 2-dimensional periodicity of the potential, according to (3) most pronounced in the repulsive part of the
potential, may cause diffraction. Let us describe incident and outgoing gas atoms by their wave vectors $\underline{k}_i$
and $\underline{k}_f$ and the components $\underline{K}_i$ and $\underline{K}_f$ of these parallel to
the surface. Then diffraction implies

$$k_i^2 = k_f^2 \qquad (5)$$

and

$$\underline{K}_f(\underline{G}) = \underline{K}_i + \underline{G} \qquad (6)$$

for energy and momentum respectively.

From equations (5) and (6) the spatial position of
diffracted beams is determined. An experimental example
is shown in fig. 3 for hydrogen atoms scattered from a
(water covered) KCl(001) plane /7/. Besides a broadening of the peaks with increasing order of diffraction
as a consequence of a finite $\Delta\lambda$ in the beam it can be
clearly noticed that e.g. the peak (-2,0) exceeds the
(-1,0) peak intensity and that the beam (-4,0) is de-
finitely missing. In a simple picture this is very
similar to diffraction from an optical grating where
it is well known that the intensities in the different
orders of diffraction depend strongly on the detailed
structure within one lattice constant (transmission function). In our case the relative intensities depend on
the corrugation of the repulsive potential wall (compare
fig.2), which may be extracted from a comparison of the
experimental results to an appropriate model calculation
/7/.

As in the case of electron, neutron or X-ray diffraction the elastic intensity I_G in diffracted atomic
beams is diminished by inelastic processes (energy
transfer from or to lattice phonons) taken into account
by the Debye-Waller-factor e^{-2W}:

$$I_{\underline{G}} = \left| A_{\underline{G}} \right|^2 \cdot e^{-2W} = \left| A_{\underline{G}} \right|^2 \cdot \exp\left\{ -\left\langle (\underline{\Delta k} \cdot \underline{u})^2 \right\rangle \right\}$$

where $\underline{\Delta k} = \underline{k}_i - \underline{k}_f$, $\underline{u}$ = instantaneous displacement of surface atom from equilibrium position, $\langle \ldots \ldots \rangle$ = thermal average, $\left| A_{\underline{G}} \right|^2$ = intensity of diffracted beam for perfectly rigid lattice.

Within the Debye-model for solids one gets for the specular beam ($\underline{G}$= (0,0)) in the high temperature limit $T_{SF} \gtrsim \Theta_D$:

$$e^{-2W} = \exp\left\{ -\frac{24m_g \cdot E_i \cdot T_{SF}}{k_B \cdot m_{SF} \cdot \Theta_D^2} \cdot \left(\cos^2\Theta_i + \frac{D}{E_i} \right) \right\} \quad (8)$$

m_g, m_{SF} = mass of gasatom and surface atom respectively, E_i = kinetic energy of incident gas atom, T_{SF} = surface temperature, Θ_D = characteristic surface Debye temperature for oszillations perpendicular to the surface, k_B = Boltzmann constant, Θ_i = angle of incidence relative to surface normal.

It should be noted that with the term $\frac{D}{E}$ in (8) the accelaration of the incoming particle in the attractive part of the potential has been explicitly taken into account, resulting from the fact that D is comparable to E_i for thermal atoms and that the repulsive collision is localised in the surface whereas the preceding acceleration accounts to a much larger crystal region. The functional dependence of (8) on D can be used to check or to determine the potential depth D, but usually from such a measurement the surface Debye temperature is extracted. The experimental situation is shown in fig. 4 where velocity selected hydrogen atoms have been scattered from a (clean) NaF(001) surface /26/. The surface Debye temperature in this kind of experiment turns out to be smaller than the bulk value as also found in LEED.

2.3   Bound Surface States and Resonant Transitions

Quantum mechanically, the attractive, finite range interaction potential provides bound states with discrete binding energies for the gasatom at the crystal surface. For a Morse potential these binding energies are given

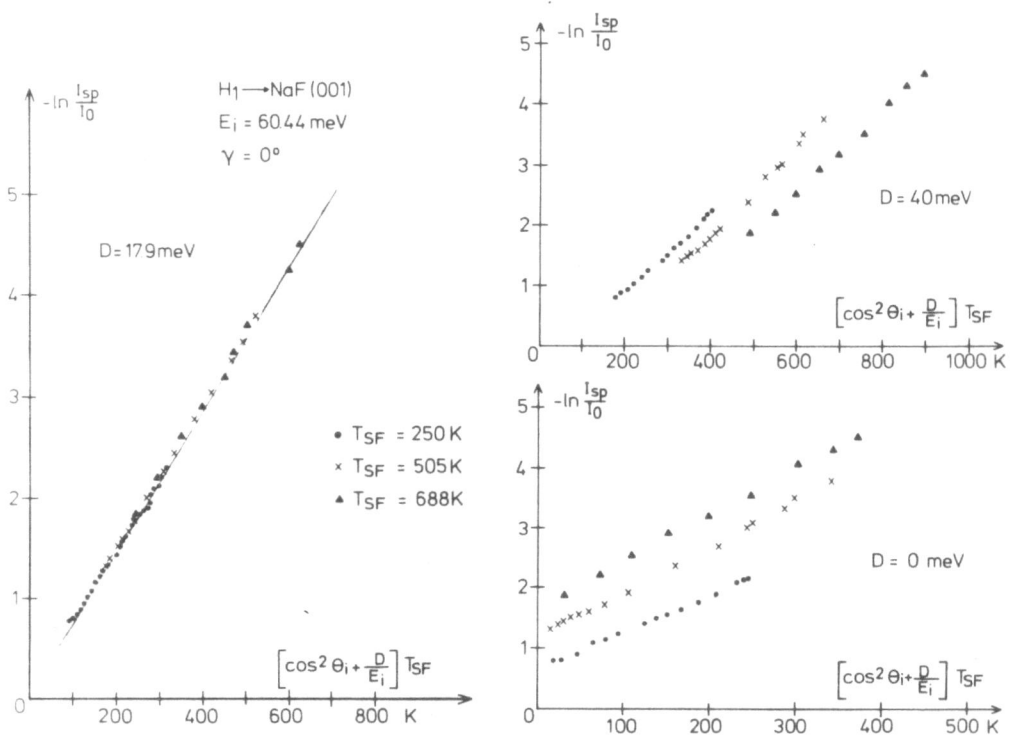

Fig.4   Logarithmic plot of specular intensity (normalized
to incident intensity) versus the product of surface temperature times momentum transfer (perpendicular to the surface). Left side: exact attractive potential D = 17.9 meV included, all data
points arrange to one straight line defining a
surface Debye temperature. Right side: Plots without or with wrong attractive potential not defining a unique surface Debye temperature.

by formula (4). These states are bound with respect to
the direction normal to the surface but not localized
parallel to the surface. Therefore, elastic processes are
possible, where an incoming gas atom enters a bound state
undergoing diffraction in a manner whereby the usual con-
servations of energy and momentum apply and the final
momentum of the gas atom is parallel to the surface. As
both binding energy and momentum change in diffraction
are discrete, the process described above will only be
possible for certain kinematic conditions of the impinging gas atoms ("resonant transition to a bound state" or
"selective adsorption"). The resonance condition can be
easily derived: From equations (5) and (6) the final
kinetic energy of a gas atom perpendicular to the surface is obtained as

$$E_{\perp,\underline{G}} = \frac{\hbar}{2m} \; (k_i^2 - K_f^2 \, (\underline{G})) \tag{9}$$

which may become negative. In this case, (9) represents
a possible diffraction process only, if

$$E_{\perp,\underline{G}} = E_j < 0 \tag{10}$$

For the specular beam at fixed angle of incidence Θ_i and
fixed wavelength λ of the incoming gas atoms the resonance can be obtained by rotating the crystal around its
surface normal until at certain values of the azimuth
angle $\gamma = \gamma_{min}$ the following condition is fulfilled

$$E_j = h^2 / (2m_g \lambda^2) \cdot \left\{ \cos^2\Theta_i - (\lambda/d)^2 (m^2 + n^2) \right.$$

$$\left. -2(\lambda/d) . \sin \Theta_i . (m \cdot \cos \gamma_{min} + n \cdot \sin \gamma_{min}) \right\} .$$

Here E_j is the binding energy of the state with quantum
number j, h = Planck's constant, m_g = mass of gas atom,
d = lattice constant, and (m,n) = order of the reciprocal
lattice vector involved. (By rotating the crystal around
its surface normal one changes the angle between $\underline{K}_i$ and
the reciprocal lattice vectors $\underline{G}_{(m,n)}$ which of course are
fixed to the lattice).

Resonant transitions to bound states produce minima
in the elastic intensity of the specular beam, as atoms
which have gone into a bound state have either high probability to suffer inelastic processes, or may undergo a

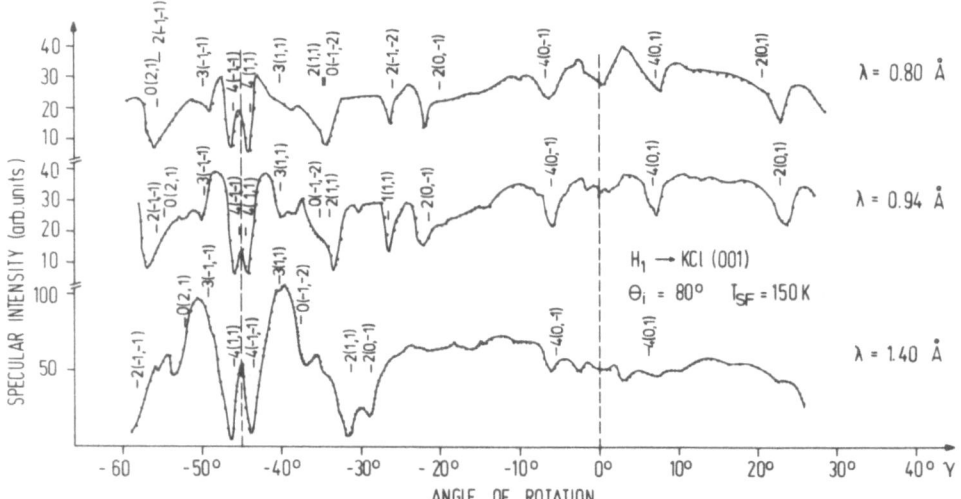

Fig.5 Selective adsorption minima in the specular inten-
       sity of atomic hydrogen, scattered from a water
       covered KCl(001) surface at T_{SF} = 150 K; crystal
       rotation γ around the surface normal. Velocity
       selected beams of wave lengths λ = 0.80, 0.94 and
       1.40 Å at an angle of incidence Θ_i = 80 have
       been used. Resonant transitions to bound states
       with quantum number j via a reciprocal lattice
       vector (m,n) are marked by j(m,n).

second diffraction by which these atoms reappear in another diffracted peak producing maxima there. Fig. 5
shows a typical measurement of the specular intensity as
a function of the rotational angle γ, here we observe
minima (as in most of the other observable diffracted
beams). Fig. 6 demonstrates, that in certain beams also
maxima may occur at resonance conditions. No assumption
concerning form and depth of the potential are necessary,
the correct labelling of the binding energies is achieved
by repeated measurements under varying conditions (other
Θ_i, λ_i), by the isotopic effect when interchanging hydrogen and deuterium, and by the symmetry relations inherent
in the resonance condition. A careful analysis of a large
number of similar measurements yields the bound state
energies E_i and finally by a fitting procedure the depth
D and the reciprocal range χ of the Morsepotential. The
quality of such an investigation is demonstrated in fig.
7 where atomic hydrogen and deuterium have been scattered
from clean LiF(001) and NaF(001) surfaces /6/.

2.4   Band Structure Effects or Channel Coupling

Recently Chow and Thompson /27/ discussed theoretically effects in gas-surface diffraction caused by strong
coupling of bound state channels to diffracted channels
and of bound state channels to one another. An example
for the first type of coupling is contained in the appearance of maxima at resonance condition as shown in
fig. 6. In this case the (0,1) beam was strongly coupled
to the bound state channel $E_{j=0}$ (1,0) via the v_{11} term
in the potential series (1). Or with other words: atoms
which have entered the bound state $E_{j=0}$ by a diffraction
of type (1,0), undergo a second diffraction of type
(-1,1) and reappear in the (0,1) beam.

Another example of strong coupling where two bound
states are involved will be discussed now. The minima in
the scattered intensity distribution as produced by
resonant transitions of atoms to bound surface states
have already been explained. In certain cases these minima showed a double minimum structure which could not
simply be attributed to bound state resonances according
to equ. (11). This is demonstrated in fig. 8 where the
arrows show two reciprocal lattice vectors (0,-1) and
(-1,-2) to fulfill approximately the resonance condition
with the bound state of energy E_o in the γ-range investigated. But especially at coincidence of the two
arrows at the same resonance angle γ_{res} we observe two

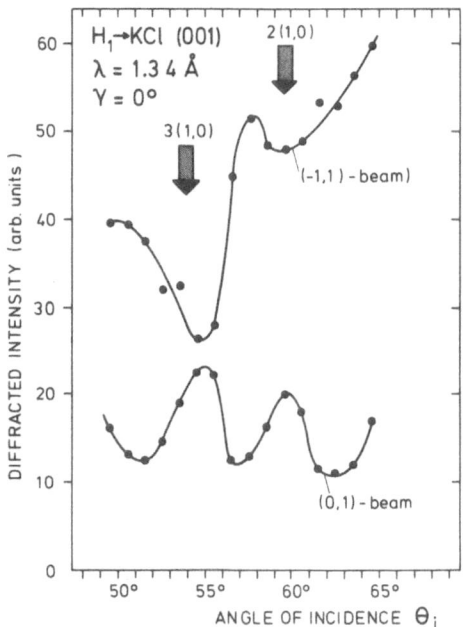

Fig.6   Diffracted intensities in the (0,1) and (-1,1)
        beams for atomic hydrogen on a KCl(001) surface
        as a function of the angle of incidence. Labelling
        j(m,n) of resonant transitions as in fig.3, showing
        up as minima in the (-1,1) beam and as maxima in
        the (0,1) beam.

minima at both sides of γ_{res}. Following the ideas of Chow and Thompson /27/, we explain this effect by energy splitting of admixed degenerate bound states. The two resonant bound states $\varphi_{0(0,-1)}$ and $\varphi_{0(-1,-2)}$ with the same total energy at $\gamma = \gamma_{res}: E_{0(0,-1)} = E_{0(-1,-2)} = E_i$ are mixed because of the perturbation from the periodic term $v_{(0,1)-(-1,-2)} = v_{11}$. For the new mixed states ψ_a, ψ_b we get from perturbation theory the new energy eigenvalues

$$E_{a,b} = \frac{1}{2} \left\{ E_{0(0,-1)} + E_0(-1,-2) \pm \left(\left[E_{0(0,-1)} - E_0(-1,-2) \right]^2 + 4 H_{12}^2 \right)^{1/2} \right\} \qquad (12)$$

with

$$H_{12} = \beta_{11} D \cdot \exp(\chi^2 \hbar / m \omega)$$

and

$$\omega = 2(D - |E_0|)/\hbar$$

(here the assumption of harmonic oscillator wavefunctions for the z-motion of the bound atoms was made). In the new resonance conditions

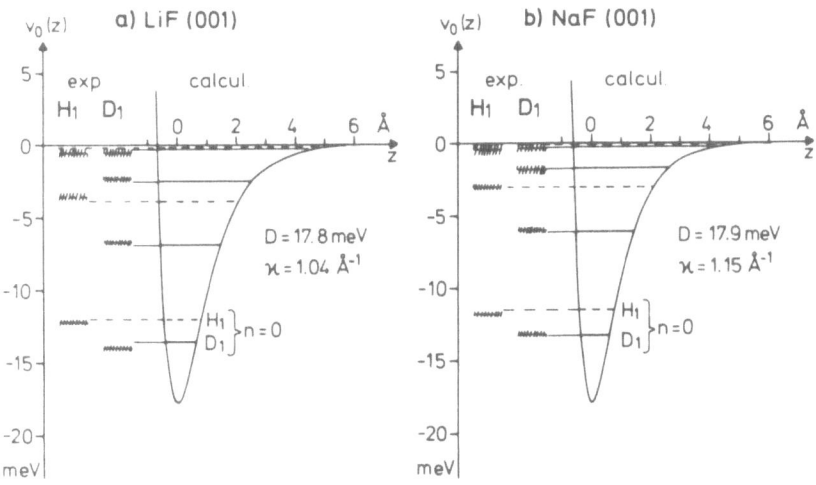

Fig.7   Actual gas atom-crystal surface potential for atomic hydrogen and deuterium on LiF and NaF with bound state energies, experimentally and calculated from a Morse-potential with depth D and reciprocal range χ as indicated.

$$E_a(\gamma_{a,min}) = E_i \quad \text{and} \quad E_b(\gamma_{b,min}) = E_i$$

the quantities D $=17.9$ meV, $\chi = 1.15\text{Å}^{-1}$ and $E_0 = -13.3$ meV are known from ref. /6/, so that the only unknown parameter β_{11} could be determined from the observed splitting to be $\beta_{11} = 0.02$. Fig. 9 shows the curves of the new resonance angles $\gamma_{a,min}$ and $\gamma_{b,min}$ as calculated with this parameter in the investigated range of E_i. These calculated curves fit quite well to the observed angles of minima. The value of $\beta_{11} = 0,02$ compares quite well to our results achieved previously with other methods /6/. A similar minima splitting in the diffraction of He from NaF(001) has been reported recently by Liva, Derry, and Frankl /17/. From a closer look to fig. 9 or formula (12)

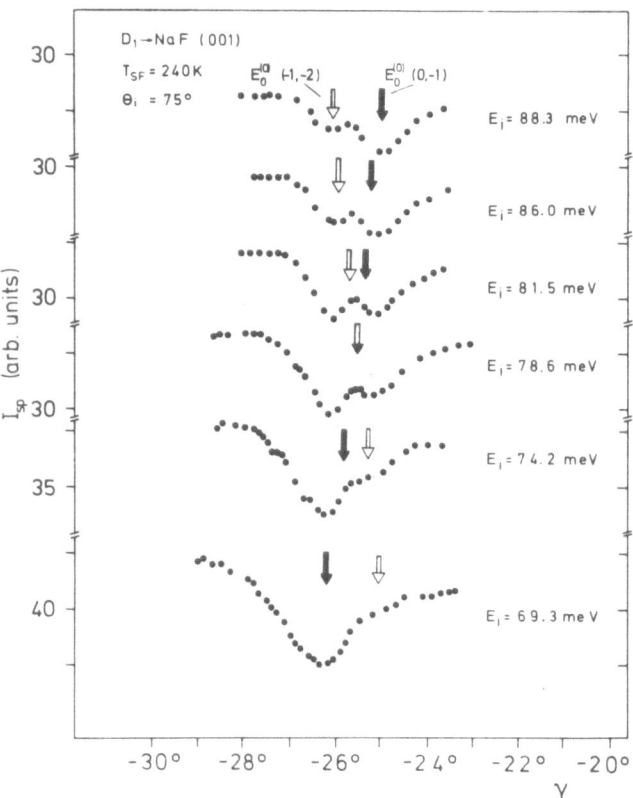

Fig.8   Splitting of bound state resonance minima in the
        specular intensity of atomic deuterium scattered
        from NaF(001). The contributing bound channels
        are indicated by arrows.

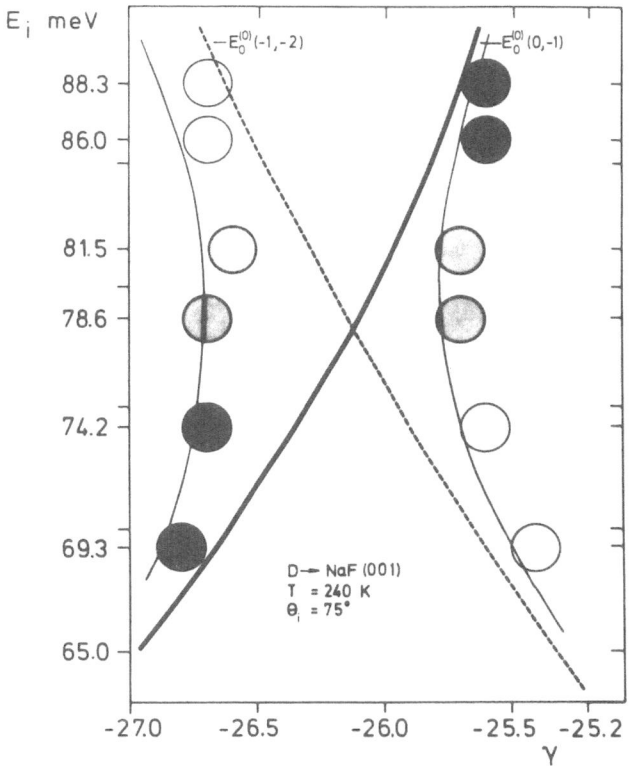

Fig.9 Calculated dependence of incident energy E_i on
       azimuthal angle γ under bound state resonance
       conditions for independent bound channels (cross-
       ing lines) and for admixed bound channels with
       ß=0.02. The disks represent the experimentally
       observed angles of resonance minima (the darkness
       indicates the depth of the minimum).

it becomes clear that the influence of the periodic
potential term on the wave functions and energy eigenvalues is to produce a forbidden energy gap, the wave
function of an atom in a bound surface state (where it
is moving parallel to the surface) becomes a Bloch wave.
Therefore, this observed behaviour is a (two dimensional)
band structur effect. The most important feature is the
fact, that by it the strength of the most important
higher order term in the potential series can be determined quantitatively (compare formula (1) and fig 1c).

2.5   Inelastic Processes

There are two special inelastic processes in molecular beam scattering which could be noticed in experiments without energy analysis of the scattered molecules
(this being difficult).
i) In diffraction experiments of light molecules as H_2,
D_2 from crystal surfaces it was found, that by the elastic interaction with the surface translational energy
of the molecules may be transferred into rotational
energy and vice versa. Energy and momentum conservation
in this case read

$$k_f^2 = k_i^2 \pm \frac{2m}{\hbar} \cdot \Delta E_{rot} \qquad \text{and}$$

$$\underline{K}_f = \underline{K}_i + \underline{G}$$

As a consequence, additional diffracted beams are observed according to the quantised rotational energies
and the selection rules involved. Experimental work can
be found in /15/ and /16/, a quantum theoretical treatment in /28/. The practical importance of these features
however is low. Of more importance is the interaction
of atoms with surface phonons: indeed in atomic diffraction it has been possible to observe creation and annihilation of surface phonons and to deduce the dispersion
relation /29/. These experiments were done in the system
He/LiF(001). Let us consider a surface Raleigh mode with
quasi momentum $\underline{q} = (\underline{Q}, q_z = 0)$ and quantum energy $\hbar \omega_q$.
Then we have the following equations for energy and
momentum of the scattered atoms:

$$k_f^2 = k_i^2 \pm \hbar \omega_q \cdot \left(\frac{2m}{\hbar^2} \right) = \left| k^{\pm} \right|^2 \quad \text{and}$$

$$\underline{K}_f = \underline{K}_i + \underline{G} + \underline{Q} \; ,$$

where + stands for phonon annihilation and - for phonon
creation. In the corresponding Ewald construction for $\underline{k}_f$
it can be easily demonstrated that diffraction connect-
ed with a one-phonon-interaction produces additional
features in the vicinity of diffracted beams, spatially
(and energetically) separated from the diffracted beams.
Scattered atoms with preceding phonon creation are also
spatially separated from those after phonon annihilation.
Also phonon interaction can be connected to the resonant
transition to bound surface states, again producing
additional structure in the scattering distribution which
can by analysed to yield the surface phonon dispersion
relation /30/. A more direct method for studying phonon
interaction in atomic beam scattering of course would
require energy analysis of the scattered atoms.

2.6   Diffraction from Clean Metal Surfaces

    For a long time until very recently, diffraction of
atoms from clean metal surfaces had only been observed in
the system He/W(112), see /31,32/. This surface shows
a structure of rows and graves and diffracted beams of
order (0,0) and ($\pm$1,0) could be found if the plane of
incidence was adjusted perpendicular to the rows. Recently diffraction of helium from an ordered stepped Cu(117)
surface was also observed /33/. Bragg diffraction from
randomly stepped Cu(100), arising from interference
between beams from planes separated by atomic steps perpendicular to the surface has also been found /34/. Yet
for smooth metal surfaces there is only the system
He/Ag(111) for which sharp diffracted beams of very low
intensity could be observed /35,36/. Missing diffraction
from metal surfaces can be explained by too low periodic
parts in the gasatom - metal surface interaction potential as large contributions to this arise from the conduction electrons of the metal the spatial distribution
of which being softly smeared out at the surface.

2.7   The Coherence Length of Atoms in a Thermal Beam

    It is clear that diffraction from structures can
only be observed if the coherence length of the waves

is larger than the lattice constant of the structure,
formulas for the coherence length therefore are given
for electron microscopy e.g. in /37/, for LEED e.g. in
/38/ and the corresponding ideas of course have to be
applied to the diffraction of atoms. But there are mis-
interpretations sometimes and therefore for the sake of
clearness some comments seem to be necessary.

From a look to the intensity of atomic particles
which are used in experiments it becomes clear that these
particles are well separated in space and time and therefore cannot interfere with each other. Consequently the
probabilities for diffraction depend on the internal
structure of each individual wave packet. According to
the Heisenberg uncertainty principle

$$\Delta k \cdot \Delta l \gtrsim 2\pi$$

for each of the 3 dimensions of the wave packet, the spatial extent of this represents the coherence length which
is related to the uncertainty of the wave-number. The
two coordinates perpendicular to the direction of propagation can be regarded as equivalent, we therefore have
to worry about $\Delta k_{\parallel}$, the component parallel to the direction
of propagation, and $\Delta k_{\perp}$, the component perpendicular to
that. By use of the relation

$$k^2 = \frac{2m}{\hbar^2} \cdot E \quad ,$$

$\Delta k_{\parallel}$ is usually expressed as $$\Delta k_{\parallel} = \frac{\pi \cdot \Delta E}{\lambda \cdot E}$$
from which one has

$$\Delta l_{\parallel} \approx \frac{2\lambda E}{\Delta E}$$

But $\Delta E/E$ does not represent the thermal energy spread of
the beam of atoms, molecules or electrons: quantum theoretically ΔE is determined by the uncertainty in phase
space as related to the last collision process which
emitts the particle into the beam direction. The corresponding $\Delta E/E$ for atoms is much better defined than by
the maxwellian velocity distribution. It does not matter
in this context whether or not a beam containes very slow
and very fast particles: they cannot and they do not have
to interfere with each other. A precise determination of
$\Delta E/E$ for the fundamental creation process however remains difficult.

Let us now consider $\Delta k_{\perp}$. When a point at the diffrac-
ting surface is excited by an incoming wave packet, the
uncertainty $\Delta k_{\perp}$ is determined by the angle under which
the opening of the beam source (radius r_s) is viewed by

the surface at the distance d downstream($\beta_s \approx r_s/d$ for
$d \gg r_s$). (Here it is important to notice that β_s is not the
angular divergence of the beam). With this geometry in
mind one easily finds

$$\Delta k_\perp \approx 2\beta_s \cdot (k + \Delta k_\parallel) \quad \text{and therefore}$$

$$\Delta \ell_\perp \approx \frac{\lambda}{2\beta_s \cdot (1 + \frac{\Delta E}{2E})}$$

For arbitrary angle of incidence θ_i (from the surface
normal) one has to sum the projections of $\Delta \ell_\parallel$ and $\Delta \ell_\perp$
to the surface, resulting in

$$\Delta \ell_{coh} \approx \Delta \ell_\perp \cdot \cos \theta_i + \Delta \ell_\parallel \cdot \sin \theta_i \quad .$$

(The relevant data for the atomic hydrogen beam at Er-
langen /5/ are: $\lambda = 1\text{Å}$, $\beta_s = \frac{1mm}{0.5m} = 2.10^{-3}$. With $\frac{\Delta E}{2E} < 10^{-2}$
the coherence length is always larger than 100 Å).
Concerning the coherence length there is not much difference between a nozzle beam and a thermal beam (emerging
from a Knudsen cell).

3. EXPERIMENTAL METHODS IN ATOMIC BEAM SCATTERING FROM
SOLID SURFACES

3.1   Atomic Beam Apparatus with Surface Analysis
Equipment

The progress in UHV technique and surface analysis
methods is utilized for well defined, reliable atomic
beam scattering experiments. Many laboratories therefore have completed or are setting up atomic beam devices
with scattering chambers which also allow in situ surface
analysis. Such a scattering chamber is shown in fig. 10
and described in more detail in /39/. In addition to an
entrance for an atomic beam, it contains a gas inlet
system, a four grid LEED-Augerelectronspectroscopy system,
a quadrupole mass filter for secondary ion mass spectroscopy and gas analysis, and electron- and ion guns, the
sample may be heated or cooled (LN$_2$), shifted (3 dimensions) and rotated (2 axis). Other facilities as e.g.
liquid helium cryostat /35/, in situ crystal cleavage
/17,7/, or extended drift path for time of flight analysis /40/ of the scattered atoms, have also been used but
in general are difficult to combine with surface analysis methods.

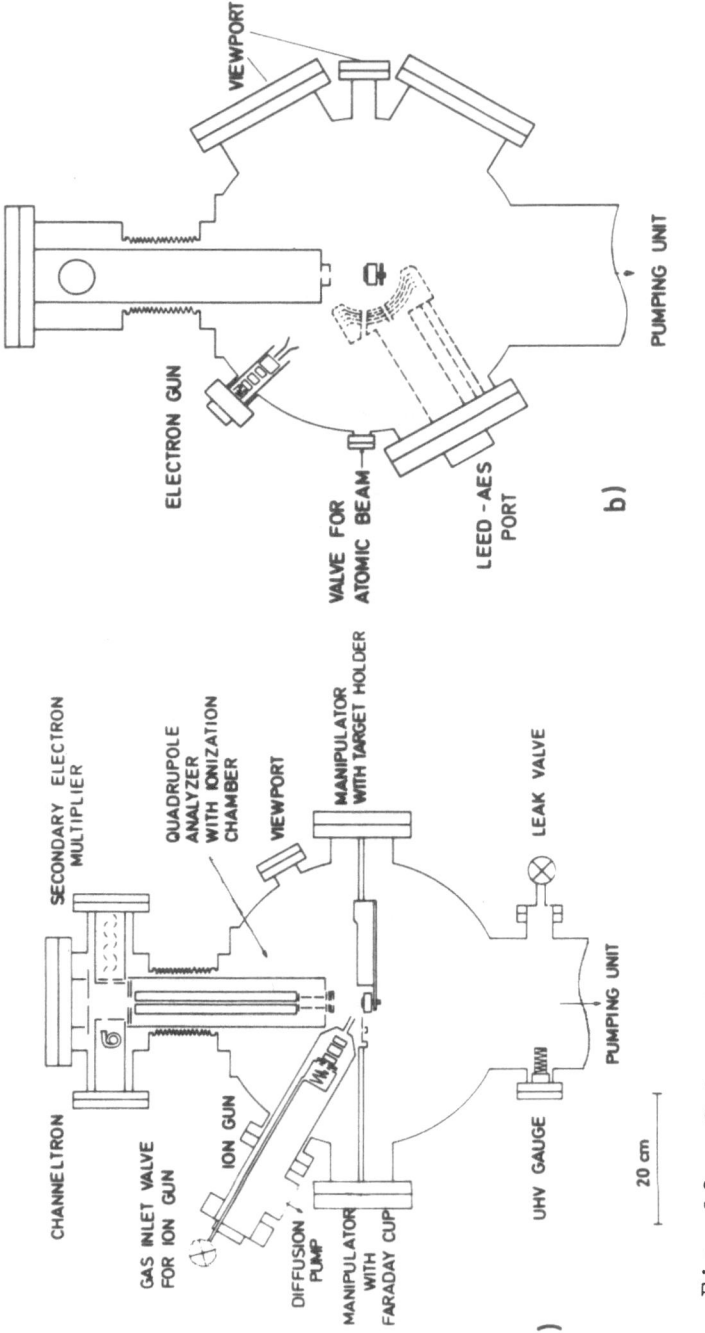

Fig. 10. Schematic views of the spherical UHV chamber constructed for in situ SIMS, LEED, AES, and atomic beam scattering analysis on solid surfaces: (a) plane with SIMS elements, perpendicular to the atomic beam axis; (b) plane perpendicular to (a), containing the atomic beam axis.

3.2   Atomic Beam Sources

     The simplest (and oldest) device for producing an
atomic (or molecular) beam which can be used for all
stable compounds of sufficiently high vapour pressure
at temperatures below~3000K, is a cell with an orifice
(Knudsen cell) towards the vacuum system and one or more
diaphragms downstream. The emerging beam is thermal
(with respect to the temperature of the Knudsen cell),
i.e. it has a maxwellian velocity distribution modified
by an additional factor v (= particle velocity) which
takes account of the probability for particles to pass
through the orifice of the cell. The pressure in the cell
is maintained at a level where the mean free path of the
particles is smaller than the dimensions of the orifice,
therefore the temperature of this directly determines
the beam temperature, a fact which can be utilized in
experiments /41/. It is also possible to excite or dissociate the particles in the cell before they are emerging.
Another source for atomic or molecular beams is the
supersonic beam (nozzle beam). Here the cell is operated
at higher pressures and by the use of a skimmer, rather
complex gas dynamical effects cooperate to produce beams
with particle velocities larger than "thermal" and a
velocity distribution which is narrower than maxwellian,
the beam intensities achieved are often higher (up to a
factor of 100) than in thermal beams, but e.g. until now
it is not possible to produce a supersonic beam of atomic
hydrogen. A useful summary of the relevant parameters and
formulas for supersonic beams together with references
to the literature may be found in /42/.

3.3   Atomic Beam Detection

     The detection of low energy neutral atoms or mole-
cules is not trivial, especially not for low intensities.
Methods that have been used include chemical reactions
(with change of colour), ionisation of alkali atoms on
hot filaments (Langmuir-Taylor detector), transfer of
thermal energy to bolometers (at low temperature), iono-
sorption of chemically active atoms or radicals to semiconductor surfaces with subsequent change in electric
conductivity /45,44/ and ionisation gauges in the stag-
nation mode (ionisation gauge with entrance channel
where beam particles enter without friction but then are
made diffusive). A very versatile method is given by

ionisation and subsequent mass spectrometry, especially if
identification of particles or quick response (as in TOF-
measurements) is required. A more detailed summary with
further references can be found in /43/.

3.4   Surfaces, Preparation, and Properties

The preparation of the single crystals and their
surfaces of course follows the usual standards of surface physics. As pointed out in 3.1, the modern surface
analysis methods enable reliable statements on the cleanliness of the surface under study.
For example it could be unambiguously shown that heat
annealed cleavage planes of LiF(001) - which are often
used in experiments since 1929 - are extremely inert
with respect to the adsorption of impurities and water
even in modest vacua up to 10^{-6} torr, these surfaces remain clean for long times /39/. If for other surfaces
adsorption occurs, the mechanism (e.g. nucleation at
certain active sites) and the amount (e.g. one monolayer)
can be explored /39,46/. It is clear that these results
are important for the interpretation of the scattering
experiments.

4.   Conclusions

This review is intended to cover the present status
in the field of elastic scattering of thermal, neutral
atoms or molecules from crystal surfaces. For shortness
only examples for the most interesting and important
features could be presented with stress on the experimental work in this area· Reference to theoretical efforts only was made when necessary (a discussion of
theoretical aspects is found in /23,42/). It should be
noted however that the restriction to elastic scattering
excluded the discussion of very important aspects of
molecular beam scattering which are more closely related
to the fundamental problems of surface reactions, residence times, adsorption and desorption kinetics and
heterogenious catalysis where (modulated) molecular beam
techniques with phase sensitive detection are becoming
widely used.

5.   ACKNOWLEDGEMENTS

Many of the examples shown were taken from the work
of the group at Erlangen. This work was financially supported by the Deutsche Forschungsgemeinschaft (Grant Nr.
Wi 350) and benefitted from the contributions of many
students and permanent collaborators, especially Dr.
H. Hoinkes and Dipl.-Phys. H. Kaarmann.

REFERENCES

/1/   I. Estermann and O. Stern, Z.Phys. $\underline{61}$ (1930) 95
/2/   O. Stern, Die Naturwissenschaften $\underline{17}$ (1929) 391
/3/   T.H. Johnson, Phys.Rev. $\underline{35}$ (1930) 1299
/4/   T.H. Johnson, Phys.Rev. $\underline{37}$ (1931) 847
/5/   H. Hoinkes, H. Nahr, and H.Wilsch; Surface Sci. $\underline{30}$
      (1972) 363
/6/   H.-U. Finzel, H. Frank, H. Hoinkes, M. Luschka,
      H. Nahr, H. Wilsch, and H. Wonka, Surface Sci. $\underline{49}$
      (1975) 577
/7/   H. Frank, H. Hoinkes, and H. Wilsch, Surface
      Sci. $\underline{63}$ (1977) 121
/8/   G. Boato, P. Cantini, and L. Mattera, Surface
      Sci. $\underline{55}$ (1976) 141
/9/   J.A. Meyers and D.R. Frankl, Surface Sci. $\underline{51}$
      (1975) 61
/10/ H. Hoinkes, H. Nahr, and H. Wilsch, J. Phys. C
      (Solid State Phys.) $\underline{5}$ (1972) L143
/11/ A. Tsuchida, Surface Sci. $\underline{52}$ (1975) 685
/12/ J.L. Beeby, J.Phys. C (Solid State Phys.) $\underline{4}$
      (1971) L359
/13/ H. Hoinkes, H. Nahr, and H. Wilsch, Surface Sci. $\underline{33}$
      (1972) 516 and Surface Sci. $\underline{40}$ (1973) 457
/14/ H. Hoinkes, H.-U. Finzel, H. Frank, H. Nahr, and
      H. Wilsch, Proc. of the 9th Int. Symp. on Rarefied
      Gas Dynamics 1974, DFVLR Press Porz-Wahn, Ed. M.
      Becher and F. Fiebig, Vol.II, E7-1
/15/ G. Boato, P. Cantini, and L. Mattera, Japan. J.
      Appl.Phys. Suppl. 2, Pt.2 (1974) 553
/16/ R.G. Rowe and G. Ehrlich, J. Chem. Phys. $\underline{62}$ (1975)
      735 and J. Chem. Phys. $\underline{63}$ (1975) 4648
/17/ M.P. Liva, G. Derry and D.R. Frankl, Phys. Rev.
      Lett. $\underline{37}$ (1976) 1413
/18/ H. Hoinkes, L. Greiner, and H. Wilsch, Proc. 7th
      Intern. Vac. Congr. and 3rd Intern. Conf. Solid
      Surfaces (Vienna 1977)
      U. Wonka, Master Thesis, Erlangen 1973

/19/ H. Frank, H. Hoinkes, and H. Wilsch, Surface
     Sci. $\underline{64}$ (1977) 362
/20/ J.P. Toennies, Appl. Phys. $\underline{3}$ (1974) 91
/21/ G.A. Somorjai and S.B. Brumbach, CRC Crit.Rev. in
     Sol. State Sci. $\underline{4}$ (1974) 429
/22/ W.H. Weinberg, Adv. in Colloid and Interface Sci. $\underline{4}$
     (1975) 301
/23/ F.O. Goodman and H.Y. Wachman, Dynamics of Gas-
     Surface Scattering, Academic Press New York 1976
/24/ H. Wilsch, Vakuum Technik $\underline{24}$ (1975) Heft 2, p. 43
/25/ A. Tsuchida, Surface Sci. $\underline{46}$ (1974) 611
/26/ H. Wilsch, H.-U. Finzel, H. Frank, H. Hoinkes, and
     H. Nahr, Japan. J. Appl.Phys. Suppl. 2, Pt.2
     (1974) 567
/27/ H. Chow and E.D. Thompson, Surface Sci. $\underline{59}$
     (1976) 225
/28/ U. Garibaldi, A.C. Levi, R. Spadacini, and G.E.
     Tommei, Surface Sci. $\underline{55}$ (1976) 40
/29/ B.R. Williams, J. Chem. Phys. $\underline{55}$ (1971) 3220
     B.F. Mason and B.R. Williams, J. Chem. Phys. $\underline{61}$
     (1974) 2765
/30/ P. Cantini, G.P. Felcher, and R. Tatarek, Phys.
     Rev.Lett. $\underline{37}$ (1976) 606 and Surface Sci. $\underline{63}$
     (1977) 104
/31/ D.V. Tendulkar and R.E. Stickney, Surface Sci. $\underline{27}$
     (1971) 516
/32/ A.G. Stoll, Jr. and R.P. Merrill, Surface Sci. $\underline{40}$
     (1973) 405
     A.G. Stoll, Jr., J.-J. Ehrhardt, and R.P. Merrill,
     J. Chem.Phys. $\underline{64}$ (1976) 34
/33/ J. Lapujoulade and Y. Lejay, to be published
     (Surface Sci.)
/34/ J. Lapujoulade and Y. Lejay, to be published
/35/ G. Boato, P. Cantini, and R. Tatarek, J.Phys. F:
     Metal Physics, $\underline{6}$ (1976) L237
/36/ J.M. Horne and D.R. Miller, to be published
/37/ H.D. Heidenreich, "Fundamentals of Transmission
     Electron Microscopy", Interscience, New York
     1961, p. 97 ff
/38/ J.B. Pendry, "Low Energy Electron Diffraction",
     Academic Press, London 1977, p. 5 ff.
/39/ J. Estel, H. Hoinkes, H. Kaarmann, H. Nahr, and
     H. Wilsch, Surface Sci. $\underline{54}$ (1976) 393
/40/ S.S. Fisher and J.R. Bledsoe, J. Vac. Sci. Technol.
     $\underline{9}$ (1972) 814
/41/ H. Wilsch, J. Chem. Phys. $\underline{56}$ (1972) 1412
/42/ M.W. Cole and D.R. Frankl, to be published

/43/ H. Pauly and J.P. Toennies, in "Methods of
     Experimental Physics" 7A, edited by B. Bederson
     and W.L. Fite, Academic Press, New York 1968,
      p. 267
/44/ K. Haberrecker, E. Mollwo, H. Schreiber, H. Hoinkes,
     H. Nahr, P. Lindner, and H. Wilsch, Nucl. Instr.
     Meth. $\underline{57}$ (1967) 22
/45/ H. Nahr, H. Hoinkes, and H. Wilsch, J. Chem.
     Phys. $\underline{54}$ (1971) 3022
/46/ H. Kaarmann, H. Hoinkes, and H. Wilsch, J. Chem.
     Phys. $\underline{66}$ (1977) 4572)

REACTIVE SCATTERING

M. Cavallini

Laboratori Richerche di Base, SNAMPROGETTI SpA

c.p. 15,   00015 Monterotondo (Rome), Italy

INTRODUCTION

The investigation of gas–solid surface interactions is
presently being approached via a considerable and rapidly
increasing number of experimental methods and techniques. These
methods are derived from different disciplines, and are
distinguishable by the nature of the information they can provide.

The static properties of solid surfaces, such as surface
structure and composition, are normally determined by means of
such well established techniques as AES, SIMS, LEED, ESCA, and
in general electron, ion, and infrared spectroscopy. On the
other hand, an increasing demand for information about dynamic
properties, particularly the kinetic aspects of adsorption, has
exerted pressure for an application to surface chemistry of
techniques traditionally applied in other fields. One of the
most powerful new techniques for this chemical investigation is
molecular beam reactive surface scattering.

We shall describe the application of this method, particularly
as it pertains to bimolecular reactive scattering on surfaces,
and especially in connection with the study of heterogeneous
catalytic reactions. Molecular beam reactive scattering has been
reviewed by several authors in the recent past [1-5] and therefore
the present paper will only deal with representative reactions
which illustrate some of the significant aspects of this field.
Particular attention will be paid to the oxidation of carbon
monoxide on palladium and platinum, since many authors have

considered it a subject of special interest. In addition, a
decisive reason for selecting this particular reaction is that
a significant contribution to its understanding comes precisely
from molecular beam experiments.

MOLECULAR BEAM AND SURFACE REACTION

The molecular beam scattering experiment with a solid surface,
shown schematically in Figure 1, can be performed to get quite
a bit of information concerning different aspects of gas-surface
interactions from different points of view. Depending on the
amount of energy exchanged in the interaction, it is possible to
follow either physical or chemical processes occurring on solid
surfaces. We generally consider as physical any interaction
involving less than about 5 kcal/mole and as chemical any process
in the upper range below a few hundred kcal/mole. Normally real
processes on surfaces are rather complicated since they are the
result of an overlap of elementary steps, each making its own
energetic contribution to the interaction.

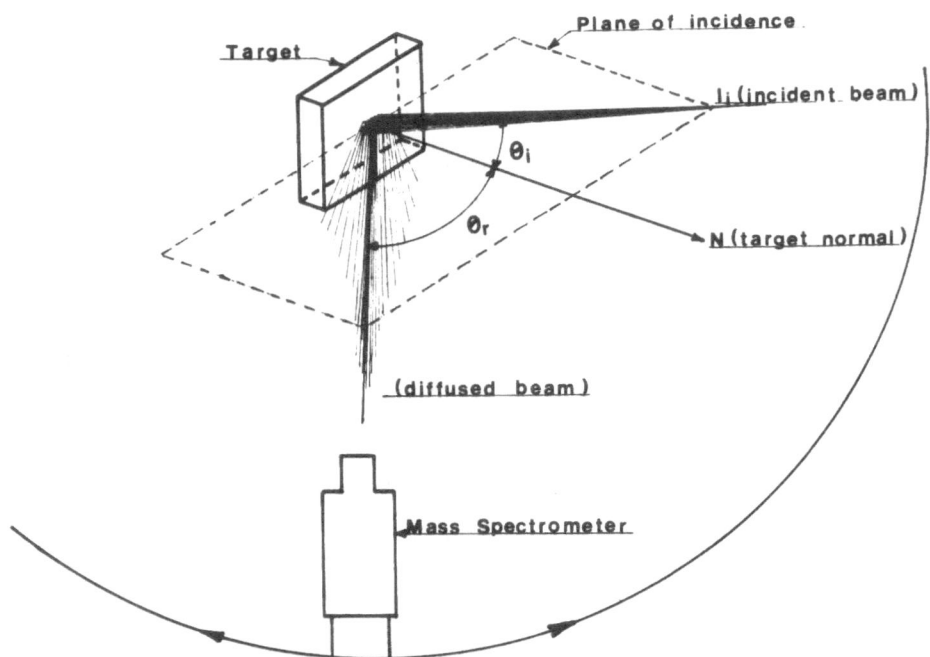

Figure 1. Schematic representation of molecular beam surface
scattering.

The aim of a molecular beam scattering experiment is the identification of these different contributions, this object being greatly simplified by the independent control of many parameters concerning the physical state of the incoming molecules that the molecular beam method furnishes. In fact, it is relatively easy in a beam to control the angle of incidence, the incident molecular flux, their average velocity, and their velocity distribution; in many cases, it is also feasible to select a specific internal energy state such as vibration, rotation, and total spin.

After the interaction with the sample surface, the products desorbed in a given solid angle can be analysed in terms of mass and angular distribution by a quadrupole mass spectrometer rotating around the scattering center. Normally, the incoming beam is chopped or modulated. In this case the time of flight of the incident molecules between the chopper and the quadrupole ionizing source is determined by phase shift measurements or by signal averaging. From these delay measurements it is possible to determine the residence time of the molecules on the surface.

All these results are usually obtained for different experimental variables, which, for example, may include surface structure, surface coverage, surface temperature, particle velocity, the angle of incidence, and the scattering angle. In many cases, all of this information uniquely determines the interaction mechanism and allows understanding of the simplest surface chemical processes. Compared to other traditional techniques like adsorption kinetics or microreactor catalysis, the molecular beam method has a limitation in the maximum number of interactions per unit time (equivalent pressure on the surface). On the other hand, it has the fundamental advantage of preventing multiple interactions with the sample surface or successive interactions of the products with the surroundings after emission. The interaction process can be examined step by step. The modification of the surface structure and composition during the reaction may also be studied by means of several different surface sensitive techniques which require UHV since the molecular beam method is compatible.

Since we shall confine our consideration to chemical applications of molecular beams, we must separate, as far as possible, each elementary contribution to the overall process. In order to get a more simplified view of the problem, we can distinguish between two groups of surface reactions.

I)   The reactions taking place between the incoming molecule and a surface atom, giving products in which surface species are present. In this group we include chemisorption, yielding a surface compound as a product. Examples of this group are:

$$A_{surface} + B_{gas} \longrightarrow AB_{gas}$$

$$A_{surface} + B_{gas} \longrightarrow AB_{surface}$$

II) Catalytic reactions are, in contrast, those in which a particular reaction can occur only under the influence of surface atoms that, however, are not among the reactants and products. Surface induced dissociation or ionization, for instance, considered only with respect to the final product, are included together in this group with more specific catalytic processes.

$$A_{2\ gas} + B_{2\ gas} \longrightarrow 2AB_{gas}$$

$$A_{2\ gas} \longrightarrow A_{gas} + A^{+}_{gas} + e$$

$$A_{2\ gas} \longrightarrow A_{gas} + A_{gas}$$

Obviously, intermediate steps of this second group of reactions can belong to the first group. In particular, adsorption is the elementary precursor step in every kind of non elastic interaction. (Let us consider the simple case of surface atom recombination following the dissociaton before reemission from the surface.) Let us consider the simple case where a molecule is dissociated on the surface and then recombines before being emitted into the gas phase. If elementary steps of this reaction are to be investigated, it would be very difficult to do so by traditional kinetic techniques. With the molecular beam technique, hydrogen deuterium exchange on platinum, for example, has been studied by Bernasek and Somorjai [6]. They have been able to establish the fundamental role of surface structure by observing the residence time of HD on the surface; they could exclude an activation energy for adsorption and identify the rate determining step below 700°K as the diffusion of D_2 molecules on the surface. This example shows how much data is typically obtainable in molecular beam reactive scattering. However, this also indicates that if a more detailed understanding of gas surface interactions is to be achieved then the introduction of a molecular beam facility to an already complicated set of surface analysis apparatus seems to be in order.

CARBON MONOXIDE OXIDATION OF METALS

The oxidation of CO can be catalyzed by several group VIII metals and in particular by Ni, Pd and Pt. Previous investigators have clearly indicated that this reaction proceeds with a very low specificity and does not need a particular surface arrangement to occur [9,11,16]. In spite of this observation concerning specificity, evidence for the strong dependence of activity on different experimental conditions is generally reported [9-19]. Promotion or inhibition of a reaction rate can be induced, for example, by the chemical composition of the uppermost layer. The presence of a considerable number of surface defects, like steps or kinks, strongly improves the reaction rate [3], while the poisoning of active sites by one of the reactants hinders chemisorption at the other one [13,11,16]. Absence of specificity and yet strong influence of experimental parameters on the reaction rate seem to be conflicting features.

Consistency is achieved by assuming that the reaction controlling mechanism is not unique. This means that different reaction paths exist which lead to the same final product. Since the two basic mechanisms that a surface reaction can follow are the Langumir-Hinshelwood (L.H.) or the Eley-Rideal (E.R.), many efforts have been made to get a better understanding of their mutual influence on the CO oxidation process. The following formulation of these two reaction mechanisms, in a simplified form, takes into account only the more important steps and neglects, for example, the desorption of oxygen in molecular form and diffusion into the bulk. These approximations are allowed in the temperature range normally used in the study of this reaction.

The Langmuir-Hinshelwood mechanism proceeds as follows:

$$CO_{gas} \underset{k_2}{\overset{k_1}{\rightleftharpoons}} CO_{ad}$$

$$O_{2\,gas} \xrightarrow{k_3} 2O_{ad}$$

$$CO_{ad} + O_{ad} \xrightarrow{k_4} CO_{2\,gas}$$

The Eley-Rideal mechanism, instead, consists of two steps:

$$O_{2\,gas} \xrightarrow{k_3} 2O_{ad}$$

$$CO_{gas} + O_{ad} \xrightarrow{k_5} CO_{2\,gas}$$

The large variety of techniques used up to now for this investigation may be divided into those following changes in the gas phase reaction products and those which are sensitive to the variations on the metal surface. Belonging to the second group is the infrared absorption technique applied by Eischens and Pliskin [7] to supported platinum and nickel and by Heyne and Tompkins [8] who also performed work function measurements. Park and Farnsworth [9] combined work function and LEED to follow the reaction on a nickel single crystal. More recently Bonzel and Ku [10] studied Pt(110) by LEED and AES, finding a change in the diffraction pattern from a (1×2) into a (1×1) due to CO adsorption in the temperature range 320-500°K. A pattern change was not observed for very low exposures to oxygen, but independent evidence of oxygen chemisorption was obtained. A complete set of LEED, AES, work function and mass spectrometry measurements on palladium (111) (100) (110) planes and on polycrystalline wire have been published by Ertl and Koch [11]. They found that no significant differences exist among different surface structures, especially with respect to the initial adsorption energy which has a value of about 35 kcal/mole. A more complex behavior was found for oxygen chemisorption, giving rise to a (2×2) structure on (100) and (111) with a binding energy of 60 kcal/mole, but on the (110) plane a whole series of ordered surface structures appear with increasing coverage and consequently a decreasing heat of adsorption from 80 kcal/mole down to 48 kcal/mole.

Other original approaches include the electrical resistance measurements on Pd and Ni films made by Kawasaky [12] and the beam scattering experiments of Bernasek and Smorjai [13] who investigated the transfer of translational kinetic energy from the incoming particle to the vibrational modes of absorbed CO molecules. This energy transfer is responsible for the strong dependence of sticking probability on the state of cleanliness of the surface and consequently on CO surface coverage. Some of these papers, which are mainly devoted to structural aspects and static surface properties, also make significant contributions to the gaseous products analytic approach for understanding basic reaction mechanisms. One of the earliest contributions to the study of reaction products was that of Stephen [14]. By simple volumetric techniques he first observed what he termed an induction period when O_2 reacted with preadsorbed CO on Pd. A confirmation of this result, together with an improved explanation of the phenomenon, was given by Park [15] who used CO and Oxygen-18 for his volumetric measurements of CO_2^{16} and $CO^{16}O^{18}$ produced on Pd films. This work established a number of restrictions on possible reaction mechanisms. It was possible to distinguish between the reactions that occur on a CO saturated surface and those taking place after oxygen adsorption. A comparison of these two situations strongly favors a model in which CO_2 is

formed on an oxygen covered palladium surface with the CO supplied
either by surface diffusion or directly from the gas phase.
Bonzel and Ku [16], by observing the CO_2 partial pressure versus
time and surface temperature, were able to establish which one
of the two mechanisms prevails under different experimental
conditions. In the temperature range between 300 and 500°K, they
observed an induction period between the time when the CO pressure
was reduced and the time that the production of CO_2 was increased.
They suggest that the reaction mechanism was of the L.H. type.
When CO was admitted to an O_2 saturated surface at temperatures
higher than 450°K, there was no induction period but, rather, a
very rapid increase in CO_2 partial pressure was observed. This
behavior clearly favors the E.R. mechanism. This was the state
of the art until the first contribution from molecular beam
reaction scattering appeared.

In 1974 Palmer and Smith [17] published a paper in which
angular, temperature, and pressure dependences were studied using
a CO beam interacting with oxygen on an epitaxially grown platinum
(111) face. The angular dependence of CO_2 intensity was found
not to follow the Knudsen law, but rather the analytical function
$\cos^d\theta_r$ where θ is the emission angle with respect to the surface
normal. A best fit of the data gave d=6. There was also evidence
that the angle at which the CO_2 was emitted from the surface
depended upon the angle at which the CO beam was incident upon
the surface. The temperature dependence of the reaction rate
was in good agreement with the Bonzel and Ku [16] results. For
surface temperatures above 525°K, however, they found a much
smaller decrease in reaction rate. They concluded, on the basis
of the CO pressure dependence, that the L.H. mechanism dominates
the reaction and that CO is the principal diffusing species. The
apparatus of Palmer and Smith was not UHV and furthermore only
CO could be supplied as a molecular beam while O_2 had to be
introduced as background pressure. The opposite experimental
arrangement (O_2 beam and CO background) was not successful,
because the CO background pressure required in order to have
measurable CO_2 formation was much lower than the ambient 10^{-7}
torr. More recently a molecular beam study was reported by Pacia,
Cassuto, Pentenero and Weber [18]. They used an UHV apparatus
in connection with a three stage differentially pumped supersonic
beam and introduced the other reactant via a leak valve directly
in the main chamber. In such a way they were able to perform
experiments using alternatively CO, O_2 or even mixed beams on a
Pt ribbon. They measured what they called the "reactive sticking
probability," defined as the probability that a molecule reacts
during a simple collision with the surface. They did not chop
the beam but intercepted it with a sequence of flags before and
after the surface interaction. By optimizing the parameters in
their kinetic model, they were able to explain the observed

different regimes in the range of low and high temperatures. The
general conclusions are partially in contrast with those of
previous papers. In fact they emphasize the existence of a unique
L.H. mechanism above 600°K, while they find that below this
temperature the E.R. mechanism will only compete in the case of
high carbon-monoxide surface coverage. This is one of the first
applications of molecular beams to the investigation of a surface
reaction with the specific aim of isolating the elementary steps
of a complex reaction mechanism. Their results give rather
complete information concerning the kinetic properties and the
dynamical aspects of the reaction, but we must stress the difficulties they met in determining the real surface coverage especially
at lower temperatures. This was due to the absence in their
apparatus of an in situ complementary technique expressly devoted
to the determination of surface chemical composition.

Almost contemporary to the present review, Engel and Ertl
presented a paper [19] at the 7th Int. Vacuum Conference (Vienna,
1977) concerning the use of molecular beam relaxation spectroscopy
[20] applied to CO oxidation on Pd(111). Evidence for the L.H.
mechanism was clearly found while they excluded the existence of
the E.R. mechanism under their experimental conditions. They
conclude, in fact, that in the complete temperature range in
which the reaction takes place, the only mechanism consistent
with their data is that occurring between adsorbed CO molecules
and adsorbed oxygen atoms. In order to emphasize that a detailed
and complete investigation of this matter is only possible with
a multi-technique approach like that of these authors, a brief
presentation of their experimental arrangement is necessary.
Their apparatus includes AES, LEED and He scattering for the
determination of cleanliness and perfection of crystalline
structure. A rotable quadrupole mass spectrometer detects
molecules coming from the surface in the scattering plane. UHV
was maintained while the nozzle beam was impinging on the surface
with a flux equivalent to a pressure of 10^{-7} torr. Phase shifts
between the primary and the scattered beam were determined with
lock-in techniques since the molecule transit time was negligible.
With the use of these surface spectroscopies and a simultaneous
evaluation of residence times by means of phase measurements,
they obtained a quite large set of data concerning precursor
states of CO oxidation and a clear distinction between different
possible elementary processes. In fact, for CO on Pd(111) they
measured the sticking coefficient versus coverage and angle of
incidence at different surface temperatures.

CO is diffusely scattered at all temperatures accessible in
the experiment (300°K-1000°K). Below 500°K scattered CO is
completely demodulated, indicating that the adsorption time
experienced by all incident molecules is larger than the

modulation frequency. If the coverage increases above the
saturation value, a subsequent increase in the modulated signal
is observed since the sticking coefficient approaches zero and
the binding energy decreases. The change in intensity of the
chopped scattered CO between 500°K and 625°K is due to the
residence time associated with the desorption process. The
desorption energy of 34 kcal/mole obtained from these data is in
good agreement with that found in previous papers. While the CO
angular distribution follows the cosine law to a good
approximation, oxygen has a coverage dependent angular
distribution. For small coverages the distribution is almost
specular; the cosine distribution becomes dominant when the
coverage approaches a quarter of a monolayer. This behavior
indicates that adsorption of oxygen must be activated, and that
for low coverage only a few O_2 molecules desorb from a trapped
precursor state.

As other authors have done, Engel and Ertl set up two
experimental arrangements, one with CO as the modulated beam and
the other with O_2 modulated and CO present as background pressure.
In the first condition, the equations governing the two mechanisms
give different results for the phase shift of the CO_2 signal.
No temperature variation is expected with the E.R. mechanism,
while the L.H. predicts a temperature dependent phase lag
corresponding to an apparent activation energy equal to the CO
desorption energy. These last predictions are, in fact, very
well confirmed by their phase shift data that yield, in perfect
agreement with L.H., an activation energy of 33 kcal/mole. A
best fit of reaction rate versus temperature measurements gives
a confirmation of L.H. as the dominant mechanism. A second set
of equations can be used to predict the results when O_2 is
modulated. In this case the phase lag should increase with
temperature for L.H. and decrease for E.R. Their data show a
clear agreement with the L.H. hypotheses giving an activation
energy for the reaction of 23.5 kcal/mole. This paper concludes
that the E.R. mechanism is strongly disfavored as the explanation
for CO oxidation reaction of Pd(111). However, it cannot be
excluded that a different surface configuration would alter the
conclusions because of the determinant role played by surface
species concentration on the reaction rate.

MOLECULAR BEAM APPARATUS FOR CATALYTIC REACTION STUDIES

In order to study bimolecular catalytic reaction mechanisms,
a quite sophisticated experimental appratus has been developed
at SNAMPROGETTI Monterotondo laboratories [21,22]. It was given
the name "Twin molecular beam SSIMS apparatus," since two
identical supersonic jets can interact, even simultaneously, with
the sample surface while surface composition can be chemically

analysed by an in situ secondary ion mass spectrometer normally
used in a static mode of operation.

This line of approach to surface reaction kinetics and the
specific performance of this apparatus are such that the formation
of reaction products is directly observed, while the simultaneous
growth of surface intermediate compounds, appearing as adsorbed
species, can be followed by SIMS. By linearly increasing the
sample temperature, thermal programmed desorption can be performed
as well. The required initial value and time derivative of the
temperature can be easily set on the control unit of a temperature
regulating power amplifier. In this measurement, the quadrupole
mass spectrometer in front of the sample is able to follow either
the neutral species desorbed from the surface or the disappearance
of secondary ion peaks due to the evaporation of the surface
compound. Thereby, the desorption energy can be evaluated, and
moreover an easier assignment of the SIMS peak is obtained. A
very close relation, in fact, exists between the composition of
surface adsorbates and the peak intensity distribution in the
mass spectra, originating either from the ionization of thermally
desorbed products or from the more complicated secondary emission
mechanism.

As shown in Figure 2, the sample is placed at the center of
a hexagonal prismatic UHV chamber. Full rotation and translation
of the sample are performed by a manipulator located on the bottom
flange. The sample temperature can be controlled or programmed
from 77°K up to 900°K. Inside the principal chamber is a second
chamber made of copper which is normally maintained at liquid
nitrogen temperature and which surrounds the sample. This shield
has the double function of screening the sample from background
residual gas and creating a vacuum separation between the
modulation chamber (designated 2 in Figure 2) and the reaction
chamber (designated 3). There is a complete separation between
the first chamber in which beams are produced and the second one
in which their modulation occurs. Moreover, differential pumping
is also performed between the second, third and fourth chambers,
respectively. In fact, the interaction chamber is differentially
cryopumped by the same liquid He trap that evacuates the second
one. Its measured pumping speed is of the order of 2×10^4 lit/sec.
This is necessary in connection with the use of a supersonic jet.
The ultimate vacuum in the second chamber when molecular beams
are not present is in the range of 10^{-10} torr.

Figure 3 gives a more detailed representation of the top view
of the apparatus. At both sides of the detection chamber the
two molecular jets are produced in two independent oil pumped
systems whose separation from the UHV section is absolute. In
these chambers, just behind the skimmers, two vacuum stepping

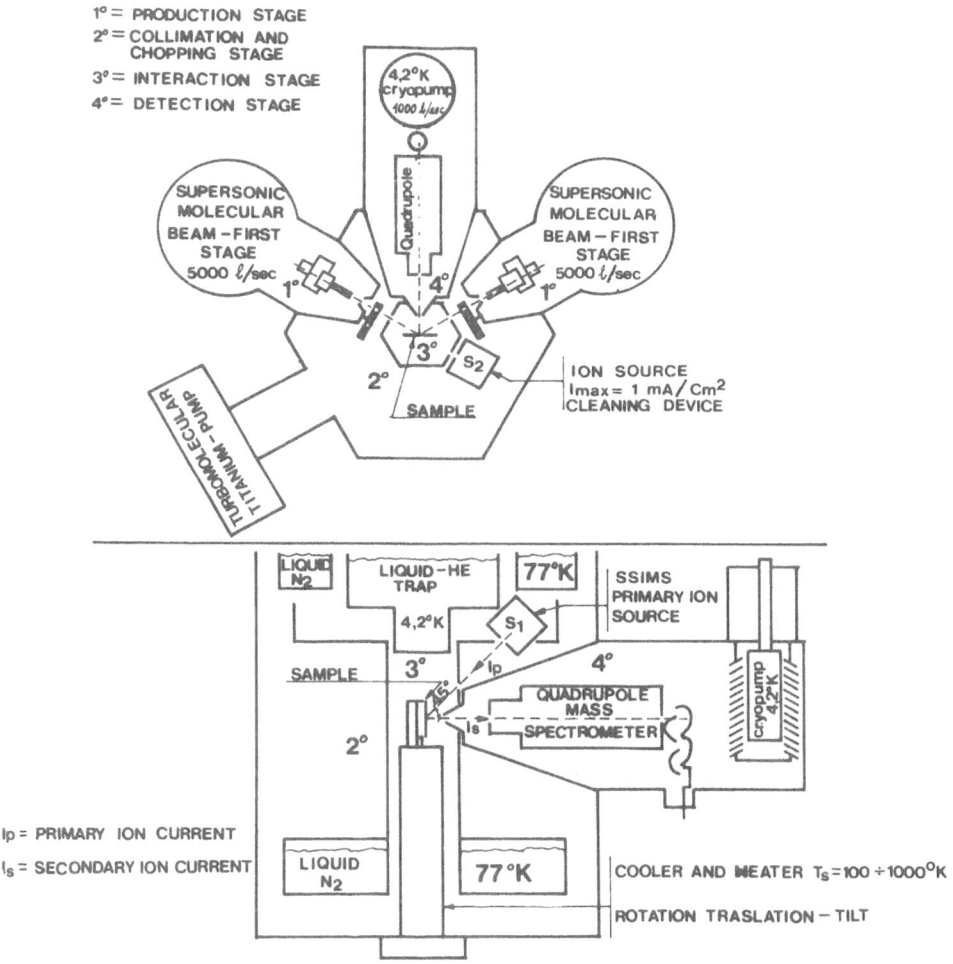

Figure 2. Top view and vertical section of "Twin molecular beam –
SSIMS" apparatus operating at SNAMPROGETTI Lab. – Monterotondo,
Rome, Italy.

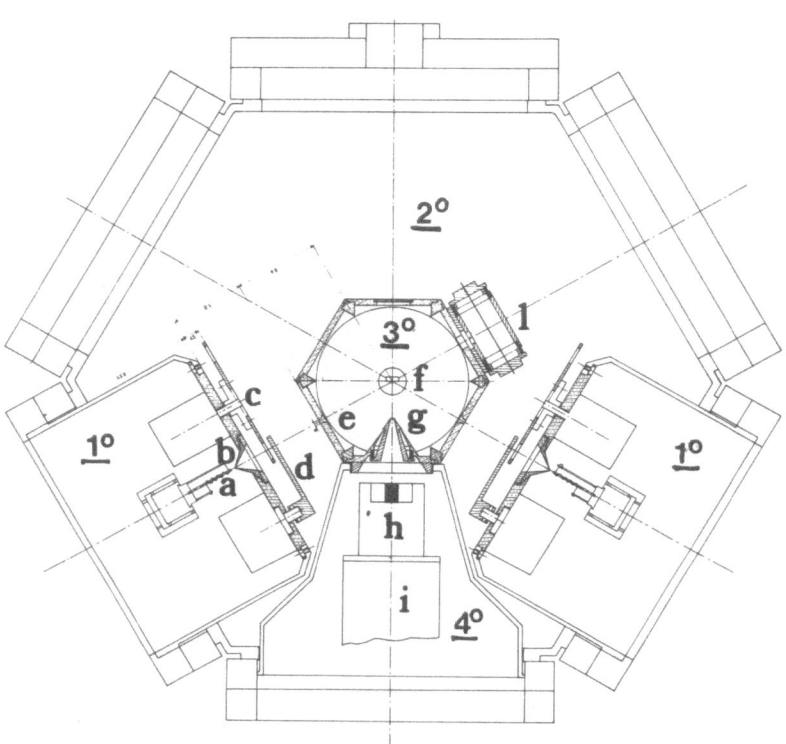

Figure 3.   Detailed top view
             1° First Stage   BEAM PRODUCTION
             2° Second Stage   BEAM MODULATION AND INTERRUPTION
             3° Third State   INTERACTION WITH SURFACE
             4° Fourth Stage   MASS DETECTION
                               a   Nozzle
                               b   Skimmer
                               c   Chopper
                               d   Shutter
                               e   Beam Collimator
                               f   Sample
                               g   Secondary Ion Extractor
                               h   Quadrupole Ion Source
                               i   Quadrupole Mass Filter
                               l   Sputtering Cleaner

motors are mounted in a position so that their axes protrude into
the modulation chamber for transmitting rotation to the chopper
wheels. By means of the stepping motors, the molecular beam
modulation can be performed in a very large variety of impulse
sequences. For example, it is possible in repetitive modulation
to change both the period and the duty cycle, that is to interpose
an arbitrarily long pause between two beam pulses. A
pseudo-random pulse sequence can equally well be obtained thus
making possible a cross-correlation treatment of the detector
output. When the sample surface is positioned in front of the
detection entrance, the angles of incidence of the two beams are
both equal to sixty degrees with respect to the surface normal,
while the primary ion particles of the SSIMS beam strike the
sample from above 45 degrees out of molecular beam plane.

A 3/4 inch quadrupole mass spectrometer, located just behind
the liquid nitrogen collimator, can be operated to detect either
neutral molecules coming off the surface or sputtered secondary
positive or negative ions. This strategic quadrupole position
meets the detection needs of both SSIMS and the molecular beam
technique. The two methods of operation differ simply by the
quadrupole ion source electrode voltages and by the filament
switch position. However, in the SSIMS detection mode the
quadrupole source can be used for energy selection as well as
analysis of secondary ions thus strongly improving the peak shape
in SSIMS spectra. The power of the SSIMS-MOLECULAR BEAM
combination is illustrated in Figure 4 which shows the SSIMS
spectrum from a Ag(111) surface containing a monolayer of adsorbed
water. The water was deposited from a seeded argon supersonic
beam. The spectrum is so rich because water-silver clusters are
sputtered from the surface.

A large quantity of information concerning adsorption was
obtained [23] by investigating the beam intensity as a function
of coverage and by conducting thermal desorption experiments.
The following section will be devoted to the exposition and
discussion of the results obtained by the Twin molecular beam -
SSIMS apparatus in the investigation of CO oxidation on
polycrystalline palladium.

TWIN MOLECULAR BEAM - SSIMS APPARATUS APPLIED
TO CARBON MONOXIDE OXIDATION OF PALLADIUM

It can easily be inferred from the review of papers concerning
CO oxidation on metals extensively discussed in the previous
section that, in spite of the very large effort, final answers
to several questions posed by this reaction are not yet given.
A clear understanding of the elementary contributions to the
overall process and the role played by the basic mechanisms

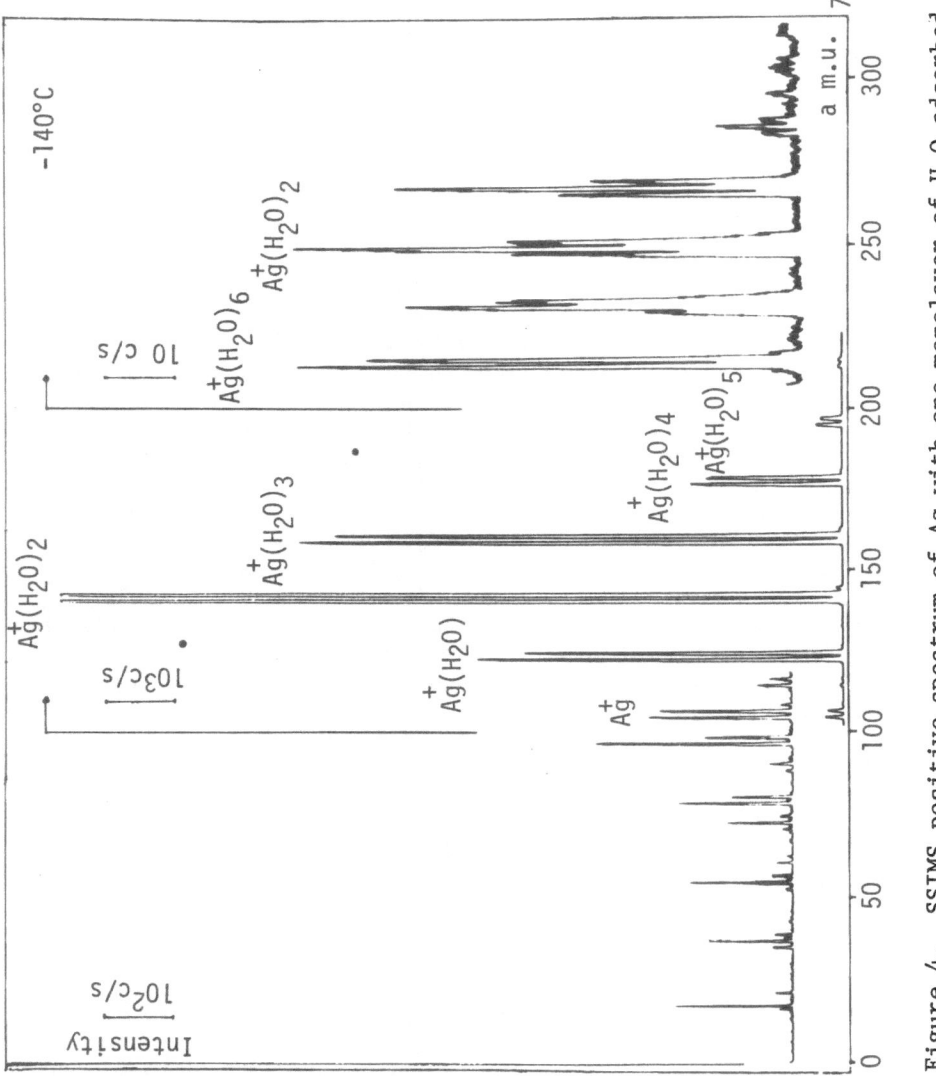

Figure 4. SSIMS positive spectrum of Ag with one monolayer of H_2O adsorbed.

occurring in this reaction have not yet been obtained. The
investigation to be described below does not pretend to be
definitive in giving a unique solution to this arduous problem.
However, we are certain that the present approach furnishes a
direct phenomenological inspection of experimental events and at
least yields an unambiguous set of background data. As we have
strongly emphasized, only a clear analysis of the process from
the two fundamental points of view, namely surface chemical
compounds and gas phase product formation (and decomposition),
gives, in our opinion, the necessary background information for
the understanding of the catalytic oxidation mechanism.

An exposition of some general considerations on surface
cleaning procedures and steady state conditions is presented
first. Response time analysis and residence time measurements
will be subsequently discussed in detail, while a comparison with
previous observations will conclude the discussion of results.
A polycrystalline 99.9% purity palladium sample was analysed by
SSIMS as soon as it was introduced into the vacuum system. The
positive and negative spectra of Figures 5 and 6 show that a
strong contamination and oxidation of the surface is still present
although a vacuum in 10^{-10} torr range was maintained, and the
surroundings were all at liquid nitrogen or Helium temperature.
A prolonged (about 30 min) ion bombardment by a 10^{-6} A/cm^2 primary
current of 1.5 keV krypton ions followed by a few days annealing
at 700°K, finally gives the typical clean SSIMS spectra presented
in Figures 7 and 8A. A few comments on some strange features of
these spectra are necessary to justify apparent contradictions
or anomalies.

First, the complete absence of Pd ion peaks in the clean
sample positive spectrum is due to the very low ionization
probability experienced in general by metal atoms sputtered from
a very clean surface. The absence of this peak confirms the
cleanliness of the surface. Metal ion formation is normally
strongly affected by surface impurities [23]. Absorbed species,
in fact, cause a localized dielectric screening of electrons
preventing ion neutralization [24]. Since the ionization
probability of alkali metals is very high compared to that of
other positive species, their presence in positive SSIMS spectra
is particularly evident. Every element or molecular species has
an ionization probability depending particularly on their
electronic configuration and, in general, on surface sample
properties. Thus, a quantitative analysis of surface
concentration does not immediately follow from peak intensities,
however, some conclusions can be made. For instance, we can
conclude that chromium, iron and nickel were only superficial
contaminants, due probably to the lamination procedures in foil

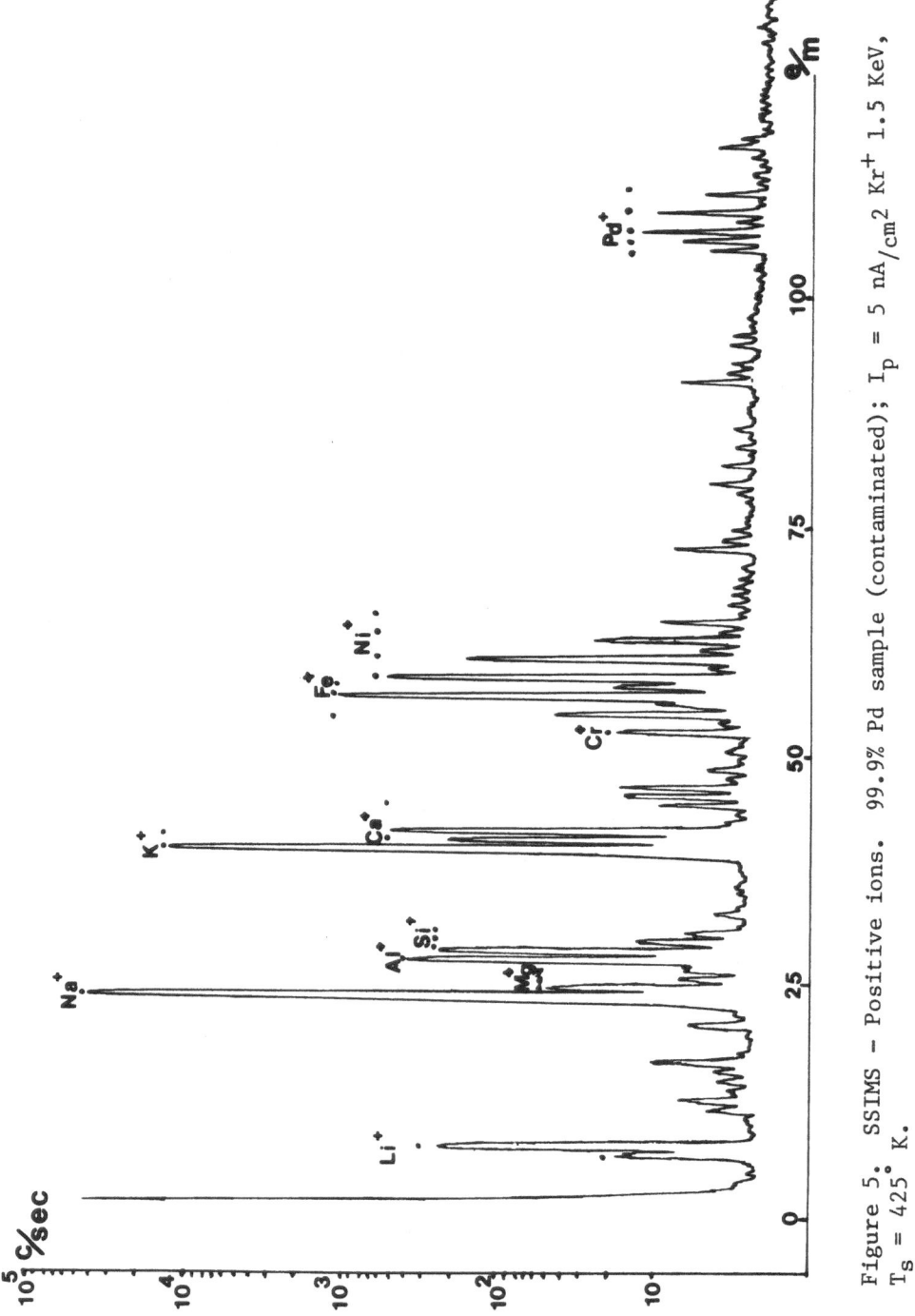

Figure 5: SSIMS - Positive ions. 99.9% Pd sample (contaminated); I_p = 5 nA/cm^2 Kr^+ 1.5 KeV, T_s = 425° K.

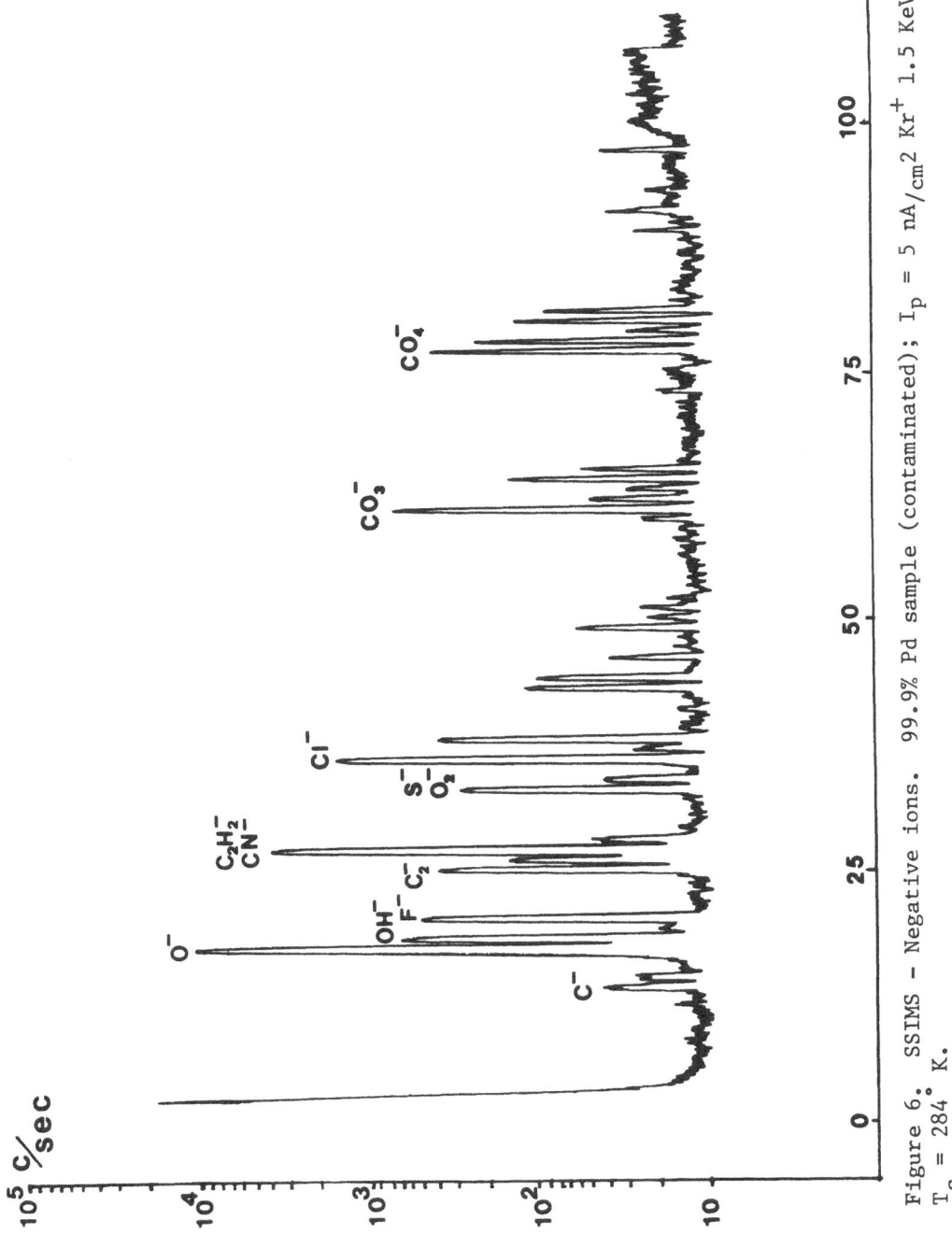

Figure 6: SSIMS - Negative ions. 99.9% Pd sample (contaminated); I_p = 5 nA/cm2 Kr$^+$ 1.5 KeV, T_s = 284° K.

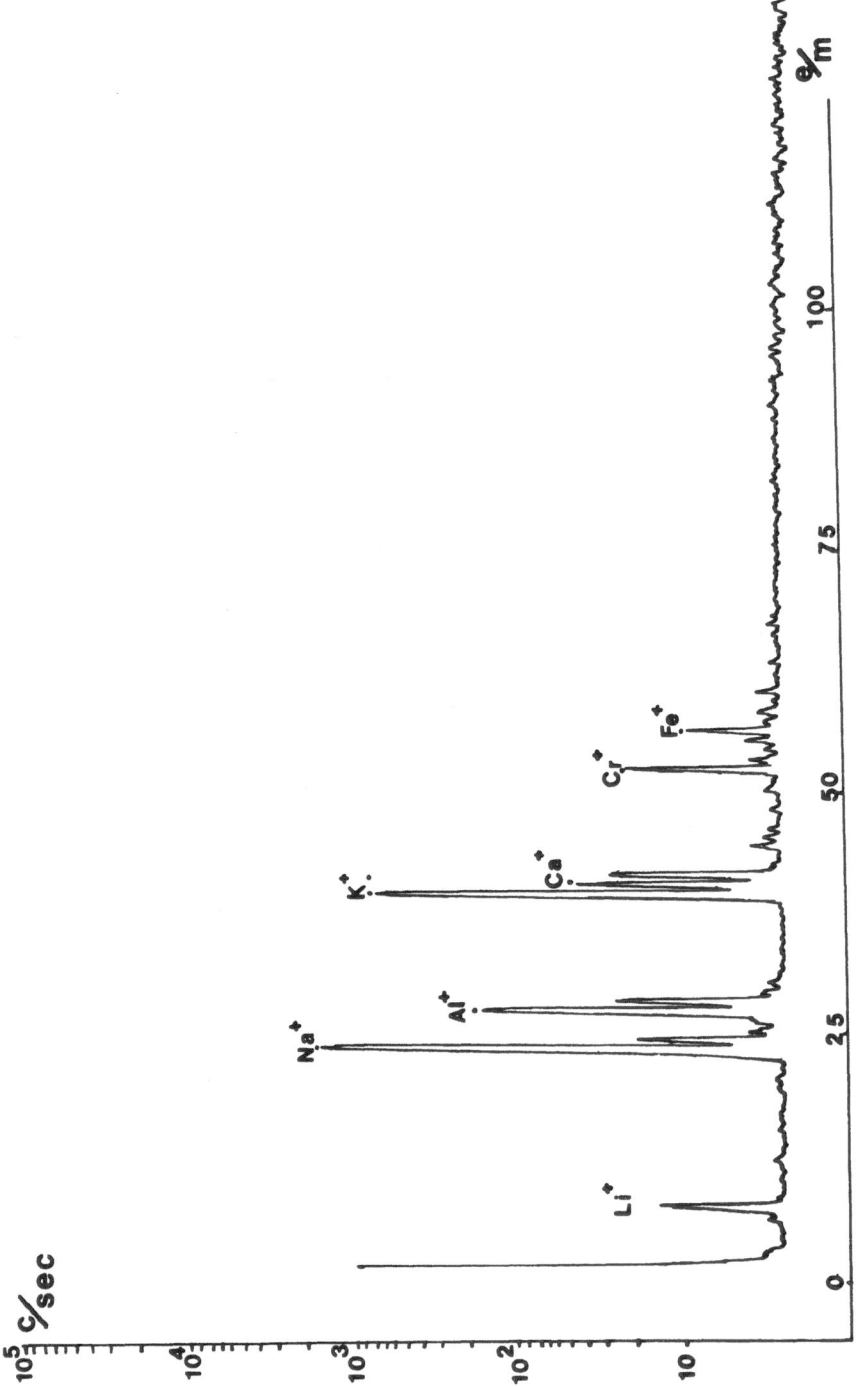

Figure 7: SSIMS – Positive ions. 99.9% Pd sample (clean); $I_p = 5$ nA/cm^2 Kr$^+$ 1.5 KeV, $T_s = 450°$ K.

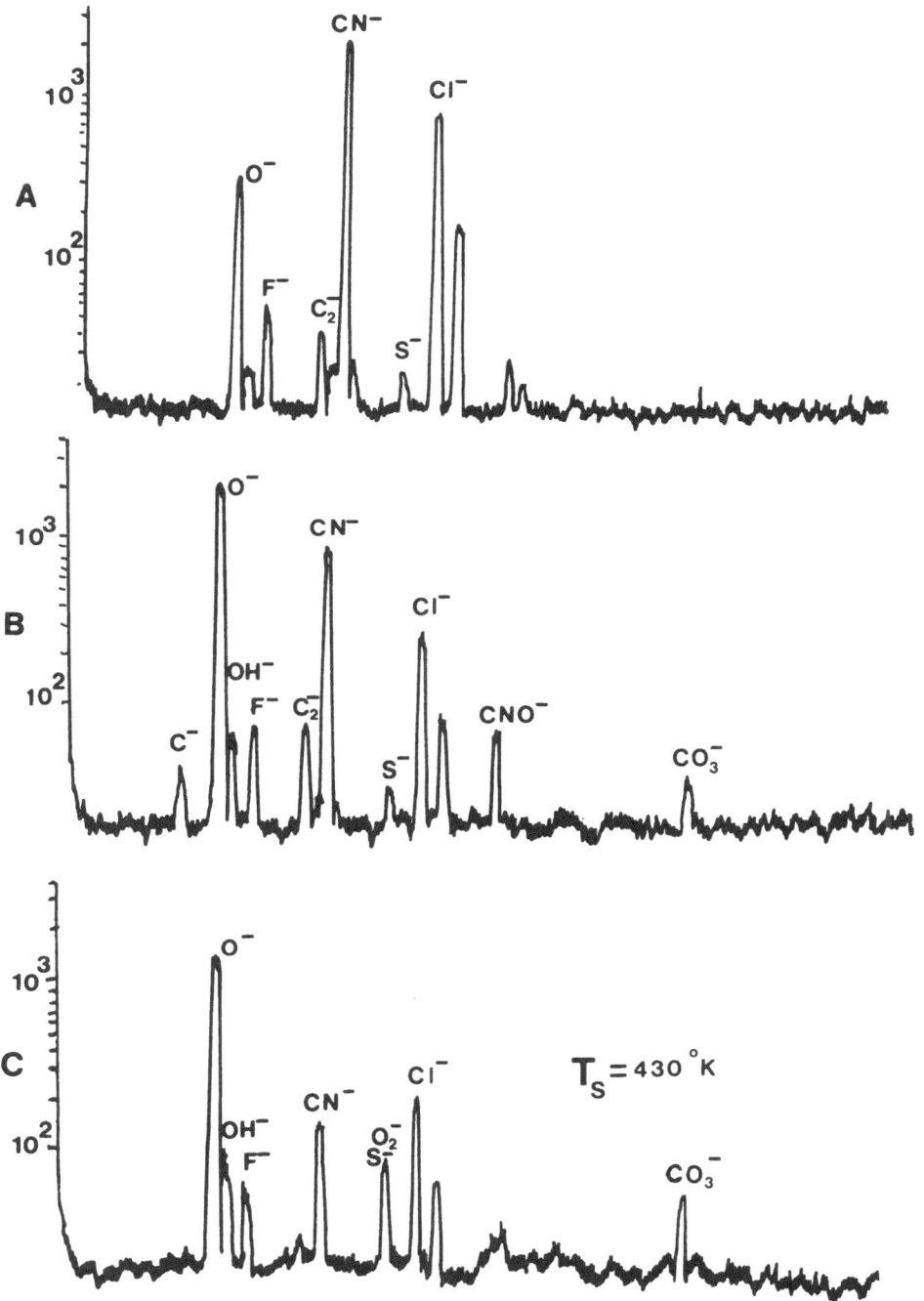

Figure 8. a) clean. SSIMS - Negative ions. 99.9% Pd;
I_p = 5 nA/cm², T_S = 350° K. b) CO beam exposed. c) O_2 beam
exposed. T_S = 430°K.

preparation. Their intensity, in fact, was seen to decrease with repeated cleaning.

Secondary negative ions are especially representative of surface adsorbates, since these species normally have high electron affinities. Accordingly, the most meaningful observations on carbon monoxide and oxygen adsorption are those of negative secondary ions. Contaminated samples yield spectra rich in hydrogen containing molecular ions coming especially from adsorbed hydrocarbon compounds. Some peaks whose origin is unknown are still present in the negative ion spectra; for example, the two groups following masses CO_3^- and CO_4^-. Although many tentative explanations are suggested in SIMS literature, no one expanation is yet generally accepted. Fortunately, the negative spectrum of a clean sample does not appear too complex, so that every variation produced by CO and O_2 exposure can be easily observed. The equilibrium coverage reached during adsorption of CO causes the appearance of the C^- ion which was completely absent in the clean surface, and also an order of magnitude increase in O^- intensity. This result is reported in Figure 8B. Figure 8C shows, on the other hand, how O_2 exposure modifies the negative secondary ion spectrum. In this case both the O^- and O_2^- peaks increased considerably provided that the sample temperature was higher than 400°K.

From these measurements we can conclude that the O^- peak is representative either of adsorbed oxygen or of carbon monoxide, while C^- and O_2^- are typical fragmentation ions corresponding to adsorbed CO and O_2, respectively. We can assume that, at low coverages, the intensity of O^- is proportional to the overall coverage of the surface and C^- to the fraction of coverage exclusively occupied by carbon monoxide. This assumption for O^- does not necessarily imply a dissociative adsorption for oxygen and CO or an inconsistency with the cooperative adsorption model. Indeed different sites are able to contribute to the same secondary ion peak intensity. In Figures 9 and 10 examples are given of adsorption curves obtained following O^- versus time, when the pressure at the clean surface was very sharply increased by lifting the shutter from in front of the beam. A soon as the molecular beam impinges on the sample surface, the mass O^- starts increasing until it reaches saturation. We have observed that CO adsorption at room temperature is irreversible. A complete desorption occurs upon heating to 530°K. We note that in both examples the final O^- intensities are equal. Indeed the two temperatures were adjusted in order to achieve this. Our data show that the CO adsorption rate is about two orders of magnitude higher than O_2 adsorption, although the temperature is lower.

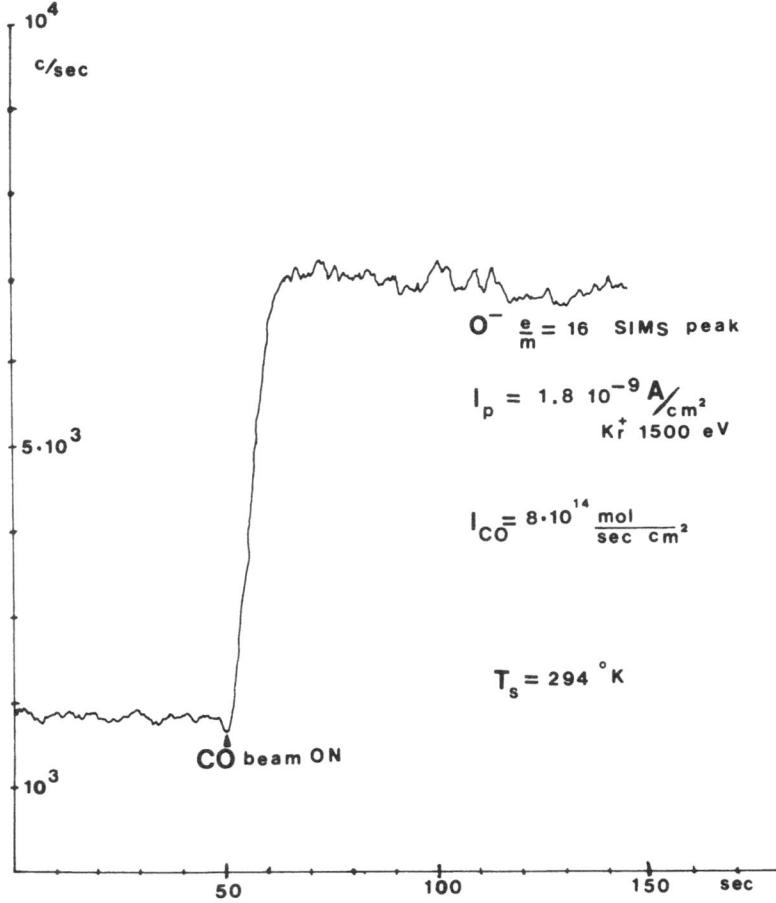

Figure 9. CO adsorption on Pd. SSIMS – molecular beam combination.

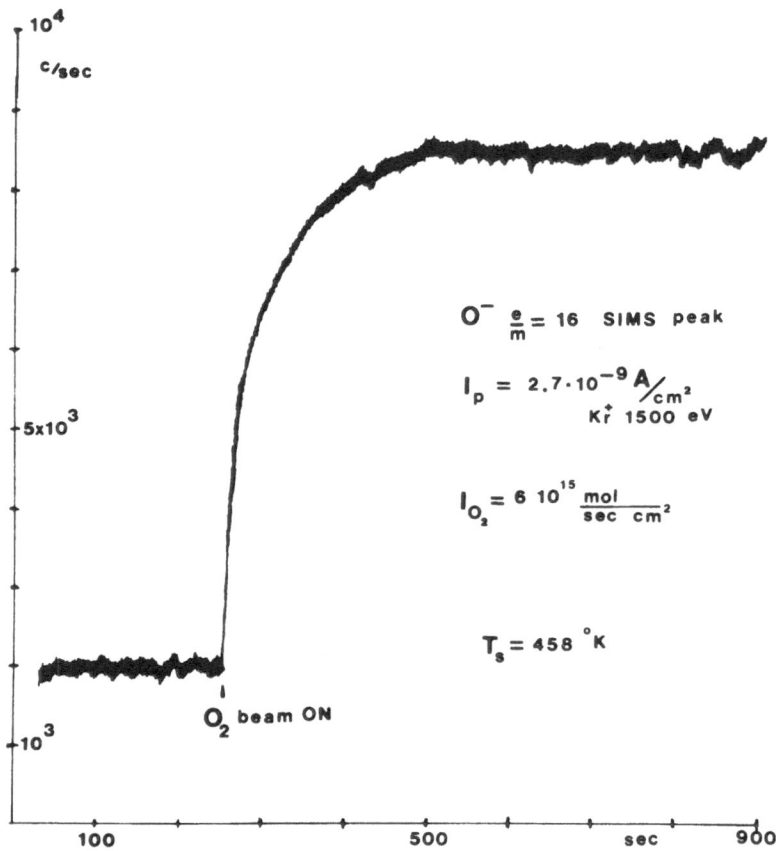

Figure 10. Oxygen adsorption on Pd. SSIMS – molecular beam combination.

The SSIMS technique used for the determination of the adsorption rate constants on the clean surface has also been applied to the palladium surface when the catalytic CO oxidation is happening. This was done in order to have a simultaneous measure of coverage and composition. Surface coverage has been measured for different steady state conditions, but in particular the variation in coverage and chemical composition have been investigated in "non-equilibrium" experiments. After a rapid pressure variation of one reactant on a catalyst surface, the restoration of new equilibria is controlled by the characteristic response time of the rate determing step. Accordingly, this sort of experiment allows the determination of the reaction parameters. We have performed this experiment at different sample temperatures. Two examples corresponding to 403°K and 580°K are presented in Figure 11, where the O^- peak intensity is recorded versus time as the O_2 and CO beams are shuttered "on" and "off". The initial condition represents the equilibrium O^- intensity when both beams are present and the reaction is taking place. According to our hypothesis, the O^- intensity is due to the cooperative contribution of adsorbed CO and oxygen. With the same technique, we have also followed C^- which depends only on CO surface coverage.

It is clear from Figure 11 that three distinct equilibrium positions are observed, especially at lower temperature. Each plateau corresponds to a particular fraction of equilibrium coverage determined by the surface reaction conditions. Let us examine the curve at 403°K. The lowest intensity of O^- is obtained when CO is off but O_2 is on. The same is valid for C^- intensity. This means that at this temperature the most important contribution to the coverage is given by CO, while the presence of O_2 on the surface determines a more efficient CO displacement, probably via the L.H. reaction mechanism event. The CO oxidation stops and the background O^- intensity is reached. The intermediate level, however, corresponds to the simultaneous presence of the two reactants. In this condition the oxidation takes place normally and the coverage assumes an equilibrium value reduced with respect to the maximum plateau by the effect of the reaction itself. The highest O^- intensity corresponds simply to the adsorption equilibrium of CO not perturbed by the reaction with oxygen. The curve at higher temperature (580°K) shows a completely reversed situation in the roles of adsorbates. The dominant contribution to the coverage comes from oxygen which causes a threefold increase in the O^- intensity. This effect is shown clearly in Figure 12 where the curve crossing point in the O^- plot corresponds to an equivalent contribution of CO and oxygen to the O^- intensity. In the range of temperatures below 480°K (crossing point), CO is the dominant adsorbed species, as also confirmed by the C^- curves. Above this temperature, adsorbed

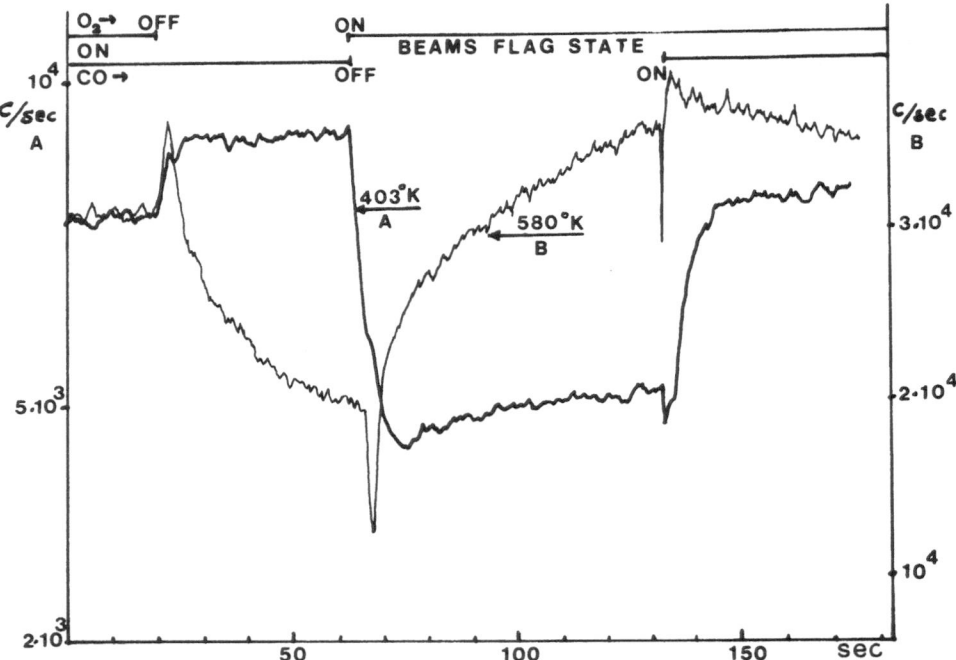

Figure 11. SSIMS O$^-$ peak versus time during the CO catalytic oxidation on Pd. On the left side, the vertical axis is the O$^-$ intensity scale for the plot at 403°K; on the right side, the scale is for the plot at 580°K. The horizontal axis at the top of the figure indicates by full lines the presence, on the sample surface, of the O$_2$ and the CO beams.

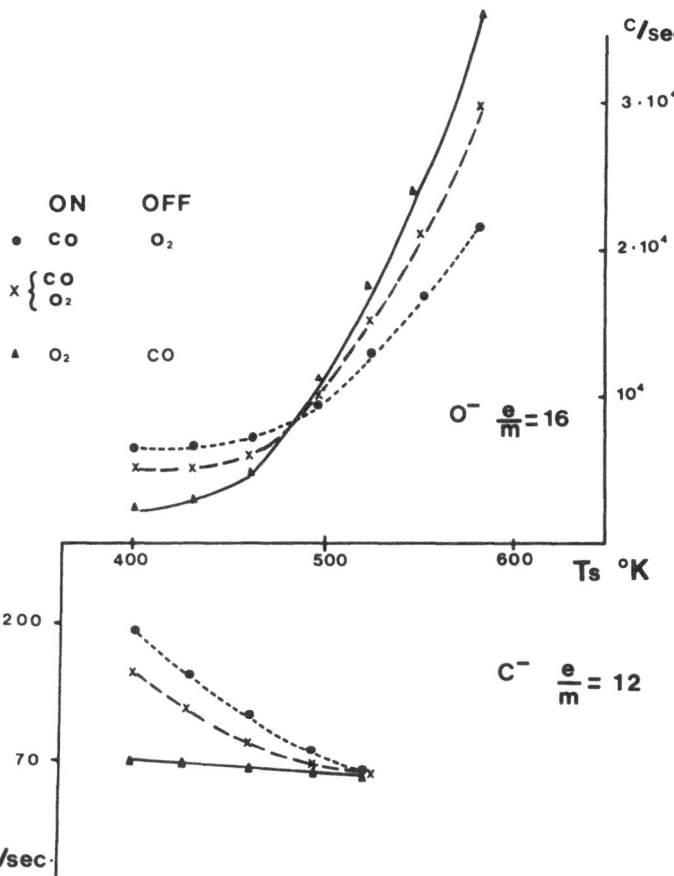

Figure 12. SSIMS peaks O⁻ C⁻ followed during CO oxidation on Pd.

oxygen occupies the majority of the surface sites until at about 520°K the disappearance of the C$^-$ peak suggests a complete absence of CO as an adsorbed species. According to the last observation, since oxygen is the only adsorbate, the L.H. mechanism cannot be operative above 520°K. This conclusion can be disproved only by assuming that CO is still present as adsorbed species above 520°K but does not contribute at all to the secondary ion emission. The latter assumption would be, in our opinion, quite unrealistic. Besides, Ertl and Koch [11] have determined by flash desorption that the position of the maximum for CO on polycrystalline Pd happens around 240°C. This is in good agreement with our result.

Let us now consider the second aspect of our experiment, the detection by mass spectrometry of gaseous reaction products. In this particular case CO_2 is the only real reaction product, although unreacted CO and O_2 can be equally well investigated. Stoichiometric calculations have, in fact, been performed by comparing the intensity of CO_2 produced per unit time with the respective lack of signal in one beam when the second one is shuttered on and vice-versa. In Figure 13 a sequence of measurements of the CO_2 signal versus time at increasing temperatures is presented. The first plot shows that no significant production of CO_2 occurs at 300°K. Raising the temperature to 335°K produces a small quantity of CO_2 which appears about 20 seconds after the interruption of the CO beam. When CO is opened again, a very rapid and intense spike of CO_2 is produced; then, with a much slower time constant, the equilibrium is restored very close to the background level. At 380°K, a non negligible steady state level of CO_2 production is obtained as well as the previously observed transients. As the temperature increases, the characteristic time of the transient which occurs after the interruption of CO is strongly reduced. This fact is clearly connected with an L.H. mechanism with a reaction rate controlled by surface diffusion of oxygen. After the CO interruption, the reaction rate is no longer hindered by the presence of adsorbed CO which poisons the surface. For sample temperatues higher than 450°K, the reaction rate tends to stabilize at a large maximum value until at 550°K the rate will decrease again; see Figure 14.

From these data combined with the previous SIMS results, we can try to give a general explanation of the observed phenomena. The initial increase of the reaction rate with temperature is a consequence of the decrease in CO coverage and the simultaneous increase in oxygen coverage until the optimum adsorption pattern for an efficient action of the L.H. mechanism is reached around

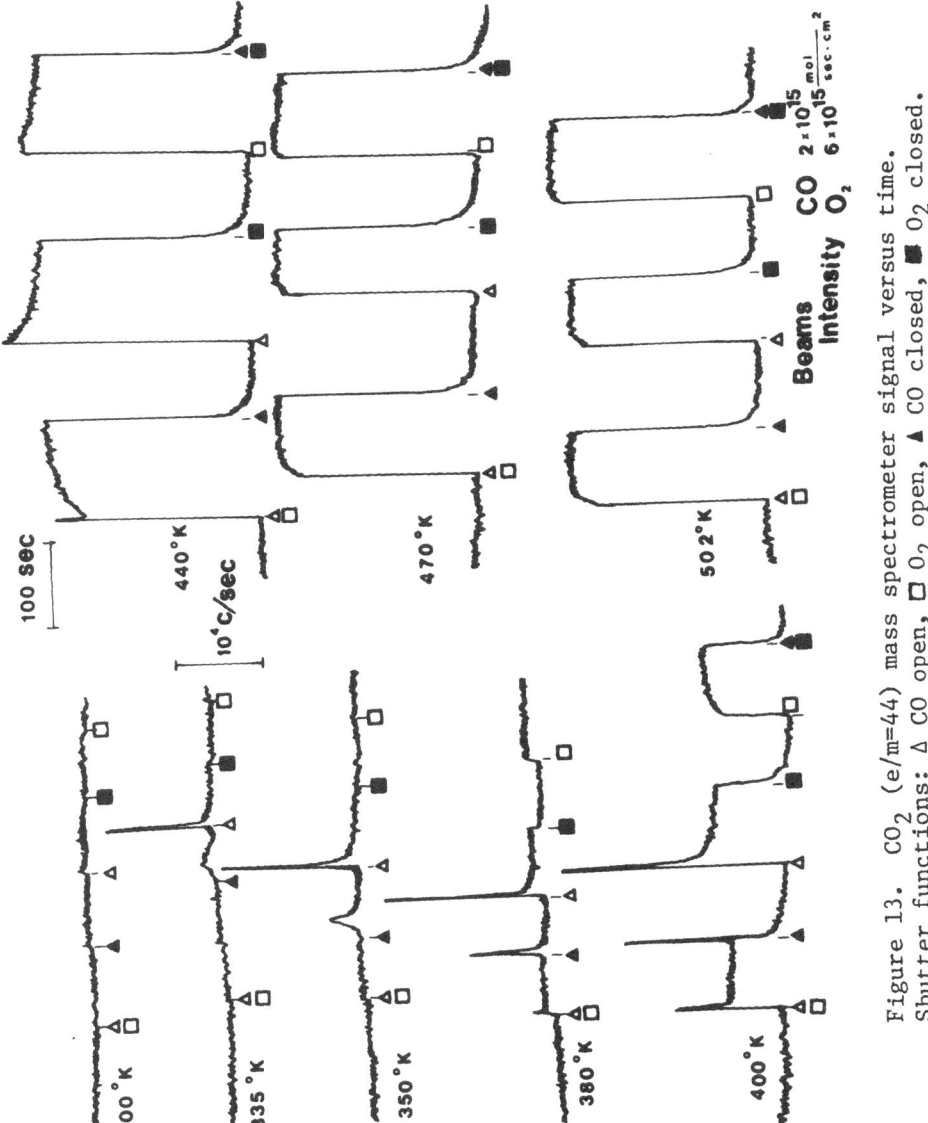

Figure 13. CO_2 (e/m=44) mass spectrometer signal versus time.
Shutter functions: Δ CO open, ☐ O_2 open, ▲ CO closed, ■ O_2 closed.

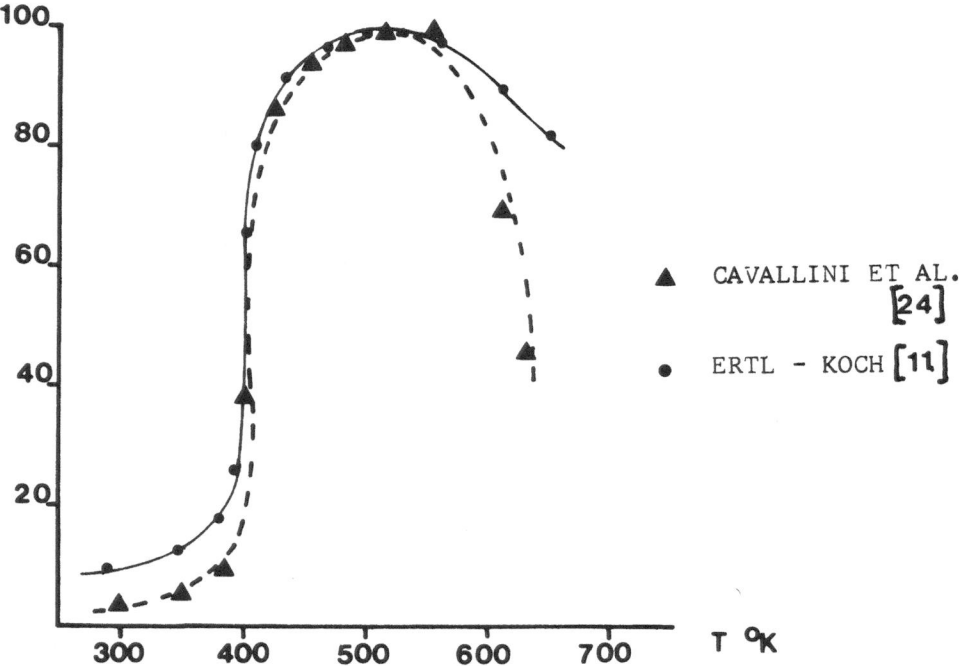

Figure 14. Steady-state rate of CO_2 formation as a function of
temperature.

450°K. In the range of higher temperatures, we found by SSIMS
that CO is completely absent as an adsorbed species, but in the
meantime we still observe that the reaction is going on. These
two pieces of experimental evidence are compatible only with an
operating E.R. mechanism. This is quite reasonable, however,
considering that at lower temperature both these mechanisms are
operative. This assumption, however, would enable us to also
explain the rapid increase in the rate of CO_2 formation observed
after the restoration of the CO beam on the surface, as is clearly
shown in the low temperature curves of Figure 13.

In connection with this interpretation, a clear distinction between the two mechanisms is reported by Bonzel and Ku in the conclusion of their paper [16]. They used the term "competitive" and "cooperative" adsorption, previously introduced by Ertl [25] to point out two opposite configurations of the adsorption state. The adsorption is, indeed, competitive. For example, the adsorption of an oxygen molecule needs two neighbouring free sites for dissociative adsorption on a surface with preadsorbed CO. This surface configuration is such that an oxygen molecule will have some difficulties in finding the correct adsorption condition because of CO poisoning. If this situation is overcome somewhere on the surface, the reaction with nearby adsorbed CO will occur and a large zone of the surface will be available for oxygen adsorption. The reaction mechanism in this case is L.H. operating only at the boundaries of the growing oxygen island. The induction time observed in the low temperature curve of Figure 13 after CO interruption can be explained by means of a competitive oxygen adsorption. The induction time is, in fact, the time necessary for the boundary reaction to reach the maximum island dimensions until a complete consumption of adsorbed CO is attained. When this process is completed, the Pd surface is partially covered with oxygen and no adsorbed CO is left on the surface. A steeply increasing CO pressure on such a surface configuration gives rise to an "instantaneous" CO_2 production that can be interpreted as a characteristic E.R. mechanism. If free sites among oxygen atoms are available for the adsorption of CO, the subsequent reaction occurs "instantaneously" between adsorbed CO and a nearby oxygen atom. This is physically indistinguishable from a process involving a CO molecule coming directly from the gas phase that reacts without experiencing even a very short adsorption time. In both instances, the reaction appears as if gaseous CO molecules would be able to establish a direct bond with an adsorbed oxygen atom. The latter CO adsorption process is called cooperative, and the subsequent mechanism cannot be considered an L.H. one even though it occurs between adsorbed molecules, since it is not limited by a surface diffusion process.

According to these considerations, some features obtained by averaging the CO_2 signal after the modulation of CO and O_2 are presented in Figures 15 and 16, respectively. We simply point out how the surface temperature strongly influences the shape of the CO_2 pulse. In the case of O_2 modulation, there is simply an integration effect due to the induction time caused by surface diffusion. In the experiments with a modulated CO beam, there is a quite rapid rise time but a much longer decay time. At 430°K, instead of a slow decay, a bump appears in correspondence with the CO interruption. This effect is caused by the excess CO that hinders the oxygen adsorption and reduces the reaction

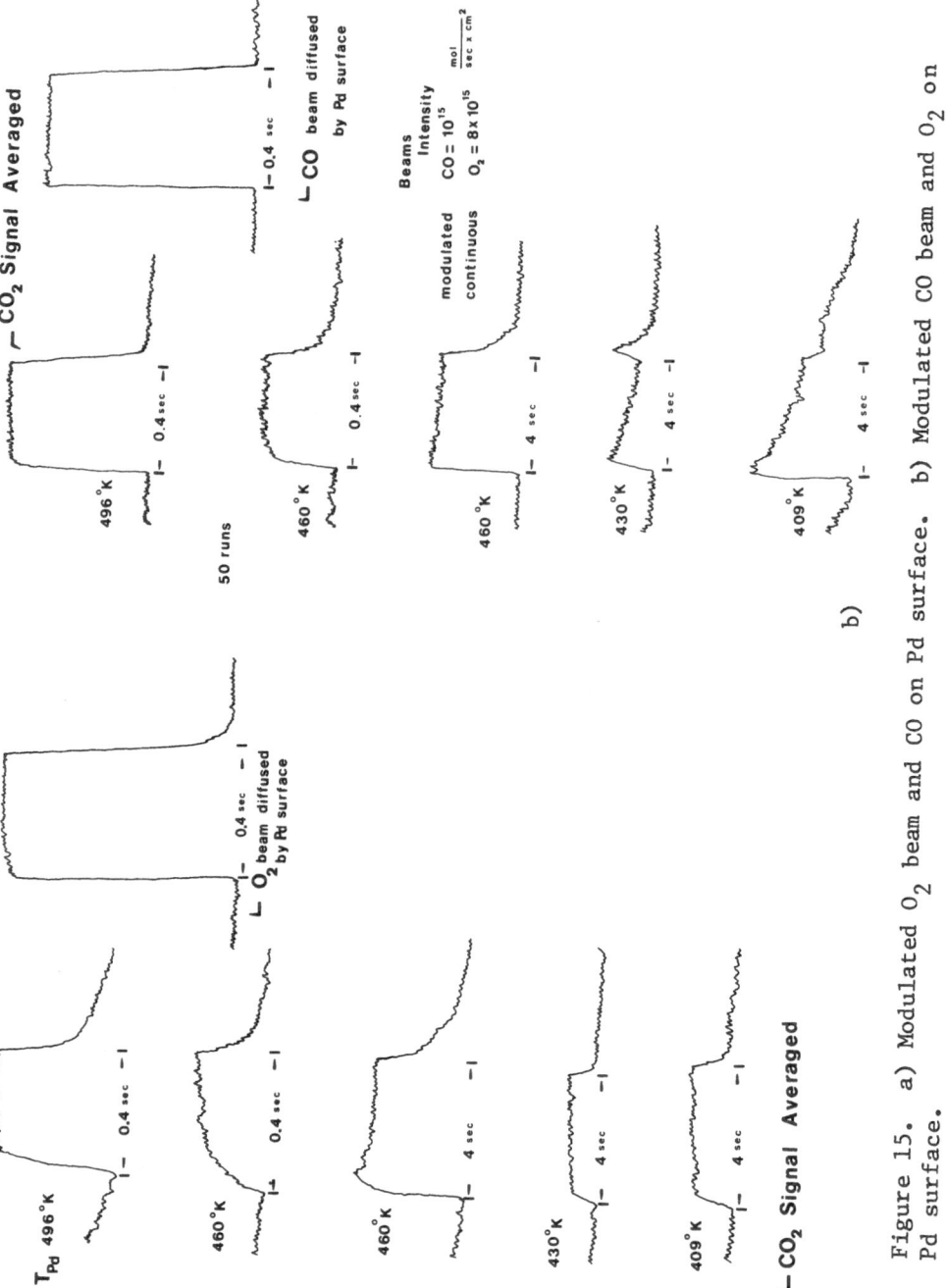

Figure 15. a) Modulated O_2 beam and CO on Pd surface. b) Modulated CO beam and O_2 on Pd surface.

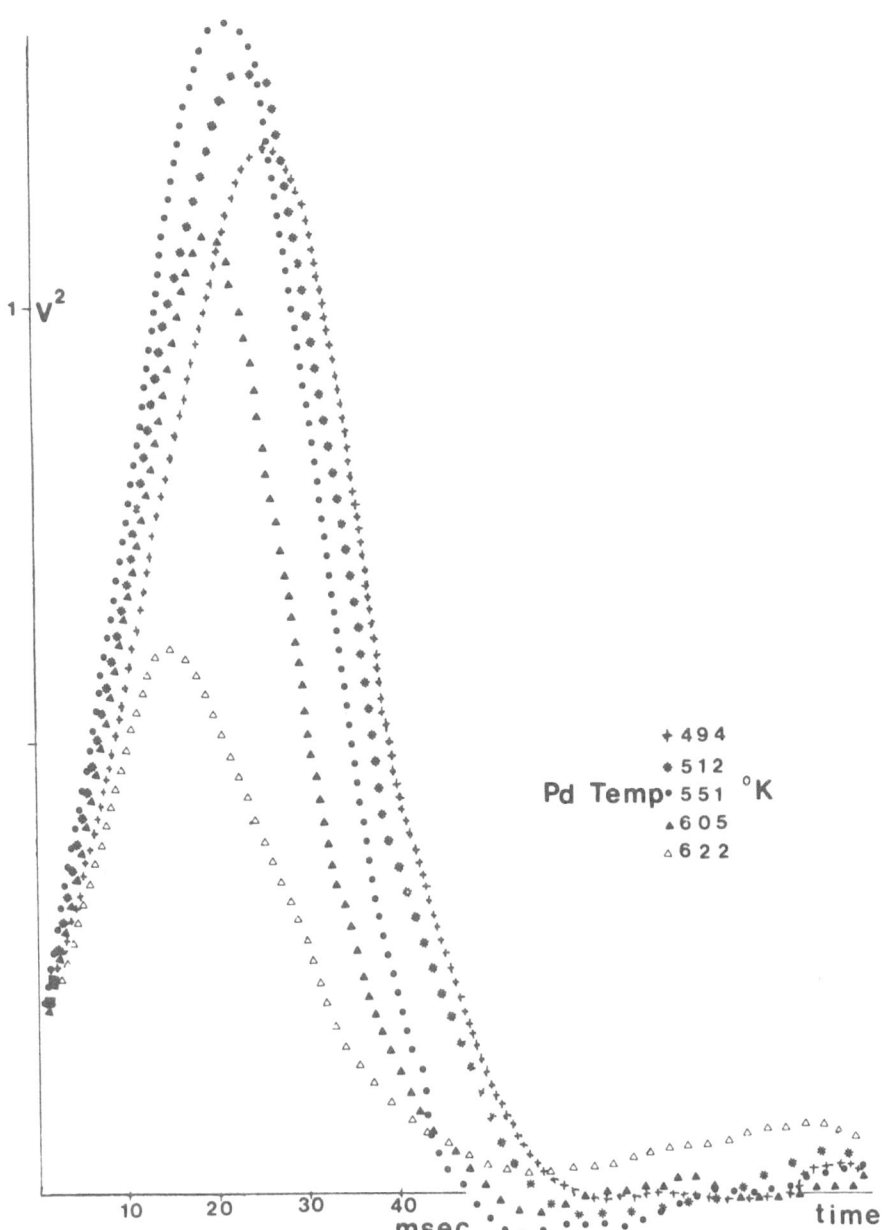

Figure 16. CO oxidation on Pd. Cross correlation: $x(t-\tau)$ = pseudo random binary sequence O_2 beam, $y(t)$ = Quadrupole MS output $\frac{e}{m}$ = 44 CO_2.

rate. When the CO pressure disappears, the oxygen starts to
react and creates free sites on the surface until all readsorbed
CO has reacted. At higher temperatures, the CO thermal desorption
is more rapid than oxygen surface diffusion. Finally, an example
of a quite sophisticated signal processing is reported, primarily
with the aim of presenting a new method of approaching the surface
chemistry in connection with the molecular beam technique. The
cross correlation of the CO_2 signal is performed by a
pseudo-random sequence of 20 msec wide impulses obtained by a
computer sorted binary random sequence controlling the stepping
motor rotation. When the randomly modulated O_2 beam reaches the
surface together with a continuous CO beam, a random sequence of
CO_2 pulses is generated. Each CO_2 pulse of the sequence undergoes
an integration process that depends on its time duration compared
with the reactant's adsorption times and the characteristic times
of other reaction rate limiting steps. The cross correlation
function represented by the integral

$$R_{xy}(\tau) = \lim_{T \to \infty} \frac{1}{T} \int_0^T Y(t)x(t-\tau)dt$$

has been measured between CO_2 spectrometer output and a reference
signal as a function of sample temperature. The results are
shown in Figure 16. The position of the maxima is temperature
dependent. In fact, an Arrenius plot gives an activation energy
of 4.5 kcal/mole that can be ascribed to the adsorption energy
of CO_2 on Pd. The latter is only one possible explanation of
these data. An independent check should be performed by sending
a CO_2 beam on Pd and measuring its residence time versus
temperature.

Finally, two important features of correlation functions
shall be recalled here. First, that the integral of this function
is proportional to the reaction rate. Second, that its Fourier
transform directly gives the power spectrum of the reaction; that
is, the characteristic times that control the reaction.

From this example, the application of molecular beam
scattering to surface chemistry studies appears to be quite
promising. This is especially true if this technique is used in
combination with other surface techniques and when the main
interest of the investigation is devoted to the dynamic character
of catalytic reactions. The conclusions presented here are the
subject of a paper in preparation [24] that will discuss a model
for the reaction and a detailed computer simulation of the results
we have presented here only briefly.

REFERENCES

1. Merrill, R. P. (1970) Catal. Rev. $\underline{4}$, 115.

2. Saltsburg, H. (1973) Ann. Rev. Phys. Chem. $\underline{24}$, 493.

3. Somorjai, G. A. (1972) Catal. Rev. $\underline{7}$, 87.

4. Palmer, L. P., and Smith, J. N. Jr. (1975) Catal. Rev. $\underline{12}$, 279.

5. Madix, R. J., to appear in "Physical Chemistry of Fast Reactions," Vol. 2, Plenum, NY.

6. Bernasek, S. L., and Somorjai, G. A. (1975) J. Chem. Phys. $\underline{62}$, 3149.

7. Eischens, R. P. and Pliskin, W. A. (1958) Advances in Catalysis $\underline{10}$, 1.

8. Heyne, H., and Tompkins, F. C. (1966) Proc. Roy. Soc. 292A, 460.

9. Park, R. L., and Farnsworth, H. E. (1964) J. Chem. Phys. $\underline{40}$, 2354.

10. Bonzel, H. P., and Ku, R. (1972) J. Vac. Sci. Technol. $\underline{9}$, 663.

11. Ertl, G., and Koch, J., Catalysis Congress $\underline{67}$, 969 (1972) Miami, Fl.

12. Kawasaki, K. et al. (1966) J. Chem. Phys. $\underline{44}$, 2313.

13. Bernasek, S. L. and Somorjai, G. A. (1974) J. Chem. Phys. $\underline{60}$, 4552.

14. Stephens, S. J. (1959) J. Phys. Chem. $\underline{63}$, 188.

15. Park, R. L., Proc. Symposium Gas Surface Interactions, pg. 295 (1966) San Diego, Ca.

16. Bonzel, H. P. and Ku, R. (1972) Surface Sci. $\underline{33}$, 91.

17. Palmer, R. L., and Smith, J. N. Jr. (1974) J. Chem. Phys. $\underline{60}$, 1453.

18. Pacia, N., Cassuto, A., Pentenero, A., and Weber, B. (1976) J. Catal. $\underline{41}$, 455.

19. Engel, T., and Ertl, G., Proc. Seventh Int. Vac. Conf.,
    pg. 1365 (1977) Vienna, Austria.

20. Schwarz, J. A., and Madix, R. J. (1974) Surface Sci. <u>46</u>, 317.

21. Cavallini, M. and Nencini, G., Proc. Ninth Int. Symp. on Rare
    Gas Dynamics E-10-1 (1974) Göttingen, Germany.

22. Cavallini, M., Cini, M., and d'Andrea, S., Proc. Sixth Int.
    Symposium on Molecular Beams (1977), Noordwijkerhout, The
    Netherlands.

23. Cavallini, M., Proc. Int. Conf. on Secondary Ion Mass
    Spectrometry (SMIS) and Ion Microprobes 1977,
    Muenster, Germany.

24. Bassi, D., Cavallini, M., and Grillo G., in preparation.

25. Ertl, G., Molecular Processes on Solid Surfaces, pg. 147,
    (1969), McGraw Hill, NY.

LOW ENERGY ION SCATTERING

U. Gerlach-Meyer and E. Hulpke

Max-Planck-Institut für Strömungsforschung

34 Göttingen, Böttingerstraße 4-8

ABSTRACT

Ion scattering at energies higher than 200 eV has been proved to be a useful tool for analyzing the structure and composition of surfaces. The additional information which can be extracted from ion scattering at energies below 20 eV will be discussed. The results of surface scattering experiments are reported which involve Li^+ ions as projectiles and a W(110) and a Ni(100)-surface as targets. High resolution measurements of the energy spectra and the spatial distributions of the backscattered ions are compared with the results of a computer simulation of the scattering process. The computer calculations are based on a scattering model which accounts for single as well as consecutive binary collisions with a chain of surface atoms, the appropriate repulsive interaction potentials, the attractive potential between the ion and the surface due to the image charge in the metal, and the thermal motion of the surface atoms.

With reasonable assumptions as to the actual values of the parameters involved, a very good theoretical prediction of the observed scattering behaviour can be achieved. The measured rainbow structure of the angular distribution is, however, not quite as well reproduced as the form of the energy spectra. This is probably due to the two-dimensional nature of the calculations.

I.   INTRODUCTION

The interaction between an atomic or molecular particle and the surface of a solid has been a very challenging problem for physicists, chemists, and engineers until today. Unfortunately such collision processes prove to be of a very complicated nature.

The difficulties in understanding the gas-surface interaction come about because these collisions are usually highly inelastic, involving multiple phonon excitation - and deexcitation processes. The scattering behaviour can therefore be very different for different collision energies. This is illustrated in Fig. 1. Depending on the magnitude of the collision time τ as compared to the vibrational period in the lattice, T, one can distinguish three different energy regimes in the gas surface interaction, the regime of quasielastic scattering (a), of inelastic scattering (b) and of impulsive scattering (c). Processes of the type a and b have been discussed in the preceding lecture. This report will be on impulsive collision processes [1]. Our goal has been to investigate the transition regime between the impulsive and the inelastic interactions.

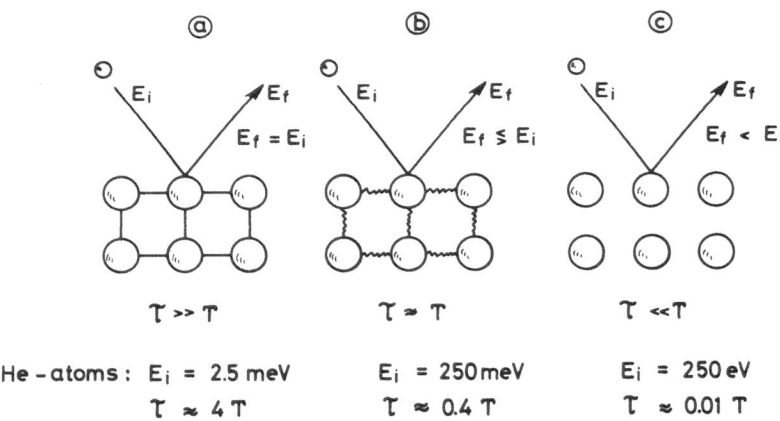

Fig. 1:   Different energy regimes in the gas-surface interaction:
a) quasielastic, b) inelastic, c) impulsive scattering

We have therefore performed surface scattering experiments at collision energies above 10 eV. For experimental reasons Li^+ - ions have been chosen as projectiles [2, 3, 4, 5, 21].

II.   EXPERIMENT

A schematic view of the experiment can be seen in Fig. 2. We have investigated the scattering of Li^+ ions from W(110) and Ni(100) surfaces at beam energies E_i between 5 and 50 eV and for incoming angles Θ_i between 10° and 70°. The scattering was observed in a plane which contains the surface normal $\underline{n}$ and the incoming beam. Using a time of flight (TOF) technique the energy spectra of the backscattered ions have been measured for a variety of outgoing angles Θ_f, ranging from $\Theta_f = 0^o$ to $\Theta_f = 85^o$. We have also measured the in-plane angular distribution for different incoming angles Θ_i.

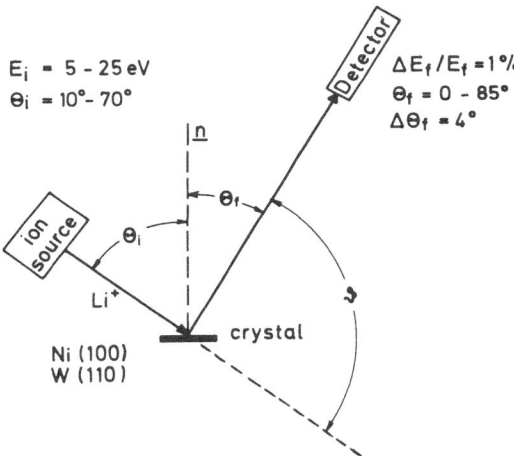

Fig. 2:   Schematic representation of the experiment.

The experiments have been performed in an ultra high vacuum system and the state of the surfaces could be controlled with in situ LEED. We have found out that the form of the measured energy spectra itself is a very sensitive indicator for possible surface contamination.

The experimental results will be presented in terms of the initial energy E_i, the final energy E_f, the incoming angle Θ_i and the outgoing angle Θ_f which are both measured with respect to the surface normal. In our discussion of the data we will also be using the deflection angle ϑ between the initial and the final momentum of the projectile. One can see from Fig. 2 that

$$\vartheta = \pi - \Theta_i - \Theta_f. \tag{1}$$

III. EXPERIMENTAL RESULTS

1.) Energy spectra

In order to facilitate the interpretation of our data we have converted the measured TOF spectra into energy spectra. Some typical energy distributions are plotted in Fig. 3. The curves (a) represent energy spectra of backscattered ions at four different initial ion energies and the narrow curves (b) show the energy distribution of the incoming beam. One can notice that all backscattered ions have lost kinetic energy and that the final energies are smeared out. The energy at the maximum of the curves (a) is roughly the average final energy and will be called E_f. Fig. 3 represents results for the Li$^+$ + W(110) scattering, the Ni-data exhibit a similar behaviour.

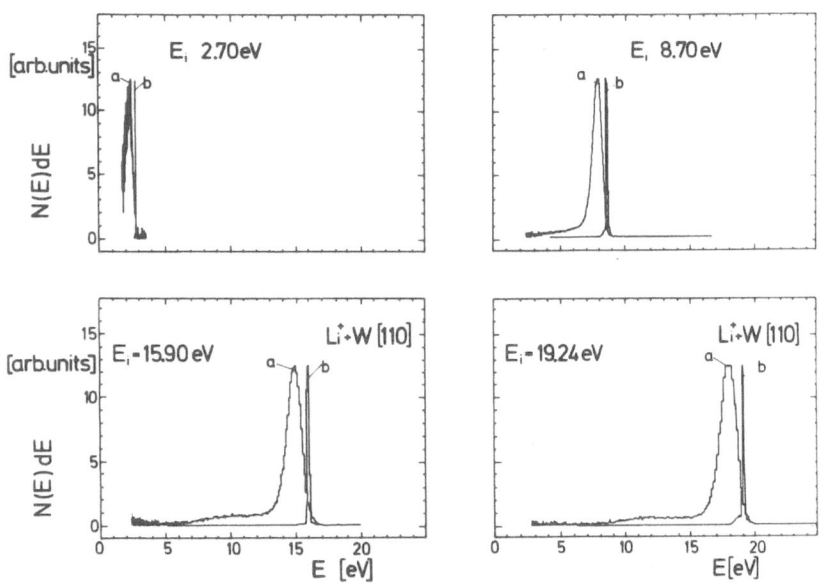

Fig. 3: Typical energy spectra of backscattered ions,
Li$^+$ + W(110) $\Theta_i = 55^o$, $\Theta_f = 64^o$.

2.)   Angular distributions

In our earlier experiments on the $Li^+ + W$ system we had observed
a rather interesting structure in the spatial distributions of the
backscattered ions [5]. We have investigated this phenomenon more
extensively for the $Li^+ + Ni$ scattering [2, 4, 21].

Some of these data are compiled in Fig. 4. The different plots show
the backscattered ion current as a function of the scattering angle
Θ_f. These curves have been obtained at an initial ion energy of
50 eV, each one for a different incoming angle Θ_i which is marked
by the arrow on the Θ_f-axis. One can notice that some of these
curves exhibit two maxima and that the positions of the maxima
cannot be correlated with the angle Θ_i in a straight forward way.
The form of the angular distributions depends also on beam energy.
Data for E_i = 10 eV are shown in Fig. 5. We see that in some cases
a more complicated structure can be found. Since the little bump
at Θ_f = 40° in the Θ_i = 30° curve proves to be a quite reproducable
feature, this curve exhibits 3 peaks.

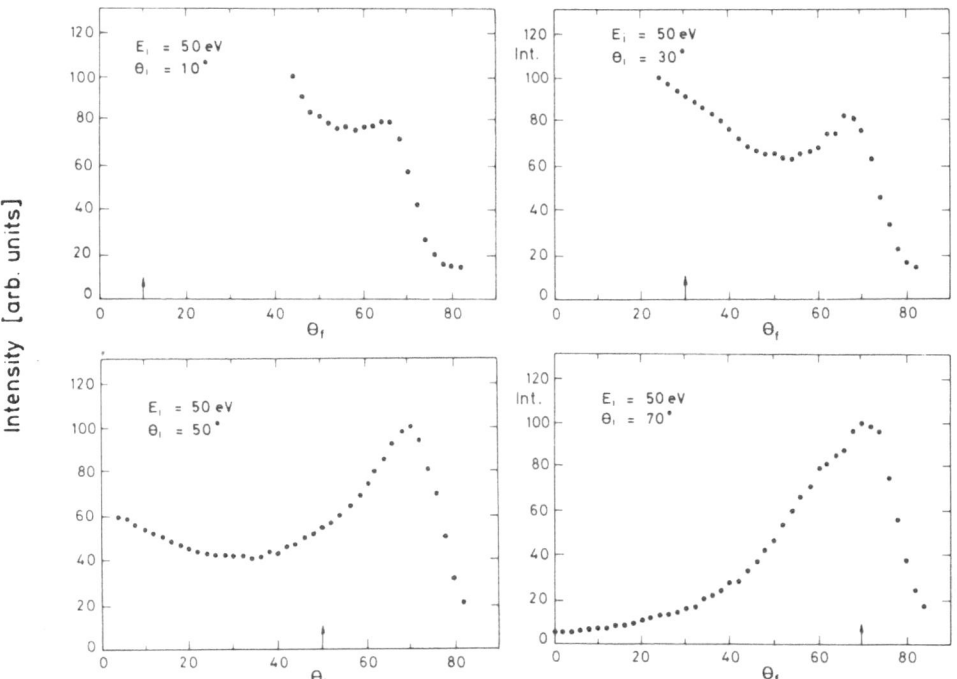

Fig. 4:   Typical spatial distributions of backscattered ions,
$Li^+ + Ni(100)$, E_i = 50 eV.

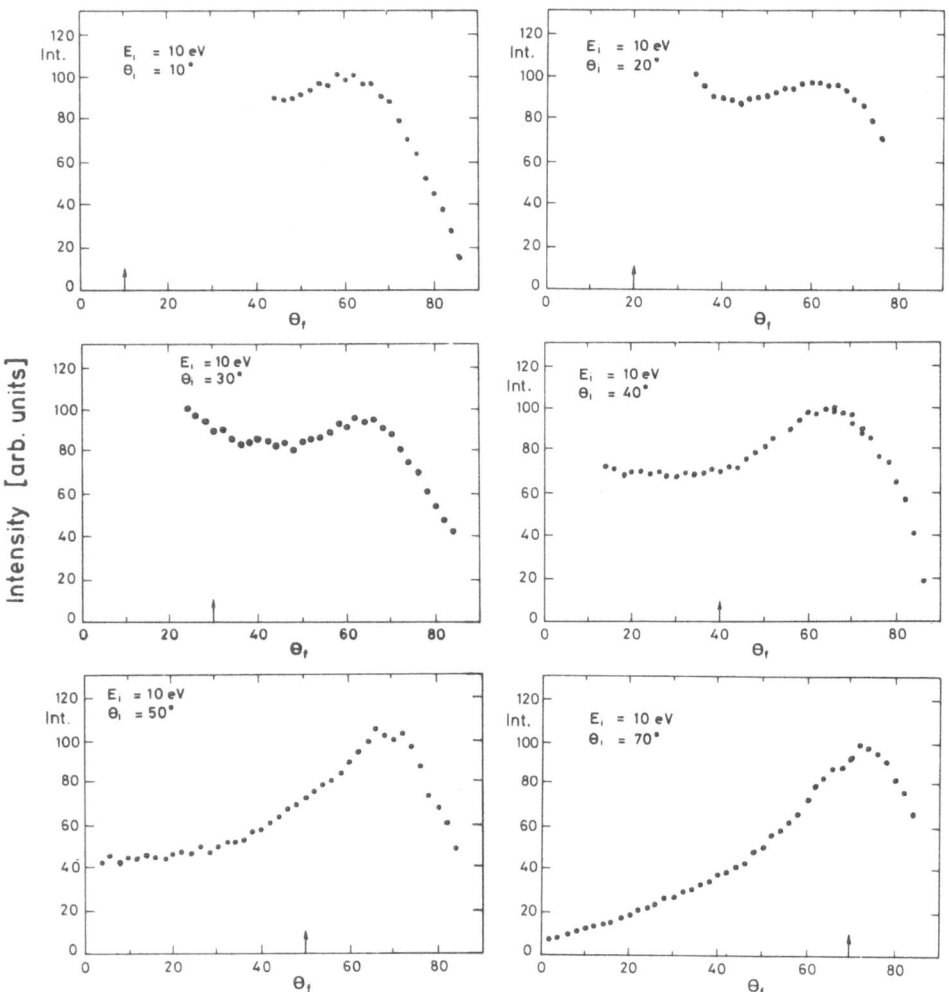

Fig. 5: Spatial distributions for Li$^+$ + Ni(100), E_i = 10 eV.

IV.   Discussion

Scattering models

In order to understand our results we have tried to compare them
with the predictions of scattering models of different degrees of
sophistication. As far as collision times τ and vibrational peri-
ods T in the surface are concerned an impulsive scattering model
should be applicable even at the unusually small ion energies in
our experiments. At 10 eV we find T ~ 40 τ

a.) Binary collisions

The simplest model of this kind is the binary collision model which postulates that the scattering be treated as a collision between the projectile and single, isolated surface atoms [6,7]. With this assumption one can calculate from energy- and momentum conservation laws the energy E_f of a projectile particle (mass m_p) which has been deflected by an angle ϑ after colliding with a surface atom (mass m_T).

$$\frac{E_f}{E_i} = \frac{1}{(1 + \mu)^2} \; [\cos \vartheta + \sqrt{\mu^2 - \sin^2 \vartheta} \;]^2 \qquad (2)$$

$$\mu = \frac{m_T}{m_p} \quad ; \; \mu > 1 \; ; \; \text{target atom initially at rest.}$$

In Fig. 6 we have tried to compare our data with this model. In order to do so we have plotted the measured ratios E_f/E_i versus the measured deflection angles $\vartheta = \pi - \Theta_i - \Theta_f$. The solid line is the prediction from formula (2). The model fails to explain our data.

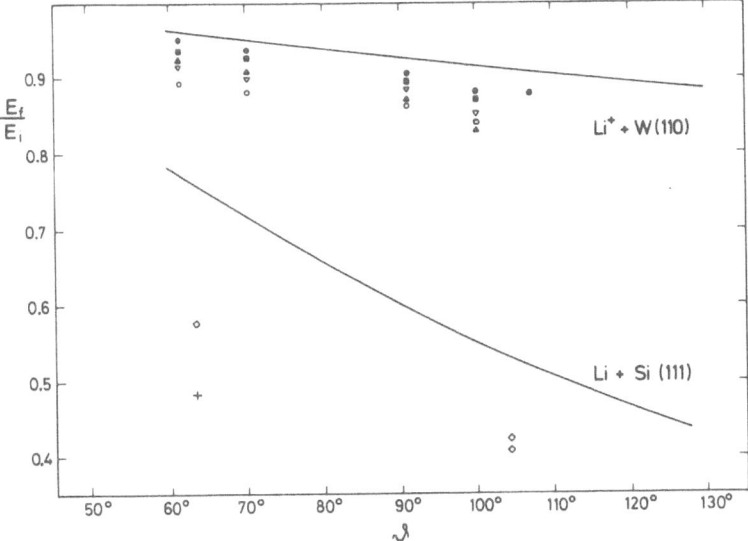

Fig. 6: Comparison of experimental data with the unmodified binary collision model, ● E_i = 19.83 eV; ■ E_i = 16.32 eV; ▲ E_i = 12.64 eV; ▼ E_i = 9.10 eV; ○ E_i = 4.65 eV; ◇ E_i = 13.80 eV; + E_i = 7.52 eV.

b.)   Modified binary collision model

We have tried to construct a modified binary collision model [5]
which takes the attractive forces between the ion and the surface
into account by means of a potential step, step hight $\mathcal{E}$, and in which
the repulsive encounter is again treated as a binary collision. Formula (2) must then be replaced by

$$\frac{(E_f + \mathcal{E})}{(E_i + \mathcal{E})} = \frac{1}{(1 + \mu)^2} \; [\cos \vartheta^* + \sqrt{\mu^2 - \sin^2 \vartheta^*}]^2. \tag{3}$$

ϑ^* denotes the deflection angle at the point where the binary collision takes place.

If we compare our energy loss data with the predictions of this
model we find a fairly nice agreement (cf. Fig. 7). The solid line
in Fig. 7 represents the theoretical curve, the symbols are data
which are plotted in a fashion that allows a comparison with
formula (3). [5]

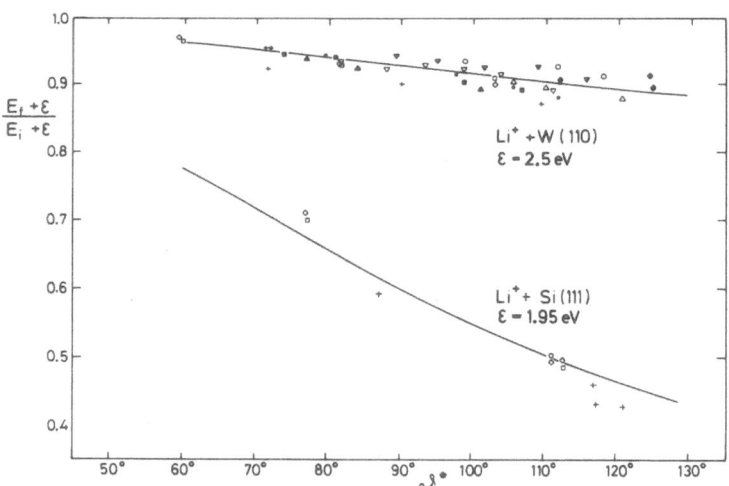

Fig. 7:   Comparison of experimental data with the modified binary
         collision model, symbols as in Fig. 6.

The binary collision model fails to explain the angular distribution.
Fig. 8a shows the differential cross section for a soft repulsive
two particle interaction (cf. [8]), Fig. 8b, c, d display measured
angular distributions for the Li$^+$ + W system. When written as a
function of $\Theta_f = \tau - \vartheta - \Theta_i$ the differential cross section looks like

the solid lines in Fig. 8b,c,d. In order to achieve a behaviour at large Θ_f as observed in the data, an artificial sharp cut off has been introduced at $\Theta_f = 75^\circ$. We will see later on that the form of the angular distribution is a consequence of the periodic structure of the surface, and that this is why it cannot be understood in terms of single collisions.

One can also try to explain the large spread in the observed energy spectra in terms of the binary collision model. In order to achieve this, the thermal motion of the surface atoms must be included in this model [5, 9-13]. In Fig. 9 measured TOF spectra for two different surface temperatures are presented which are compared with the predictions from the binary collision model denoted by the two large bars inside the curves.

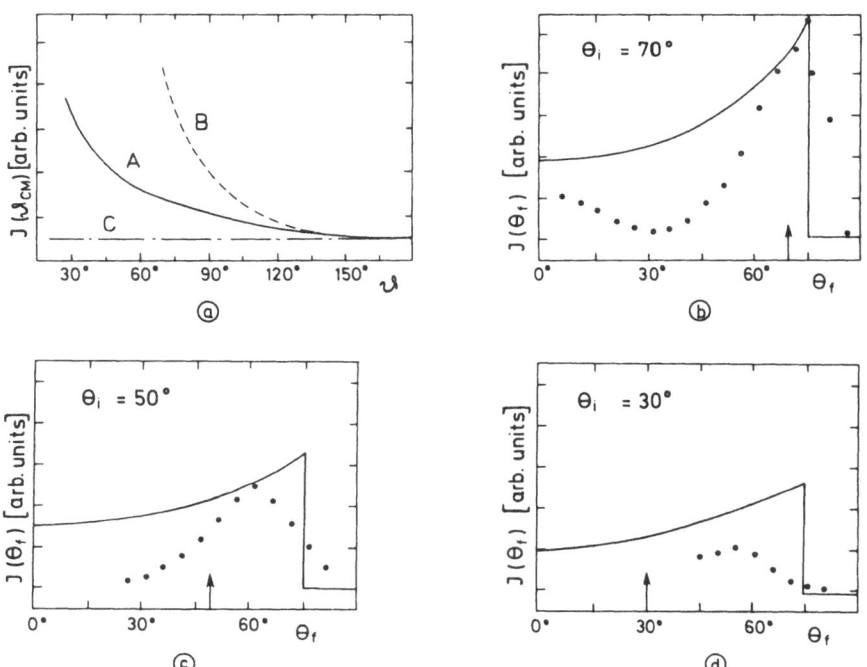

Fig. 8a: Differential cross section for two particle interaction at
           high energies:
           A: for linear deflection function (cf. [8])
           B: Coulomb interaction
           C: hard sphere interaction.

Fig. 8b,c,d: measured angular distributions, $Li^+ + W(110)$,
               $E_i = 14$ eV ($\bullet$) and binary collision model (solid line).

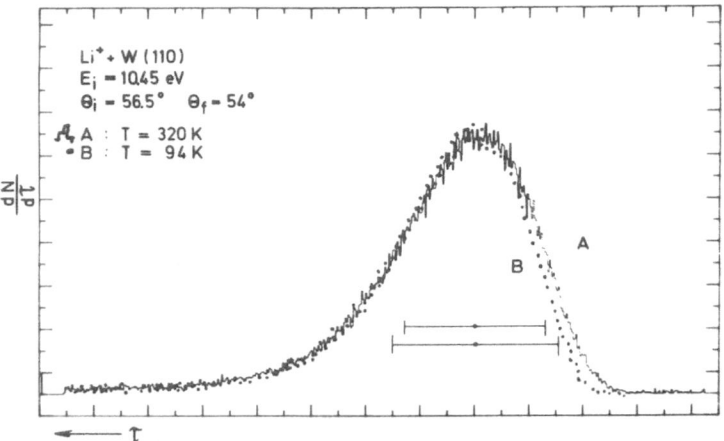

Fig. 9:   TOF spectra at different surface temperatures, bars
          denote predicted smear out (binary collision model).

We can summarize that the modified binary collision model proves
to be capable of reproducing the energy loss of the scattered ions
quite nicely, the predictions for the spread in the final energy are
but qualitative and angular distributions cannot be explained at all.
This model has the advantage of providing analytical formulae for
energy loss and energy spread. It has, however, a number of
shortcomings. The most serious shortcoming is that the predictions become more and more qualitative with decreasing mass
ratio μ. The reason for all these difficulties is that collisions with
neighbouring surface atoms have been neglected.

c.)   Double collisions

Because of the close packing of the surface atoms many projectile
particles will suffer a second encounter with a neighbouring surface atom. Let us therefore ask the question: How do double collisions influence the scattering behaviour ?

Energy losses

Fig. 10 despicts a double collision for which the recoil momenta
of the two target atoms involved are both of the same magnitude
and point in almost the same direction. The kinematics of such
a collision is quite similar to one in which the momentum $\vec{p}$
(cf. Fig. 10) is picked up by an atom of twice the mass of the target atom. Larger recoil momenta result in a smaller change in
the kinetic energy of the scattered particle. We arrive at the
surprising result that such double collisions are correlated with

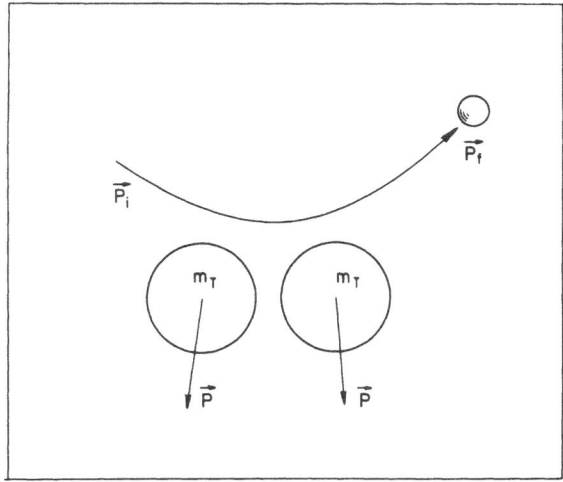

<u>Fig. 10:</u>   Collision with two atoms, symbols explained in text.

a smaller loss of kinetic energy for the projectile than a single
collision which leads to the same deflection.

Single collision rainbow scattering

The periodicity of the arrangement of the surface atoms leads to
the so called surface-rainbow scattering [14, 15]. We would like
to discuss this effect for an interaction potential which acts like
a sinusoidially corrugated hard wall. In Fig. 11 each incoming
trajectory is labeled by its impact parameter b. For each trajectory the orientation of the actual surface normal depends on the
impact parameter and the particles are specularly reflected with
respect to this actual surface normal.

We can calculate the deflection function $\Theta_f(b)$ which is plotted in
the lower part of Fig. 11. One notices that the deflection function
repeats the periodicity of the potential. One can also see that a
large range of impact parameters Δb_2 contributes to scattering
into angles at the extrema of $\Theta_f(b)$ while only very few trajectories
(Δb_1) are scattered into an equally large range of angles ($\Delta \Theta_f$)
around other Θ_f. The resulting angular distribution will therefore
exhibit two maxima, located at Θ_f for which $d\,\Theta_f/db = 0$.

Double collision rainbow scattering

If we allow for more than one collision with the repulsive potential,
the picture becomes even more complicated. We would like to

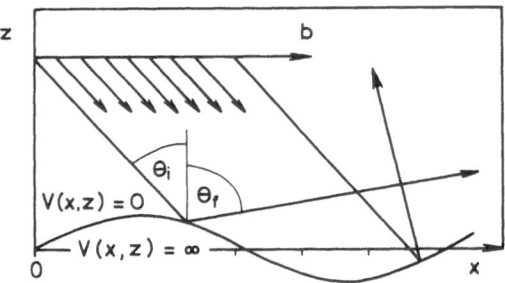

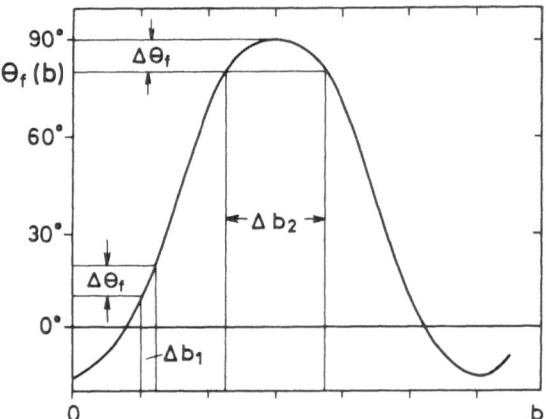

Fig. 11:   Rainbow scattering from corrugated wall, symbols explained in text.

discuss these effects for the scattering from a potential which acts
like a solid wall corrugated in zig-zag form. In this case the rainbow structure of the angular distribution degenerates into δ-func-
tions at the rainbow angles.

We notice in Fig. 12 that in addition to the single collision rainbows
(trajectory 1 and 2) a double collision rainbow (trajectory 3) can
be observed. The locations of the three rainbow angles depend on
the parameter α:

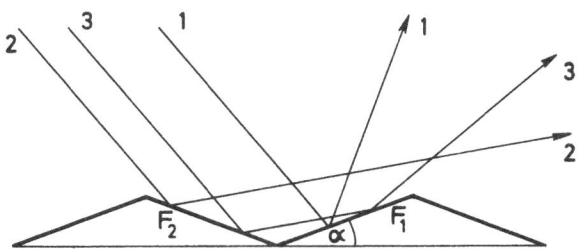

Fig. 12:   Multiple rainbow scattering, see text.

$$\Theta_f^{\,1} = \Theta_i - 2\alpha$$
$$\Theta_f^{\,2} = \Theta_i + 2\alpha$$
$$\Theta_f^{\,3} = \pi - 4\alpha - \Theta_i \,. \qquad (4)$$

If we take α as a measure for the "waviness" of the surface, we
can see that even small changes in the "waviness" of the potential
can cause considerable changes in the position of the rainbow
maxima. We note in particular that the double rainbow $\Theta_f^{\,3}$ is
twice as sensitive to changes in the surface roughness as the
single collision rainbows.

    d.)   Two dimensional computer simulation of the scattering

Scattering model

The previous discussion of the influence of single and double binary collisions on the surface scattering in the impulsive collision regime involved fairly unrealistic assumptions as to the
interaction potential between the projectile and the surface, like
the assumption of rigid, hard wall repulsive potentials or a
potential step which accounts for the attractive forces. In order
to explain our data we have constructed a scattering model that
comprises all the details which we expect to influence the scattering behaviour. These details are the following:

1.) the impulsive nature of the collision which guarantees that
   the perturbation of the surface due to the encounter itself
   does not influence the scattering behaviour;

2.) a realistic representation of the attractive forces between the
   ion and the surface which come mainly about because of
   forces between the ion and its image charge;

3.) collisions with more than one surface atom at the same time;

4.) a realistic form of the repulsive interaction potential between
   the ion and the surface atoms;

5.) the bonding of the surface atoms to their equilibrium positions,
   and last not least

6.) the influence of the thermal motion of the surface which
   results in an instantaneous displacement and a finite value
   of the initial momentum of the surface atoms.

All this can be achieved by an appropriate choice for the interaction potential. It is obvious that predictions from such a sophisticated scattering model can be obtained only through a computer
simulation of the scattering.

Since our data have been measured in the scattering plane and in
order to save computer time we have restricted ourselves to
calculating two dimensional particle trajectories, thus simulating
the scattering from a chain of surface atoms. In this case it is
reasonable to include the interaction with only two adjacent surface atoms because of the short range of the repulsive forces. As
we will see later, this restriction can prove to be too strong. It
is of course straight forward to extend the model potential by
including the interaction with more surface atoms and to perform
three dimensional trajectory calculations.

Model potential for the interaction with a chain of atoms

The model potential used in these computer calculations consists
of three parts:

1.) an attractive potential V_1 which does not change in the course
   of the collision process,

2.) a term V_2 which accounts for the interaction between the projectile and two adjacent surface atoms, the positions and mo-

menta of which are influenced by the collision and

3.) a term V_3 representing the bonding of each one of the two surface atoms to their equilibrium positions.

Let us now take a look at the actual form of these three terms and the coordinate system used. We have chosen a cartesian coordinate system which is fixed in space, the z axis perpendicular to the surface. The positions of the two surface atoms are denoted by $\underline{r}_1$ and $\underline{r}_2$, their equilibrium positions by $\underline{r}_1{}^0$ and $\underline{r}_2{}^0$ $\underline{R}$ describes the position of the projectile.

V_1 has been chosen to be a truncated Coulomb potential

$$V_1 = \frac{-e^2}{4((z-z_o)^2 + z_m{}^2)^{1/2}} \qquad z > z_o$$

$$V_1 = \frac{-e^2}{4 z_m} = \text{const.} \qquad z \leqslant z_o \qquad (5)$$

The truncation has been performed in order to avoid the unphysical singularity of the Coulomb potential.

V_2 is obtained from a superposition of two Born-Mayer potentials.

$$V_2 = D \exp (-\gamma |\underline{R} - \underline{r}_1|) + D \exp (-\gamma |\underline{R} - r_2|) \qquad (6)$$

The bonding to the equilibrium positions of both target atoms is achieved by assuming harmonic forces:

$$V_3{}^i = \frac{m_T \omega_D{}^2}{2} (\underline{r}_i - \underline{r}_i{}^0)^2 \qquad (7)$$

with $\hbar \omega_D = k_B \Theta_D$; Θ_D Debye temperature; m_T mass of the surface atoms.

In the calculations to be discussed below the following set of potential parameters has been used:

$$z_o = 1.0 \text{ Å} \qquad D = 13\,400 \text{ eV}$$
$$z_m = 0.11 \text{ Å} \qquad \gamma = 3.0 \text{ Å}^{-1} .$$

Trajectories and scattering behaviour

In order to compare our data with the predictions of the collision
model which we have just discussed, two dimensional classical
trajectories have been calculated for ions that are scattered from
the model potential $V_1 + V_2 + V_3$ by numerically solving the
equations of motion. Some typical results are shown in Fig. 13.
As explained in the previous chapter, we have chosen the coordinate x parallel and the coordinate z perpendicular to the surface. The wavy lines in this plot are equipotential lines for the
model potential, the spacing of the lines being 300 meV. The
bundle of parallel straight lines represents trajectories of the
incoming beam. One can notice a strong spatial dispersion of the
of the scattered ions.

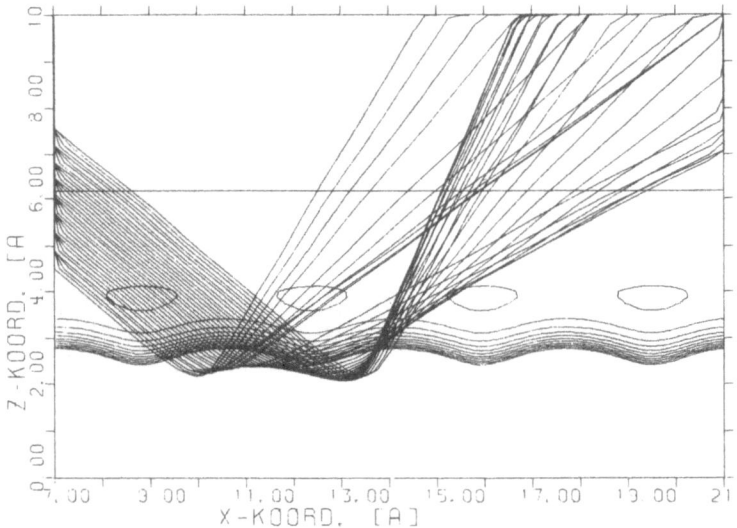

Fig. 13: Typical particle trajectories for scattering from the
model potential $V_1 + V_2 + V_3$.

For each trajectory the scattering angle Θ_f and the final energy
E_f can be determined and thus the dependence of Θ_f and E_f on the
impact parameter b. In the upper part of Fig. 14 we have plotted
the deflection function $\Theta_f(b)$ and the function $E_f(b)$ which have been
obtained from a set of trajectories starting at $\Theta_i = 30^{\circ}$ and
$E_i = 20$ eV.

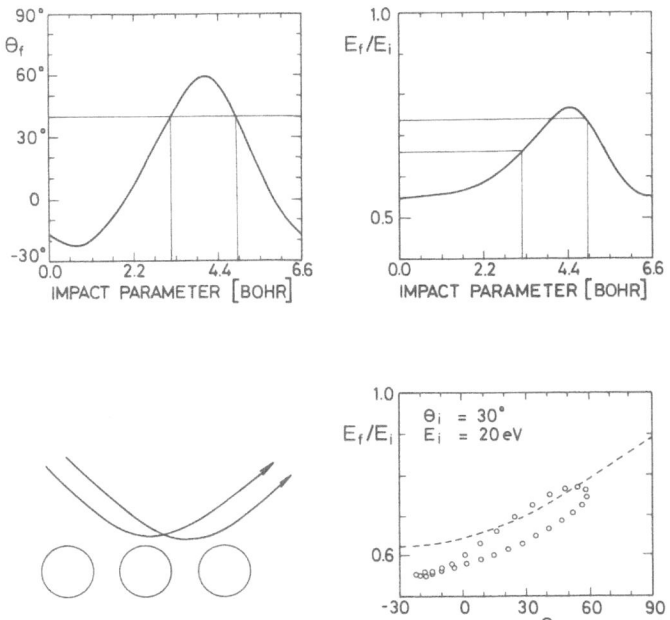

<u>Fig. 14:</u> Deflection function and energy loss from trajectory calculations.

One can notice that the deflection function displays two extrema which, as we know, leads to the phenomenon of rainbow scattering. For each angle Θ_f between the two rainbows we find two different impact parameters correlated with two trajectories that are scattered into the same outgoing angle. From the E_f/E_i plot we can see that these two trajectories end at two different final energies E_f. We find that one of these trajectories has suffered a single impulsive collision and the other one a double collision. The plot E_f/E_i versus Θ_f (cf. lower part Fig. 14) shows that all the possible values of E_f/E_i form a loop in the E_f/E_i - Θ_f plane [16-20]. The behaviour for single collisions is marked by the dotted line.

The energy spectrum associated with this energy loss loop consists of two infinitely sharp peaks. An associated angular distribution is shown in Fig. 15. We recognize the two rainbow maxima which drop off sharply beyond the rainbow angles.

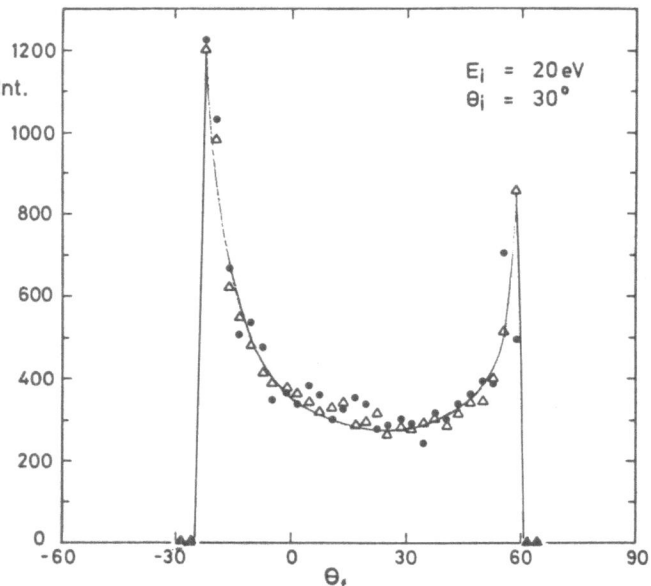

<u>Fig. 15:</u>   Calculated spatial distribution

Energy loss loops

Fig. 16 shows an energy loss loop which has been obtained from
trajectory calculations for $V_1 \equiv 0$. From the discussion above we
know that the larger value of E_f is correlated with a double collision and the smaller one with a single collision. The dotted lines
in Fig. 13 have been calculated from formula (2) for two different
mass ratios. The lower curve describes a behaviour which is
expected for single binary collisions with one target atom and the
upper curve describes this behaviour for target atoms which are
twice as heavy. One notices that the energy loss loops are located
between these two curves, which indicates that double collisions
that contribute to the upper branch of a loop behave like single
collisions with atoms of an "effective" mass somewhat smaller
than twice the actual mass of the target atom.

This observation makes it possible to explain why the single binary
collision model has been successful in predicting the scattering
for systems with large mass ratios. The reason is simply that for
large mass ratios both dotted curves in Fig. 16 are located so

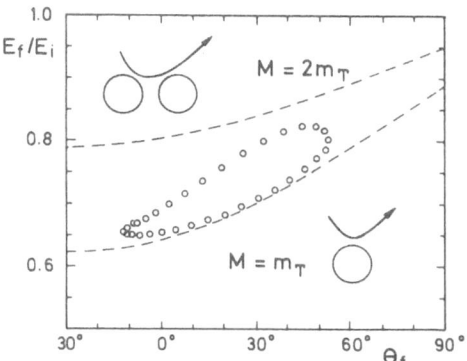

<u>Fig. 16:</u>   Energy loss "loop", M = "effective" mass of target atom,
            see text.

closely together that the failing of the single binary collision model
is less easily detected in the experiment.

Thermal averaging

The influence of the surface temperature on the scattering can be
taken into account by including finite initial displacements and
velocities of the target atoms in the computer calculations. These
initial values of the coordinates and momenta are chosen in such
a way that the phase of the oscillator is chosen at random and by
assuming that the energy is Boltzmann-distributed, i. e.

$$f(E) = \frac{1}{<E>} \exp(-E/<E>), \tag{8}$$

where E denotes the energy of the oscillator for one dimension
and $<E>$ is the mean energy per atom and dimension in a Debye
crystal [2, 21].

Results of the Monte Carlo calculations

The influence of the surface temperature on the calculated energy
spectra is shown in Fig. 17 for three different surface temperatures.
One observes an increase in smear out with increasing temperature. As a result the double peak structure disappears at sufficiently high temperatures. It is, on the other hand, possible to
regain the original structure of the energy spectra by cooling the
surface. One must, however, keep in mind that the zero point

vibration of the surface atoms sets a principle lower limit on the
initial momenta and displacements of these atoms. For the Ni
surface cooling below ~50 K no longer affects the smear out of
the spectra [2].

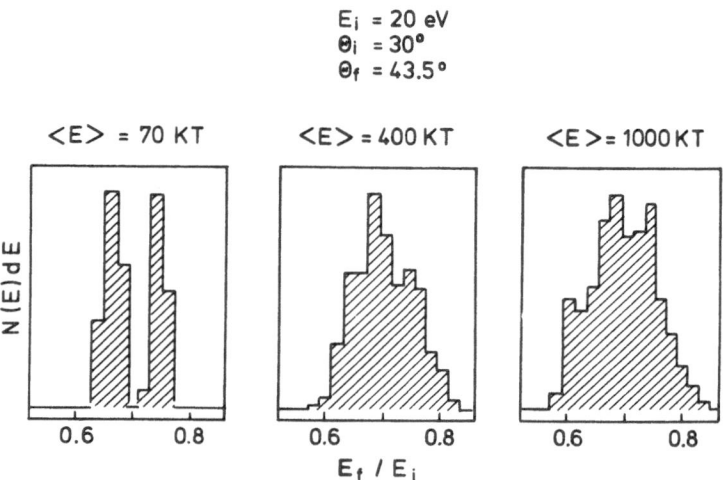

<u>Fig. 17:</u>   Influence of surface temperature on calculated energy
distributions.

In Fig. 18 we have plotted calculated angular distributions for two
different surface temperatures. With increasing temperature the
rainbow maxima become broader and less intense. If we remember
the case of the zig-zag-potential we know that changes in the
"waviness" can cause considerable shifts in the positions of the
rainbow maxima. The displacements of the oscillating atoms result
in such changes in the "waviness" of the surface and thus in a
smeared out rainbow structure.

Comparisons with the experiment
Fig. 19 shows a number of calculated energy spectra represented
as histograms for different beam energies and scattering angles.
These spectra are compared with the experimental data. One
notices that the position and the shape of the experimental spectra
is reproduced very nicely. As before, we define the energy at the
maxima of these histograms as the average final energy E_f.

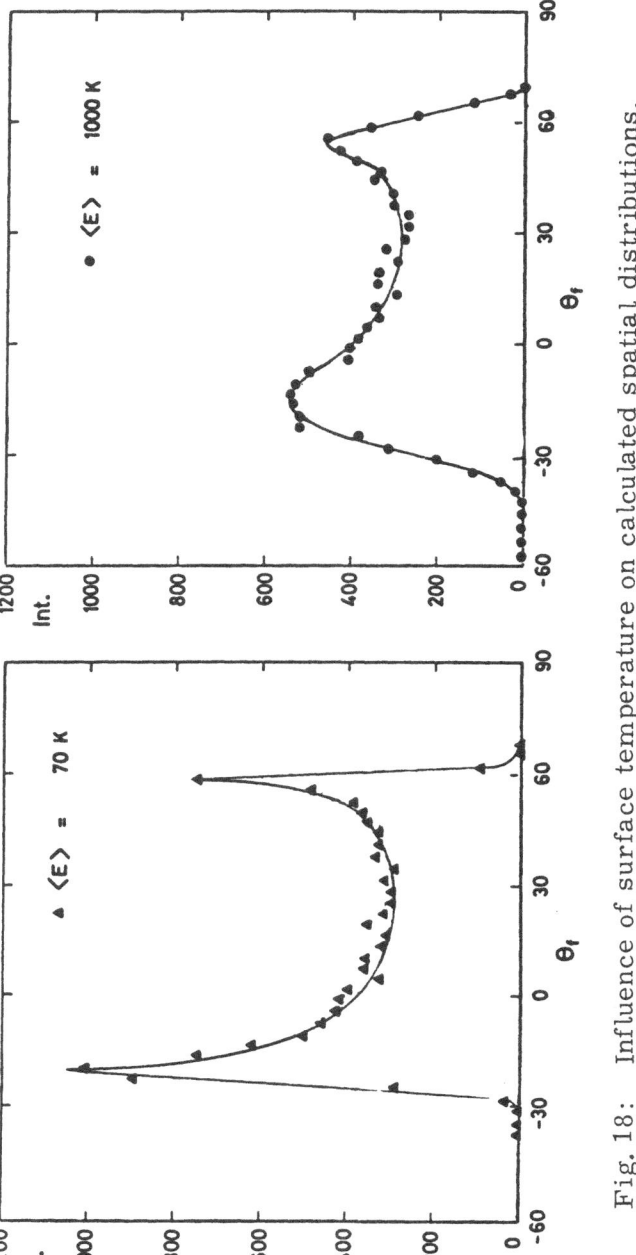

Fig. 18:   Influence of surface temperature on calculated spatial distributions.

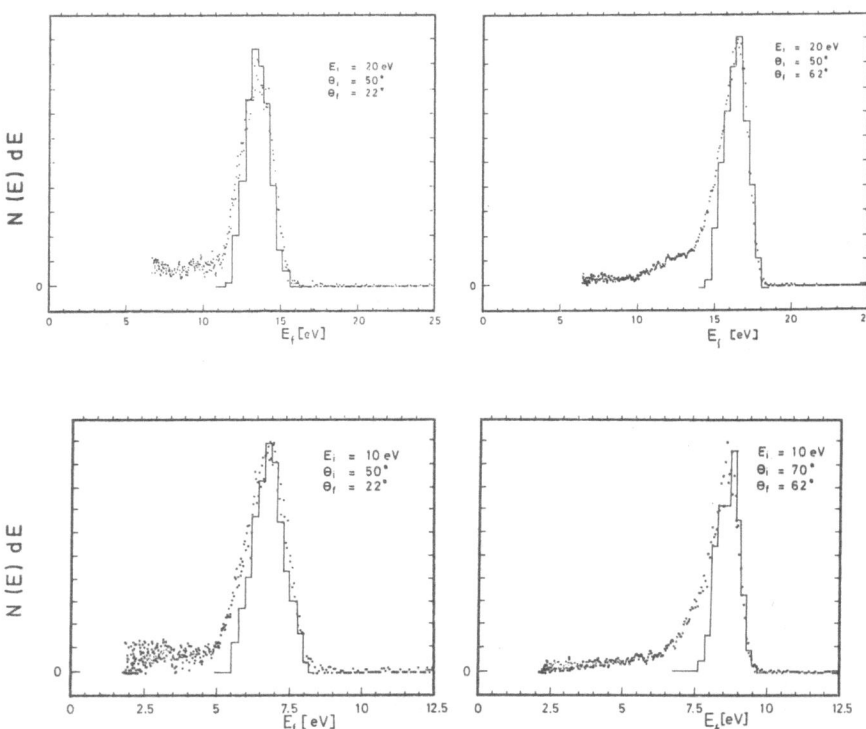

Fig. 19: Shape of calculated energy spectra (histograms) compared
with data (•) .

In Fig. 20 measured values of E_f/E_i as a function of Θ_f (represented by the symbols) are compared with the solid line which has been found from the computer calculations. Again the agreement with the experiment is a rather satisfactory one.

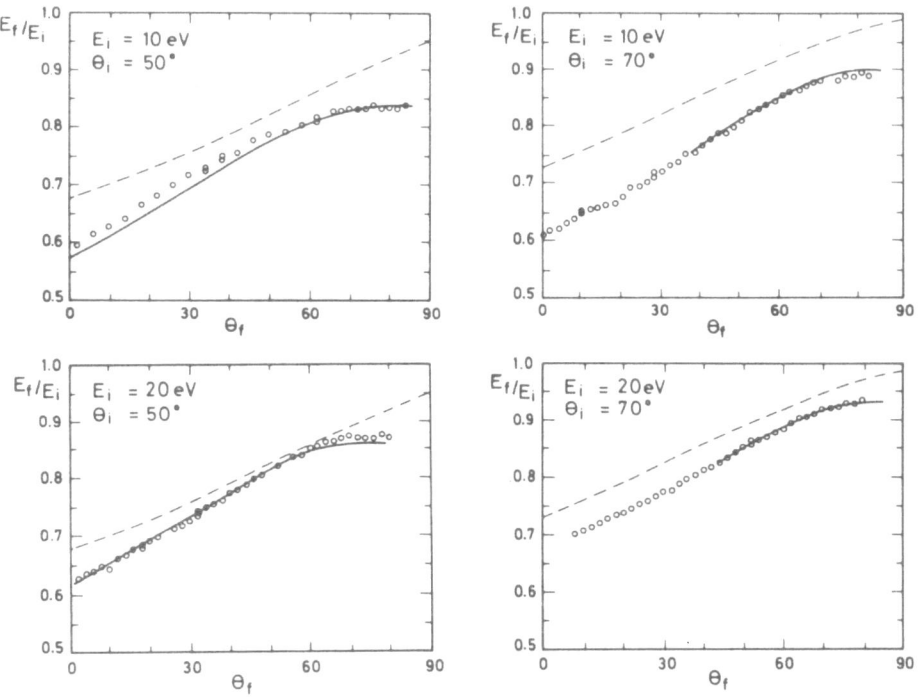

Fig. 20:   Behaviour of average loss in kinetic energy:
           o : measurements, solid line: computer simulation,
           broken line: binary collision model.

Calculated and measured angular distributions for different values of E_i and Θ_i are plotted in Fig. 21. The general features of the data (circles), particularly the rainbow structure for $\Theta_i = 30^0$ and $\Theta_i = 70^0$, are qualitatively reproduced by the calculations (crosses). The agreement between theory and experiment is, however, not nearly as good as for the energy spectra. In particular the strongly developed rainbow structure at $\Theta_f \sim 20^0$ for $\Theta_i = 50^0$ in the calculated spectra compares very poorly indeed with the data.

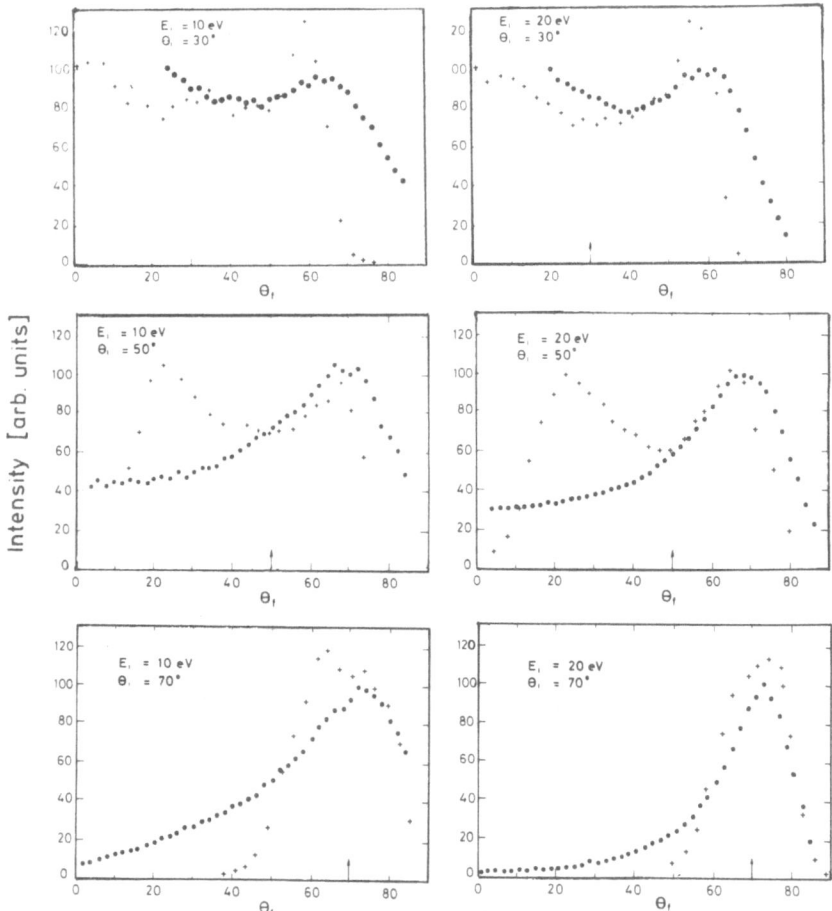

<u>Fig. 21:</u>   Comparison of measured (●) and calculated angular
distributions (+).

How can such a discrepancy come about ? We had seen before that
changes in the "waviness" of the potential influences the position
of the rainbow maxima. Such variations of the waviness can be
introduced by changing the "lattice constant", i. e. the distance
between the centers of the two surface atoms involved.

We have performed calculations for different lattice constants
which are up to 14 % smaller than for the row of atoms explored
in the experiment. The influence of averaging over this range of

lattice constants is illustrated in Fig. 22. We have found that for
$\Theta_i = 50^\circ$ the strong rainbow structure at lower $\Theta_f(\Theta_f \sim 23^\circ)$ for
the fixed lattice constant (crosses +) becomes washed out
(triangles ▲), if the average is taken over calculations for lattice
constants ranging between 3.02 and 3.52 Å. Comparing the averaged angular distribution (▲) with the experimental results (symbols (•) in the upper part of Fig. 22) we find that the agreement with
the data has been improved.

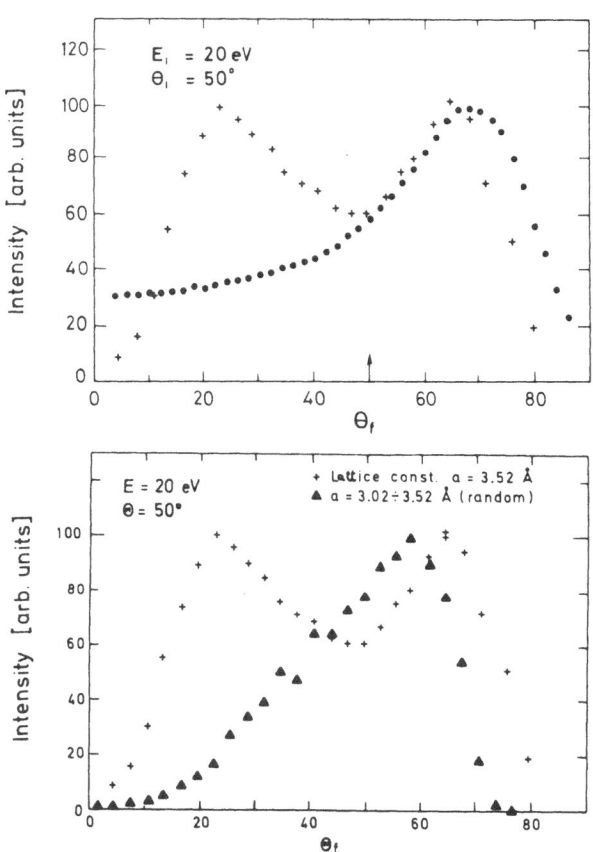

Fig. 22: Influence of different periodicities of the potential on the
          rainbow structure: (▲) angular distribution averaged
          over different lattice constants, (+) angular distribution
          for fixed lattice constant, (•) data.

We have applied the same averaging procedure to the calculation of spatial distributions for other Θ_i ($\Theta_i = 30^O$ and $\Theta_i = 70^O$) and have found that here the agreement with the data does at least not become worse. We have also found that the averaging procedure does not change the energy spectra.

The motivation for averaging the calculated distribution in the described manner has been the following:

Contributions of out of plane scattering

We suspect that the problems in reproducing the measured angular distributions with our calculations are due to the two dimensional nature of these calculations. One can for instance visualize out of plane collisions which lead to in-plane back scattering [11]. This is illustrated in Fig. 23 which shows a top view of the Ni(100) surface. The solid line despicts the intersection between the scattering plane and the surface. Our calculations include only trajectories in this scattering plane, in the experiment, however, we detect also particles with trajectories the projections of which are close to the dashed line in Fig. 23, particles which have undergone zig-zag-collision in a potential with a period that is smaller than the period of the potential in the scattering plane.

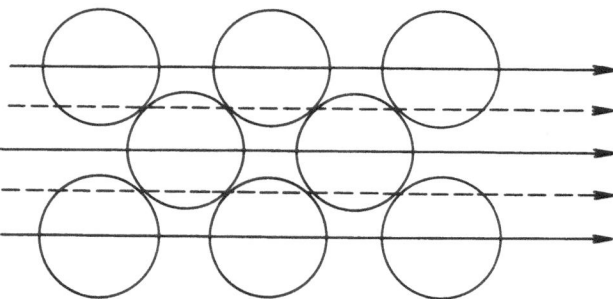

Fig. 23:   On top view of the Ni(100)-surface with projection of
           trajectories that lead to in-plane scattering, see text.

V.   Conclusions

Summarizing we can state that the impulsive collision model can be successfully applied to ion-surface scattering even at energies as low as in our experiments. In order to achieve a good agreement with the experiment one must, however, account for the influence of single and double collisions, for the influence of the surface temperature and the surface roughness and also for the attractive forces between the projectile and the surface.

The energy spectra of the backscattered ions contain mostly information about the kinematics of the collisions, i. e. about energy and momentum transfer. They are otherwise influenced only by the attractive forces between the projectile and the surface. The form of the angular distributions on the other hand is determined by the total interaction potential.

From surface scattering experiments at such low energies one can thus get information about the interaction potential between the projectile and the surface. The influence of the surface temperature on the scattering will cease, once the zero point vibration of the surface atoms is reached. Therefore it should be possible to determine the surface Debye temperature by successively lowering the surface temperature.

Future experiments will be concerned with the exploration of the scattering from cold surfaces. The calculations for the $Li^+ + Ni$ system show that it appears to be possible to resolve the two peaks in the energy spectra in experiments at sufficiently low surface temperatures. This would provide for studying single and double collisions independently in the experiment.

We would also like to study the influence of the surface structure experimentally by investigating the scattering for different orientations of the scattering plane with respect to a fixed direction in the surface. This can be achieved by rotating the crystal surface around its normal. It is also desirable to study the influence of out of plane scattering in the computer simulation by performing three dimensional trajectory calculations.

The influence of adatoms on the surface has been found to strongly influence the scattering [5]. This effect cannot yet be explained and should therefore be studied more closely.

Acknowledgements

It is a pleasure to thank Prof. J. P. Toennies for stimulating discussions. This work has been performed under the auspicies of the Sonderforschungsbereich 126.

References

Because of the great wealth of publications on ion - surface scattering this list of references is not complete. For more detailed information cf. references in [1].

[1]   E. P. Th. Suurmeijer and A. L. Boers, Surface Sci. 43 (1973) 309

[2]   E. Gerlach-Meyer, Ph. D. thesis, Göttingen (1977)

[3]   E. Hulpke and U. Gerlach-Meyer, Vakuum-Technik 25 (1976) 233

[4]   U. Gerlach-Meyer and E. Hulpke, Chemical Physics 22 (1977) 325

[5]   E. Hulpke, Surface Sci. 52 (1975) 615

[6]   E. Taglauer and W. Heiland, Appl. Phys. 9 (1976) 261

[7]   D. P. Smith, Surface Sci. 25 (1971) 171

[8]   P. Barwig, Ph. D. thesis, Bonn (1973)

[9]   D. S. Karpuzow, Surface Sci. 45 (1974) 342

[10]   M. T. Robinson and I. M. Torrens, Phys. Rev. 9 (1974) 5008

[11]   B. Poelsema, L. K. Verhey and A. L. Boers, Surface Sci 56 (1976) 445

[12]   B. Poelsema, L. K. Verhey and A. L. Boers, Surface Sci. 60 (1976) 485

[13]   W. Heiland, E. Taglauer and M. T. Robinson, Nucl. Instr. Meth. 132 (1976) 655

[14]   J. D. McClure, J. Chem. Phys. 52 (1970) 2712

[15]   A. G. Stoll and R. P. Merill, Surface Sci. 40 (1973) 405

[16]   Z. Juvela and B. Perovic, Can. J. Phys. 46 (1968) 773

[17]   V. M. Kivilis, E. S. Parilis and N. Yu. Turner, Soviet Phys. - Dokl. 12 (1967) 328

[18]   V. E. Yurasova, V. I. Shugla and D. S. Karpurzow, Can. J. Phys. 46 (1968) 759

[19]   D. G. Amour, G. Carter and A. G. Smith, Rad. Effects 3 (1970) 175

[20]   W. Heiland, H. G. Schäffler and E. Taglauer, Surface Sci. 35 (1973) 381

[21]   U. Gerlach-Meyer and E. Hulpke, to be published, 3rd International Conference on Solid Surfaces, Vienna (1977)

PHOTOELECTRON SPECTROSCOPY AND SURFACE CHEMISTRY

A. M. Bradshaw

Fritz-Haber-Institut der Max-Planck-Gesellschaft
1000 Berlin 33, Germany

D. Menzel

Institut f. Festkörperphysik der Technischen Universität
München, 8046 Garching, Germany

ABSTRACT

Photoelectron spectroscopy has fulfilled much of its early
promise as a useful complementary technique in surface science investigations. In the present paper we review certain aspects of
recent work, in particular, the application to adsorption phenomena
on single crystal surfaces. It is found that excitation with monochromatic soft X-radiation produces core level spectra which can be
used to give an overview of different chemisorption states. Because the binding energies corresponding to such spectral features
are often characteristic of the chemical environment of the atom
concerned, it is possible, for example, to differentiate between
molecular and dissociative chemisorption of molecules such as carbon
monoxide and formaldehyde adsorbed on metal surfaces. Certain
relaxation effects specific to adsorbed layers can be identified.
The use of the method in chemisorption and oxidation kinetics is
also discussed. Valence level spectra generally excited by longer
wavelength radiation give information on the molecular orbitals of
the adsorbate complex. A precise identification of such orbitals
and thus the construction of gas phase/adsorbed phase correlation
diagrams is achieved by investigating the dependence of the spectrum on energy and state of polarisation of the incoming photons
as well as on the photoelectric angular distribution. The much-
investigated CO|Ni (100) system is discussed here as an example.
The establishment of empirical orbital energy diagrams is considered
a necessary prerequisite for further progress in chemisorption
theory. Recent measurements of the adsorption of unsaturated
hydrocarbons on transition metals demonstrate the sensitivity of
UPS in fingerprinting chemical reactions on surfaces.

INTRODUCTION

The present paper concentrates on certain aspects of the application of photoelectron spectroscopy in surface studies. It is not intended to be a comprehensive review of that subject, but rather an attempt to indicate where such measurements have contributed and are continuing to contribute to our understanding of chemisorption. Most of the examples are taken from work performed under ultra-high vacuum conditions on clean metal single crystal surfaces: a deeper insight into adsorption processes on polycrystalline surfaces--a problem of paramount interest in surface chemistry and heterogeneous catalysis--must necessarily follow such investigations [1]. The conventional division of the subject into XPS (= X-ray photoelectron spectroscopy) and UPS (= ultraviolet photoelectron spectroscopy) will be avoided here, as it merely derives from the experimental choice of photon source. The increasing use of the continuum provided by synchrotron radiation makes the XPS/UPS distinction somewhat arbitrary. In this review we shall refer simply to core level and valence level spectroscopies; X-radiation is always sufficiently energetic to cause photoelectron emission from valence levels. Before considering examples of each experiment and the chemical information they supply, it is necessary to discuss in general terms the photoionisation process [2].

Let us picture electromagnetic radiation of sufficiently short wavelength incident upon a solid surface. The parameters characterising such radiation are its energy (inversely proportional to wavelength) and the orientation of its electric field vector $\vec{E}$ relative to the co-ordinate system dictated by the surface. In the case of unpolarised radiation the relevant parameter is the angle of incidence, α. The emitted photoelectron is characterised by its momentum, $\vec{p}$, or, alternatively, by its kinetic energy, E_K, together with the polar and azimuthal angles, θ and ϕ, locating $\vec{p}$ in the surface co-ordinate system. For a single crystal surface of defined orientation the crystal azimuth must also be defined. This is particularly important for discussions of ordered adlayers. The photoelectron is energy-analysed under particular conditions of θ and ϕ and registered with a suitable detection system. For detailed information on experimentation the reader is advised to consult general reviews on photoelectron spectroscopy of surfaces [3-7] as well as the relevant texts on radiation sources [8,9], electron energy analysers [10] and ultra-high vacuum [11].

In general it is not possible to describe the photoionisation event in terms of a one-electron theory, even for atoms and molecules in the gas phase. Let us consider photoelectron emission from a particular orbital in a N-electron system. The relevant final state is the system of N-1 electrons and an electron with defined kinetic energy at infinity. The appropriate energy equation is thus

$$\hbar\omega - E_K^V = E_f(\text{sys.}) - E_i(\text{sys.}) \equiv E_B^V \qquad (1)$$

where $E_i(\text{sys.})$ and $E_f(\text{sys.})$ represent the initial and final states of the system. The binding energy, E_B^V, is thus rigourously defined as the difference in total energies between the N-1 and N electron systems. Koopmans' theorem states simply that the experimental binding energy, or ionisation potential, is approximately equal to the single electron orbital energy in the Hartree-Fock (HF) approximation:

$$E_B^V \simeq -\varepsilon_j . \qquad (2)$$

An important effect leading to the breakdown of Koopmans' theorem is orbital relaxation: if the removal of the one electron changes the effective potential seen by the other electrons, the latter will readjust energetically, or "relax." The measured binding energy is then smaller than the calculated HF orbital energy by an amount termed the relaxation energy. Correlation effects in the initial and final states are also different, leading to a further contribution to the net Koopmans defect. For valence levels of molecules in the gas phase, Koopmans' theorem is often quite a reasonable approximation, as it appears that the relaxation energy partly cancels the correlation energies omitted in the HF picture [12]. The status with regard to molecules on surfaces is still not clear and further attention is given to this point in the next section.

In equation (1) E_K^V will only be equal to the measured analyser deflection voltage if the work function of sample and analyser entrance slits are identical. The problem of the necessary correction does not arise for a metal, and thus for an adsorbate layer, because the Fermi level is visible in the spectrum itself. Under these circumstances

$$E_B^V = E_B^F + \phi_S \qquad (3)$$

where E_B^F is the binding energy referred to the Fermi level and ϕ_S the sample work function. The so-called reference level problem in adsorption studies has been discussed by one of us elsewhere [5].

The application of photoelectron spectroscopy in chemisorption studies derives from its high surface sensitivity. Even incident photons with only a few eV energy have a mean penetration depth of several hundred Ångstrom in a solid, whereas the mean depth of origin of the corresponding photoelectrons given by equation (1) is at least an order of magnitude smaller. A measure of the mean depth of origin of such hot electrons and thus of the surface sensitivity of photoelectron spectroscopy is provided by the mean free path for inelastic scattering, $\lambda(E_K^F)$. This quantity is shown plotted in Figure 1a for aluminium. The curve derives

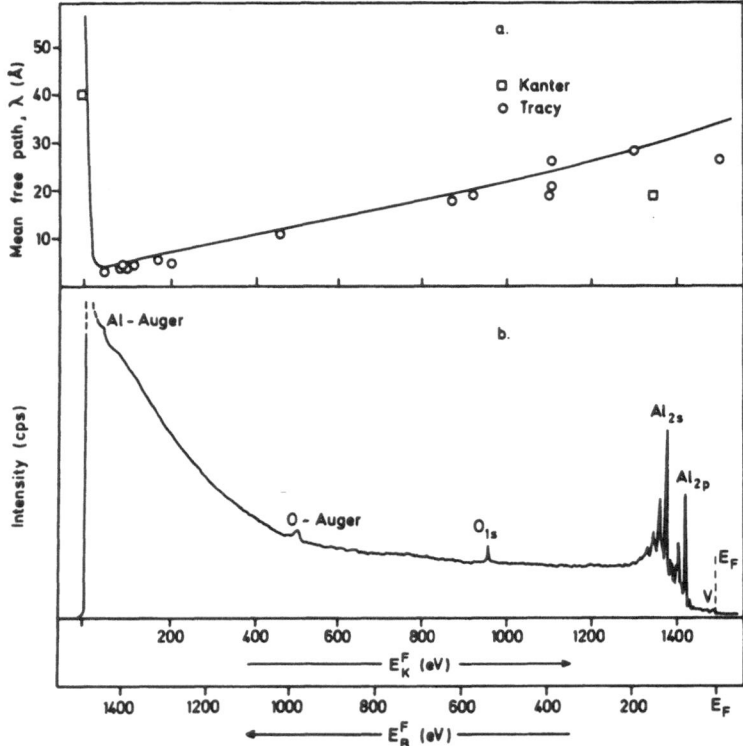

Figure 1.   a) Electron mean free paths in aluminium after
Bradshaw et al. [3].   Experimental points from Kanter [15]
and Tracy [16].   b) Typical core level spectrum of aluminium
with chemisorbed oxygen.   AlKα.   After Bradshaw et al. [25].

from theoretical analyses of scattering by single particle excitations [13] and by plasmons [14].   The measured values are taken
from the literature [15,16].   The mean free path clearly drops to
a minimum of ∿ 5 Å in aluminium between 50 and 100 eV electron
energy, a prediction which is substantiated by experiment.   Thus
for photoemission in this energy range over 70% of the "no-loss"
current (that is, the current comprising electrons whose kinetic
energy is given by equation (1) and which have not suffered an
energy loss) will originate from the first three atomic layers of
metal.   At the higher energies generally of interest in core level
spectroscopy the mean free path rises (e.g. ∿ 15 Å at 720 eV,
which would correspond to the oxygen 1s level excited by MgKα
radiation).   Although these results apply specifically to
aluminium, many authors have shown that a curve similar in form
and scale applies for a wide range of solids above 10 eV electron
energy.

In Figure 1b the core level spectrum of aluminium is plotted on the same scale as $\lambda(E_K^F)$ in Figure 1a. The strong secondary electron tail is a direct consequence of the low mean free path, or, alternatively, of the high inelastic collision cross-section at low electron energies. Other features of the spectrum are the plasmon satellite lines accompanying the 2s and 2p core levels, which may arise from both intrinsic and extrinsic processes [17], and the O_{1s} peak due to chemisorbed oxygen. It is the study of such adsorbate core levels which forms the material for the next section of this paper. The feature marked V is the aluminium valence band, which, because of a higher photoionisation cross-section and higher instrumental resolution, would usually be studied with lower energy radiation. Typical spectra for an aluminium surface with chemisorbed oxygen are shown at two different photon energies in Figure 2. The valence band extending from the sharp Fermi edge downwards derives from aluminium 3s and 3p electrons. The sharp structure at $\sim$ 7 eV is due to the oxygen 2p

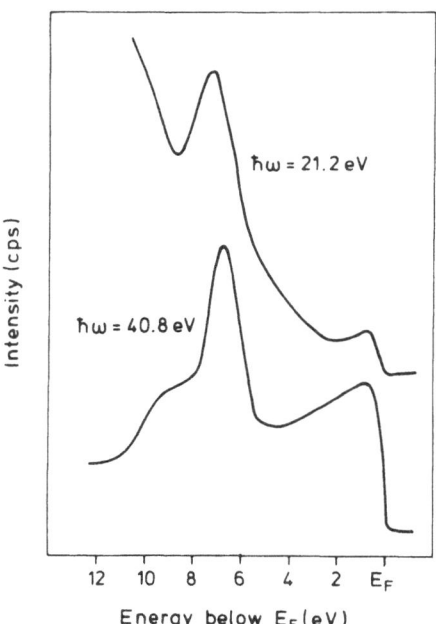

Figure 2. Valence level spectrum of oxygen adsorbed on aluminium at two different photon energies. 5L oxygen. After Flodström et al. [18] with permission.

adsorbate resonance. Later we consider such spectra and the
information on the chemisorption bond which they can provide.

CORE LEVEL SPECTROSCOPY

Core level spectroscopy of chemical compounds aims at
examining the effect of the electrostatic potential due to the
valence field on core level binding energies. Thus in a particular hydrocarbon, for example, it may be possible to distinguish between non-equivalent carbon atoms, on account of the
differing chemical environment. This is the notion of the chemical
shift. With proper consideration of the final state (i.e., inclusion of relaxation and change in correlation energy) it is
possible to quantify the effect and to derive chemical information [19]. When core level spectroscopy was first applied to surface studies some five years ago, many experimentalists, including
the present authors, hoped that similar developments would provide
detailed chemical information on adsorption phenomena. Unfortunately, this aim has not been fully realised and there is some reason
to doubt whether in fact it can be. The problem is basically one
of determining the initial state chemical shift relative to some
reference state, most sensibly in this instance, the molecule or
atom in the gas phase. The extraction of this information is
hindered firstly by another initial state effect, namely the ill-
defined electrostatic potential of the solid in the surface region
and secondly by final state effects, which are considerably more
difficult to account for than in the gas phase. These two problems
will now be briefly considered.

On bringing an atom from vacuum into a solid, the electrostatic potential changes from the vacuum potential to the average
internal electrostatic potential in the solid as the intrinsic
surface dipole layer is traversed. In order to understand this
situation we must consider with reference to Figure 3 the initial
state shift which would occur for a hypothetical non-interacting
molecule, i.e., where no chemical shift or final state effect
takes place. For photoionisation in the gas phase the kinetic
energy of the electron is given by equation (1). For the molecule
or atom in the solid the photoelectron has to overcome the electrostatic potential difference between solid and vacuum indicated by
D, the surface dipole part of the work function [20]:

$$E_B^V \text{ (met.)} = E_B^V \text{ (gas)} + D \qquad (4)$$

$$\phi_s = D - \bar{\mu} \qquad (5)$$

where $\bar{\mu}$ is the chemical potential change [21]. For the adsorbate
case [22] the situation lies probably somewhere between the two
extremes shown in Figure 3:

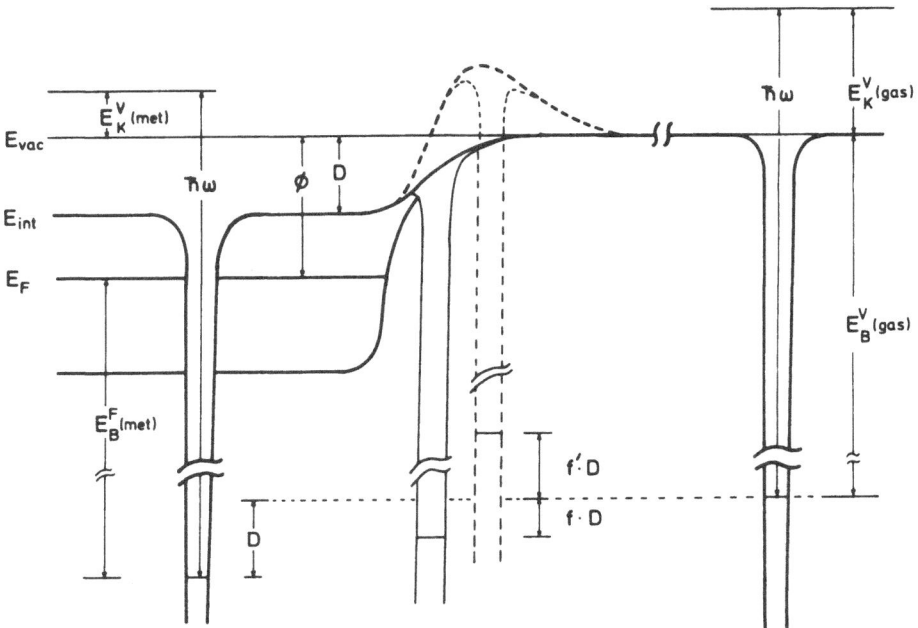

Figure 3. The influence of local electrostatic potential on core level binding energies for the two limiting cases of photoionisation in the vacuum and inside the solid, as well as in an adsorbate layer. After Menzel [5].

$$E_B^V \text{ (ads.)} = E_B^V \text{ (gas)} + f \cdot D \qquad (6)$$

$$E_B^V \text{ (ads.)} = E_B^V \text{ (gas)} - f' \cdot D \qquad (7)$$

Equation (6) will apply to a homogeneous layer or a densely packed surface, where we can expect a smooth continuous potential variation through the surface region. The situation represented by equation (7) (dashed lines in Figure 3) could apply for a rough surface, for co-adsorbed layers or for strong intermolecular interactions in the layer. In any case such an analysis will take us no further, because the values of both D and f(f') are unknown. As there seems to be no apparent way of measuring or calculating them, we are at present unable to identify the "true" chemical shift, even if we can correctly describe the final state.

As noted in the Introduction, an important contribution to the electron binding energy derives from relaxation. A corollary of final state relaxation is the presence of satellite peaks on the high binding side of the core hole peak corresponding to excited states of the N-1 electron system. The intensity of such

satellites--often called "shake-up" peaks--is related by a sum rule
to the relaxation energy. This simply means that the main peak--
normally referred to as the adiabatic or "fully relaxed" peak--
is found at an energy lower than that predicted by Koopmans'
theorem [23], but that the centre of gravity of the complete
spectral function is found at the Koopmans energy. For photoionisation in a solid or on a surface the sum rule can break down if the
sudden approximation does not hold, i.e., if there is interaction
between the excitation coupled to the hole state and the same
excitation coupled to the photoelectron. Alternatively, we might
say that there is interference between the intrinsic and extrinsic
coupling terms [24]. Relaxation effects are naturally different
in the gaseous and adsorbed phases. One such effect characteristic
of a surface is image charge screening. The photoionisation event
outside the surface of the metal results in a movement of the
conduction electrons to screen the suddenly created core hole. In
a strictly classical model this interaction results in an induced
surface charge and a corresponding interaction, or screening energy,
of $e^2/4z$, where z is the separation between the hole and the surface
of the electron gas. In a quantum model the induced image charge
can be expressed as a coherent superposition of surface plasma
excitations. As a consequence intrinsic surface plasmon satellites
are expected to appear in the spectrum if the divergence from the
sudden approximation is not too great. An example will illustrate
that such effects can indeed be observed, but that they are not so
important for screening as was hitherto supposed.

The inset of Figure 4 shows the surface plasmon satellite
which is found to accompany the O_{1s} core level peak for oxygen
adsorbed on aluminium(111) [25]. Similar observations have been
made for polycrystalline aluminium [26]. The figure itself
demonstrates the dependence of the coupling parameter (satellite
intensity relative to adiabatic peak intensity) on the electron
escape angle, θ. The intrinsic contribution to the satellite
intensity should be angular-independent, whereas the extrinsic
contribution should vary as $1|\sin \theta$. The latter is represented by
the dashed curve in Figure 4, fitted to the experimental points at
low escape angle. The experimental curve (solid line) lies under
the "extrinsic curve" at higher escape angles, indicating that the
interference term is also important. An analysis of the data [25]
revealed a surface plasmon screening energy of ~ 0.4 eV, considerably lower than expected. A comparison of the measured O_{1s} binding
energy with the best calculated value for an oxygen atom in the gas
phase [27] reveals a discrepancy of 10-11 eV, which has to be
divided between initial and final state effects. It was therefore
concluded that atomic and molecular relaxation (single particle
excitations) was mainly responsible for the screening of the core
hole. Some recent theoretical work on this problem [27] offers
another intriguing possibility: it has been shown that for a hole
in an adsorbate layer outside an electron gas the dielectric

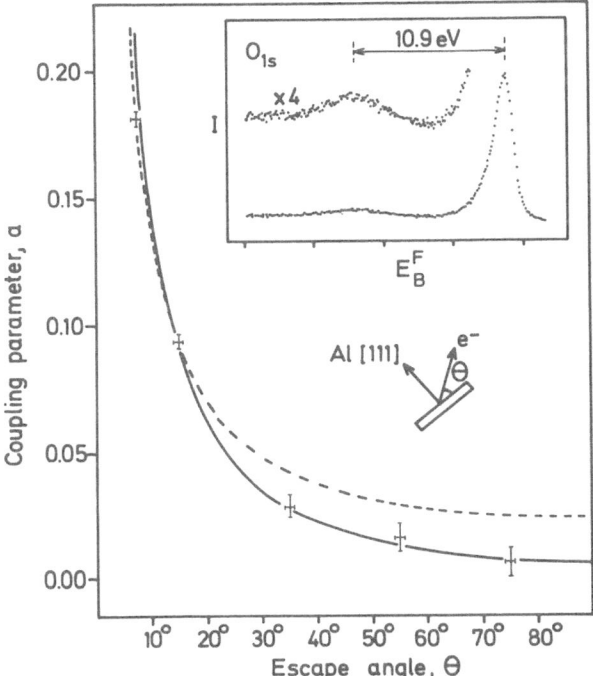

Figure 4. Angular dependence of the surface plasmon coupling parameter for O_{1s} in the system oxygen|Al(111). The solid line is a fitted curve using a theoretical expression with two parameters. The dotted line represents the extrinsic effect and is fitted at the two lowest escape angles.

function describing the spatial distribution of the screening charge must correspond to the metal and adsorbate and not just to the bare metal. In other words, not only will surface plasmons contribute to the screening but also many-body effects associated with the metal-adsorbate interaction. In particular the unoccupied valence levels of the adsorbate can be pulled down below the Fermi level due to the electrostatic potential of the suddenly turned on core hole. Screening then occurs by the filling of these levels with metal electrons. This analysis leads to a calculated screening energy of 7.8 eV for oxygen on aluminium, assuming incorporative chemisorption [27].

The explicit inclusion of the latter screening effect in an Anderson-type Hamiltonian and the calculation of the core hole spectral function [28] shows that the satellite structure depends markedly on the position and width of the unoccupied valence level.

For a broad adsorbate resonance, corresponding to the large
hopping matrix element associated with strong chemisorption, most
of the spectral weight is expected in the fully relaxed peak.
For a very narrow resonance, corresponding to the smaller hopping
matrix element typical of weak chemisorption, a considerable part
of the spectral weight is expected in the satellite structure.
In the limit of physisorption the spectral weight might even be
found entirely in the "satellite" or "unrelaxed" peak, the width of
which is given simply by the width of the adsorbate valence level.
The recent observation of strong satellite structures on core levels
in two weak chemisorption systems provide some experimental basis
for these predictions. In Figure 5 are shown the C_{1s} and O_{1s} core

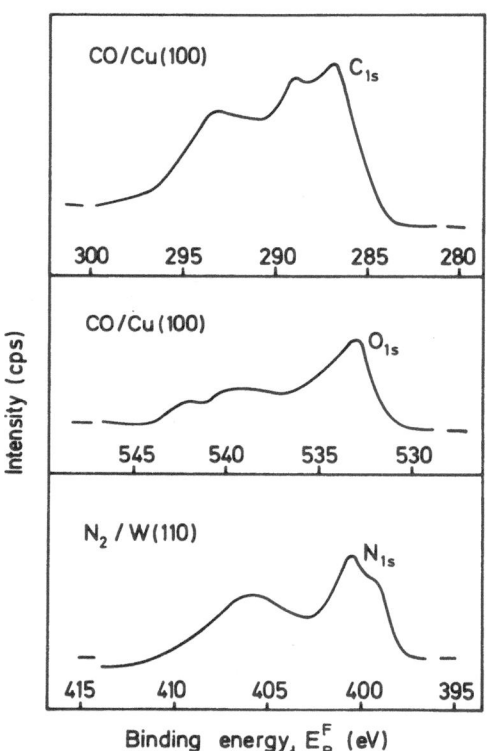

Figure 5. Strong satellite structure on adsorbate core level
peaks in two weak adsorption systems: CO|Cu(100) (after Brundle
and Wandelt [29], with permission) and N_2|W(110) (after Fuggle
and Menzel [30]).

level spectra from the CO|Cu(100) system [29] and the N_{1s} spectra from the N_2|W(110) system [30]. All three spectra show satellite structures similar to the calculated model spectra, and there is good reason to believe that the correct explanation has been found. We note that the double peaked structure on the low binding energy edge of the N_{1s} spectrum is in all probability due to an initial state chemical shift: here we are observing molecular chemisorption with the two nitrogen atoms in different chemical environments. The most favored configuration is with the nitrogen molecular axis perpendicular to the surface.

To sum up the results of this discussion on the problems of core level spectroscopy, we note that progress is being made in understanding and calculating the final state effects, but that the initial state remains a problem. Separating the chemical shift from the electrostatic potential contribution appears at the moment insoluble. This does not mean, however, that surface chemists should give up using core level spectroscopy. The technique has brought some fascinating many-body effects to light and also provided considerable chemical information of a less ambitious nature. For complex adsorption systems core level spectroscopy can be used, as we shall see below, to distinguish between different adsorption states. The differences in binding energy associated with such states can be due to a combination of initial and final state effects, but one might, perhaps, expect in this case that the initial state chemical shift will be mainly responsible. The second field of application--chemisorption and oxidation kinetics--will also be illustrated. We shall also look briefly at X-ray excited Auger spectroscopy of adsorbate layers and at the use of satellite structures for "fingerprinting" purposes.

Under experimental conditions which make the simultaneous existence of more than one surface species likely, several peaks are often observed for one core level [31]. Figure 6 shows as an example the C_{1s} spectrum under the experimental conditions which give rise to the adsorption states "virgin" (v-), α- and β-CO on W(110) [32]. This system has been partially characterised by other methods, particularly by thermal and electron-induced desorption, and demonstrates how core level spectroscopy can provide information complementary to other techniques. We note that unlike the situation in thermal desorption the different adsorption states are observed non-destructively; i.e., it is not necessary to destroy or alter in any way the composition of the adlayer. The v-CO state is formed by adsorption at low temperatures on the clean surface, β-CO by adsorption above room temperature or by heating a v-CO layer, and α-CO together with β-CO after re-exposure of a β-covered surface at low temperatures [33]. Taking the results from both the O_{1s} and the C_{1s} spectra for this system it can be concluded that the peaks obtained for v- and α-CO have binding energies

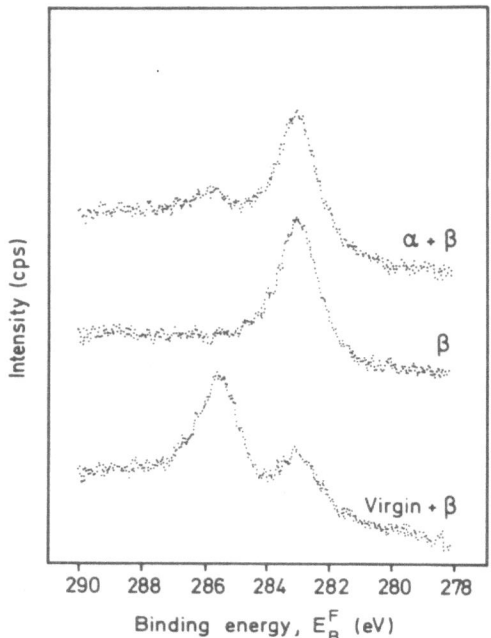

Figure 6. The C_{1s} region for three types of CO layer on W(110). After Umbach et al. [32].

similar to those found for CO adsorbed on metals where molecular adsorption is believed to take place. The O_{1s} peak for β-CO on W(110) was found to be identical to that obtained for half a mono-layer of oxygen on the same surface with respect to energy, half-width and intensity. The results thus indicate that v- and α-CO are molecularly adsorbed and that β-CO is dissociatively adsorbed. Further verification of this thesis is provided by the measurement of the O_{1s} satellite structure discussed below.

The chemisorption of formaldehyde on W(100) at low tempera-tures [34] provides a further example of this type of analysis. (There are many others, e.g. [35-39].) It is known that H_2CO dissociates into H (ads.) and CO (ads.) when low coverage adsorp-tion takes place at ~ 100 K [40]. At higher coverages, but before condensation is reached, other kinds of surface species have been postulated (for example, HCO (ads.) and CH_2 (ads.)). Figure 7

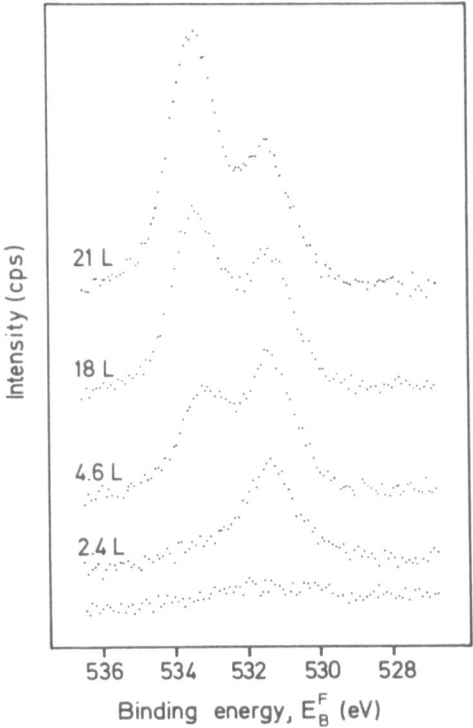

<u>Figure 7</u>. The O_{1s} region for an increasing coverage of for-
maldehyde on W(100). After Worley et al. [34] with permission.

shows the O_{1s} spectra for increasing H_2CO exposures at 100 K. The
spectrum after 2.4 L (1 L = 1 Langmuir = 1 x 10^{-6} Torr-sec.)
corresponds to $\sim$ 0.5 monolayers of adsorbate and resembles that of
v-CO on the same surface. At higher H_2CO exposures a new peak
develops at $\sim$ 2 eV higher binding energy and grows in intensity
while the v-CO peak rapidly reaches a maximum. This peak was
attributed to HCO (ads.) or H2CO (ads.). On longer exposure the
peak grows further and shifts to higher binding energies concomitant with the condensation of a thick layer of H_2CO.

An example of the use of core level spectroscopy in the investigation of chemisorption kinetics is also provided by the adsorption system CO|W(110) [41]. (Once again, there are many other examples, e.g. [42, 35, 37].) For a discussion of the problems associated with relative coverage measurements including angular effects [43] and satellites the reader is advised to consult ref. [5]. Here we simply assume that the main peak area in a given adsorbate core level spectrum is proportional to the surface concentration of the species containing that particular atom. Figure 8 shows the fractional coverage as a function of exposure, as derived from O_{1s} intensities, for the adsorption of CO on W(110) at 100 K. The CO molecules are thus in the virgin state. To illustrate the reproducibility, the results from six runs on two different crystals are shown [41]. The inset illustrates the procedure used to obtain the peak intensities.

An exciting new field of application of core level spectroscopy has recently been opened up by investigations of the oxidation

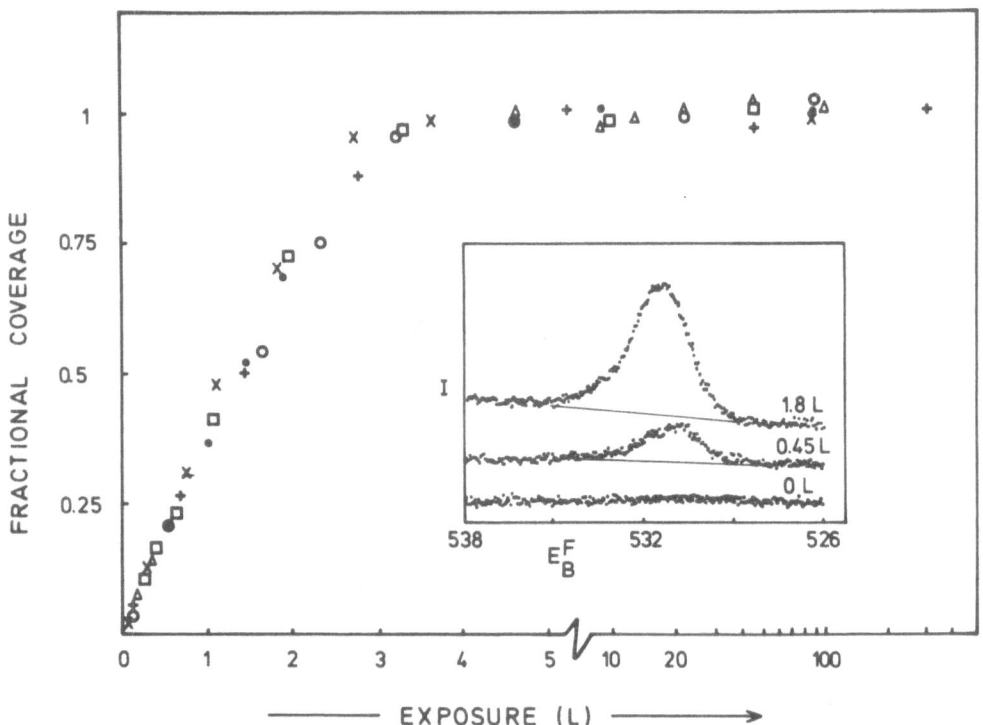

Figure 8. The kinetics of chemisorption of CO into the virgin state on W(110). After Umbach and Menzel [41].

kinetics of nickel single crystal surfaces [38]. The use of a
simple model backed up by a radiochemical calibration technique
showed that the limiting coverage at 295 and 485 K after ∿ 300 L
corresponds to the equivalent of three NiO layers. As expected
the oxide growth process was found to proceed in two distinct
steps, namely, an initial fast dissociative chemisorption followed
by nucleation and lateral oxide growth up to the limiting coverage.
As shown in Figure 9, where the corrected O_{1s} intensities are
plotted as a function of exposure, the oxide growth rates are in
the order Ni(110) > Ni(111) > Ni(100).

    The satellites discussed above in connection with relaxation
phenomena, although a hindrance for the use of core level spectroscopy in quantitative analysis, can nonetheless be used for
adsorbate characterisation. Figure 10 shows the O_{1s} spectra from
full layers of oxygen, β-CO and v-CO on W(110) [32] (the v-CO layer
contained about 30-40% β-CO in this experiment). The spectrum
from α-CO was similar to that from v-CO. Adsorbed oxygen and
β-CO give rise to satellites at 15.3 and 16.2 eV from the main peak.
The v-CO spectrum is quite different, being dominated by a peak at
6.9 eV loss energy and with considerably more spectral weight

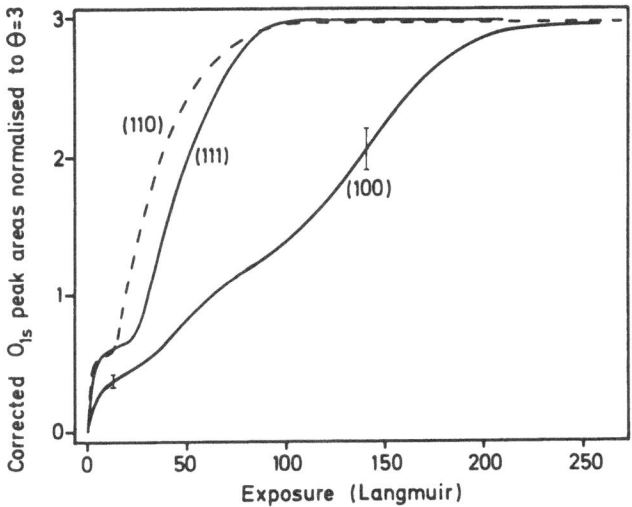

Figure 9. Oxidation kinetics for the Ni(100), (110) and (111)
surfaces at room temperature. After Norton et al. [38] with
permission.

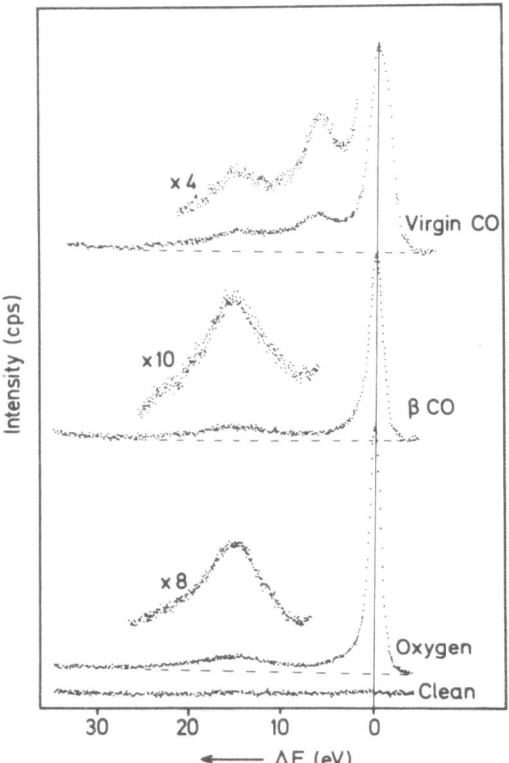

Figure 10. O_{1s} core level satellites for virgin CO, β-CO and oxygen adsorbed on W(110). After Umbach et al. [32].

occurring in the satellite structure. The v-CO spectrum exhibits similarities with the corresponding metal carbonyl and gas phase spectra as well as with the CO|Ru spectra where CO adsorption is clearly non-dissociative (see below). Metal carbonyls, for instance, show a strong satellite at ∿ 6 eV [44]. It is the strong 7 eV satellite in the v-CO spectrum which disappears on conversion to β-CO. Its origin is possibly due to screening by a single particle excitation also typical of metal carbonyls. In summary, the satellites give a further indication that v- and α-CO on W(110) are non-dissociatively adsorbed in a carbonyl-like fashion,

whilst the binding state of oxygen in β-CO is similar to that
of adsorbed oxygen.

Although the potential of high resolution N(E) mode Auger
spectra [45] for the characterisation of adsorbate layers has been
demonstrated in the literature [46,47], the advantages of X-ray
excitation have not yet been fully appreciated. As Figure 1 shows,
such spectra are the by-product of core-level spectroscopy. The
advantages of X-ray excitation are the absence of electron-beam-
induced effects and the smooth, low background. In addition, the
electron energy analysers used in core level spectroscopy have generally a higher resolution than those constructed specifically for
electron-induced Auger spectroscopy. Figure 11 shows, for purposes

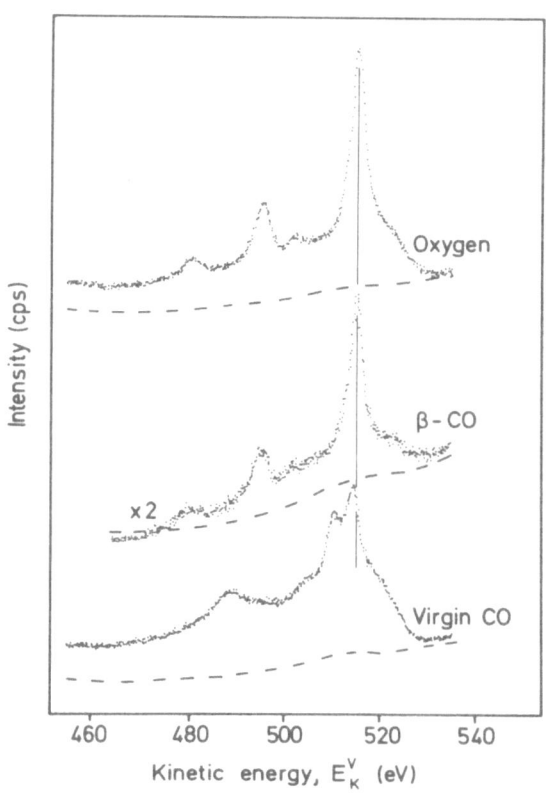

Figure 11. Oxygen KLL Auger spectra for oxygen, β-CO and
virgin CO adsorbed on Ru(001). After Fuggle et al. [48].

of illustration, the O KLL Auger spectrum for oxygen, β-CO and
"virgin"-CO adsorbed on a ruthenium (001) surface [48]. The
molecularly adsorbed ("virgin") state is normally encountered
on Ru(001) but the β-state can be produced by electron impact;
thermally activated conversion is not observed. The shape of the
"virgin" CO spectrum is very similar to that of gas-phase CO [49],
although the peaks are broadened and the whole spectrum occurs at
higher energy, as discussed elsewhere [50]. The spectrum of β-CO
differs from that of oxygen on Ru(001) only in its overall
intensity. Together with similar evidence collected from the core
level peaks and their corresponding satellites as well as from
valence level spectra, it could be shown that the conversion to
the β-state was tantamount to electron-beam-induced dissociation.
As we have mentioned above, the same conversion can be accom-
plished thermally on tungsten.

VALENCE LEVEL SPECTROSCOPY

Early in the last decade a series of experiments on molecules
in the gas phase established the importance of photoelectron
spectroscopy in the study of the chemical bond [51]. The combined
application of HF ground state calculations and the determination
of the corresponding ionisation potentials via a Green's function
approach has since enabled a large number of molecules to be
characterised [12]. The potential of photoelectron spectroscopy
in the investigation of the band structure of metals and semiconductors had also been realised for many years [52]. That the
surface sensitivity of the method was sufficiently high to probe
the valence levels of chemisorbed species was first shown de-
cisively in 1971 for the CO|Ni system [53]. Progress in the method
was helped considerably by the use of the easily realisable, highly
monochromatic radiation in the vacuum UV provided by rare gas
discharges [7] (the so-called rare gas resonance lines, e.g., He I,
$\hbar\omega$ = 21.2 eV). The use of monochromatised synchrotron radiation is
currently accelerating these developments. The aim of valence level
spectroscopy of adsorbate layers is to identify and assign correctly
the molecular energy levels, or resonances, of the adsorbate complex and to compare the results with calculations. More often than
not the latter are unavailable and where they do exist, the present
developing state of the theory of chemisorption does not guarantee
their accuracy. As an alternative we can construct a gas phase-
adsorbed phase correlation diagram for cases of molecular chemisorption, whereby we attempt to account quantitatively for the
bonding and relaxation shifts [54]. Due care must be exercised
here, because the gas phase spectrum is often itself used to
assist in the initial assignment. The third approach is to use
the spectra in a fingerprinting sense as a part of a correlated
multi-method study (as in, e.g., ref. [32]).

Very often valence level spectra of adsorbates are presented in the literature as "difference spectra," which are simply obtained by subtracting the spectrum of the clean surface from that of the surface and adsorbate. Such spectra naturally accentuate strongly the "split-off" levels, i.e., the adsorbate valence levels below the substrate valence band. (See Figure 2.) They also show that both positive and negative changes occur in the substrate valence band region, superimposed on an overall attenuation. The latter is due to the increased scattering of substrate photoelectrons as they traverse the surface region. Difference spectra also reveal changes in the secondary electron background at low kinetic energies, which can misleadingly resemble adsorbate resonances. Whilst difference spectra are useful, it is not wise to put too much reliance on them, particularly when valence levels are not entirely "split-off."

Figure 12 shows the valence level spectrum for CO adsorbed on a Ni(100) surface under defined conditions of photon incidence and electron exit [55]. The two peaks labelled P_1 and P_2 are ascribed to the valence levels of adsorbed CO and should be compared to the three levels in the corresponding gas phase spectrum (inset) obtained using the same photon energy and similar analyser resolution [56]. The two-peak structure is characteristic of molecular CO adsorption and has been observed on a wide range of metal surfaces [3]. The three gas phase peaks represent, in order of increasing ionisation potential, photoionisation of the 5σ, 1π and 4σ molecular orbitals. For the adsorbed phase we thus have an assignment problem which is not readily solved just by comparison with the gas phase spectrum. As in Figure 2, the emission under the Fermi edge derives from the substrate, here mainly from the nickel d states. The structure in this spectral region is strongly dependent on angular parameters, due firstly to the vectorial nature of the photoionisation event and secondly to the allowed momentum state of the photoelectron. Investigations of such effects on clean surfaces fall outside the scope of this article, but it is worthy of note that the symmetry considerations discussed below in connection with adsorbed layers can be used equally well in photoemission studies of bulk material [57-59]. As noted in the Introduction, the angular parameters characterising the photoelectron spectrum from a surface are the electric vector orientation (or, alternatively, for unpolarised light, the angle of incidence) and the direction of electron emission. The relative intensities of the valence levels in an adsorbate spectrum will depend on both variables, as well as on the photon energy. All three parameters can be used for the purposes of assignment and thus for constructing the desired orbital energy diagrams. In addition a certain amount of information on molecular orientation and bonding site geometry can, in principle, be derived. We will discuss first of all the class of experiments where the

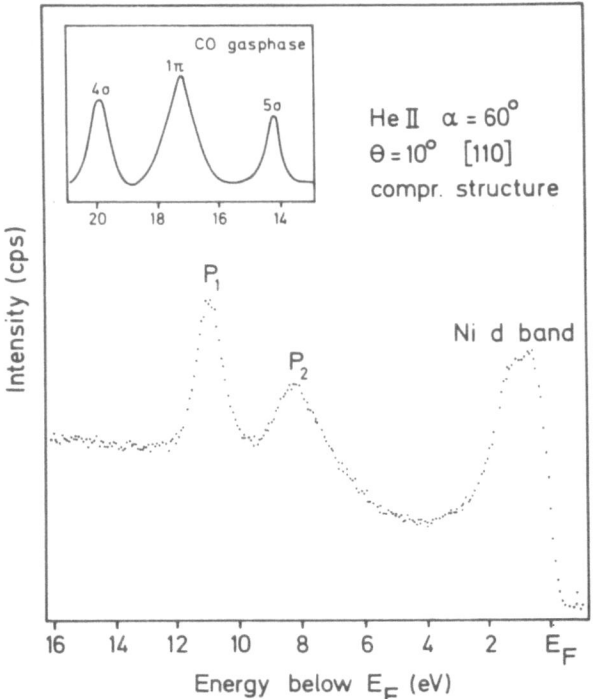

Figure 12. Valence level spectrum of CO adsorbed on Ni(100) at a coverage ~ 0.6. After Horn et al. [55].

photoelectron angular distribution is determined at constant photon energy and electric vector orientation.

A theoretical description of photoelectron angular distributions from adsorbate layers is achieved through the analysis of two effects. Because the wave functions of the occupied molecular orbitals are anisotropic, the <u>initial state</u> will contribute angular information to the photoionisation matrix element [60]. At the same time attention must be paid to the nature of the final state. In the simplest approximation the latter can be represented by a plane wave [61], under which circumstances the angular-dependent matrix element is simply proportional to the scalar product $\vec{p} \cdot \vec{A}$ modulated by the Fourier transform of the initial state wave function. For orbitals involved in bonding to the substrate, the geometry of the surface site should then be reflected in the

angular distribution [60]. Obviously the best set of initial
state wave functions would derive from a chemisorption calculation, although quite fair results have been obtained for
CO|Ru [62] with plane wave final states in the so-called oriented
molecule approximation (i.e., no interaction with the substrate:
the surface serves merely to orient the molecule). An improvement in the oriented molecule approach is achieved by specifically
including the final (continuum) state of the molecule. Such
calculations for oriented CO using the MSX$_\alpha$ method [63] have
been used to show that the CO molecular axis is normal to the
surface to within 5° [64]. The accuracy with which this particular experiment can be performed is due to the occurrence of a
strong final state resonance in the photoionisation cross
section of the 4σ molecular orbital at $\hbar\omega \sim 35$ eV. The 4σ
orbital is, as we shall show below, scarcely affected by
chemisorption. The angular distribution of the resonance is very
strongly peaked along the molecular axis and can only be excited
by the component of the vector potential parallel to the
molecular axis [65]. The second effect containing configura-
tional information and contributing to the angular dependent
matrix element is final state scattering [66-68]. Emitted
photoelectrons will scatter from the metal substrate and
from other adsorbate molecules, processes which can only effectively be handled with the multiple scattering theories
developed for LEED structure analysis [69]. The full description of the photoelectron angular distribution from adsorbate
layers and the extraction of detailed structural information
are obviously very cumbersome procedures and must treat
specifically the initial and final states in a chemisorption
calculation and the LEED problem, as well as dispersion [70].
Whilst molecular orientation may be derived in favourable
cases, the application of the technique to exact structural
analysis remains untested.

As mentioned above, the dependence of photoionisation
cross section on photon energy under fixed angular conditions
can be used for the purposes of orbital assignment. Figure
13 shows the intensity ratio of the peaks P$_1$ and P$_2$ defined
in Figure 12 for CO adsorbed on Pd [71] and Ru [72] compared
to the intensity ratio 4σ|(1π + 5σ) for gas phase CO [51,71,73].
The results indicate, and are reinforced by measurements
presented below, that P$_1$ represents 4σ and that P$_2$ represents
the superposition of 5σ and 1π. A similar situation is
encountered in the metal carbonyls [74-76]. (Note that we
continue to refer to the 4σ, 5σ and 1π orbitals in the
adsorbate molecule. They are, of course, not identical
with the corresponding orbitals in free CO, but it is convenient to retain the nomenclature.) The photon energy

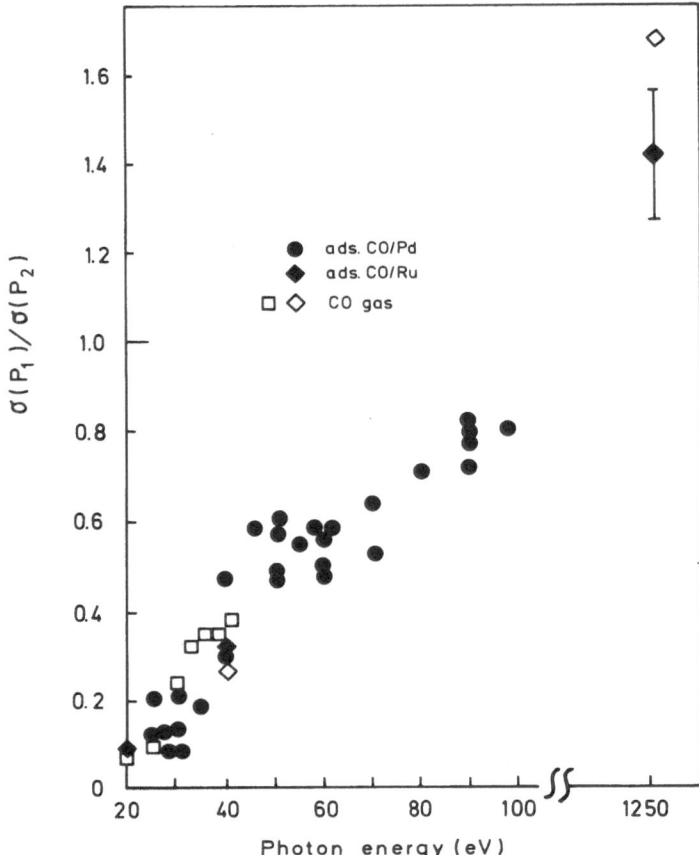

Figure 13. The intensity ratio $P_1|P_2$ for adsorbed CO on Pd
and Ru compared to the ratio $4\sigma|(1\pi + 5\sigma)$ for CO in the gas
phase. CO|Pd data from Gustafsson et al. [71], CO|Ru data
from Fuggle et al. [72] and gas phase data from Turner [51],
Gustafsson et al. [71] and Siegbahn [73].

dependence of the valence level spectrum has also been used for
the case of CO to derive information on molecular orientation [64].
This experiment again relies on the sharply defined 4σ final state
resonance at $\hbar\omega \sim 35$ eV. Figure 14 shows the experimental 4σ (P_1)
peak intensities for normal exit photoemission (p-polarised light
at $\alpha(\theta_1) = 45°$) from the CO|Ni(100) system. The solid curves are
based on SWX_α calculations in the oriented molecule approxima-

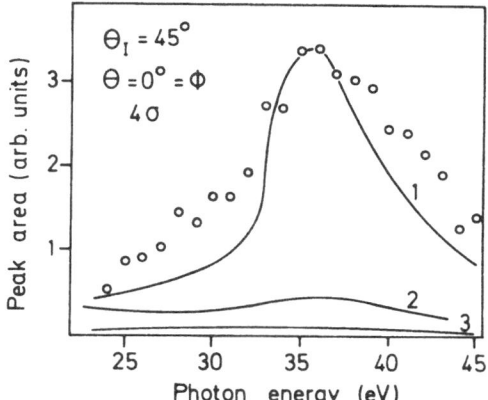

Figure 14. 4σ peak intensity in the CO|Ni(100) system as a function of photon energy. Normal emission, 45° angle of incidence and p-polarisation. After Allyn et al. [64] with permission. For further details see text.

tion [63]. (1) represents the case where the CO axis is normal to the surface with carbon end down, (2) the CO axis normal to the surface with oxygen atom down, and (3) the CO axis parallel to the surface. Clearly the best agreement is obtained for case (1), although the exact prediction of the position of the resonance is probably coincidence: agreement with the corresponding gas phase data is not so good [63].

The third technique which can be used for assignment purposes is based on the variation of electric field vector orientation of the incident light. The absorption of radiation in the photo-ionisation experiment is governed by dipole selection rules, that is, by the symmetry properties of the integrand in matrix elements of the type $\langle\psi_f|\vec{A}\cdot\vec{p}|\psi_i\rangle$. For a transition to be allowed the product, or some term in the product, of the three functions must be invariant to all symmetry operations of the molecule. In the language of group theory, it must be or contain a basis for the totally symmetric representation of the molecular point group. This is only the case if the representation of the direct product of any two of the functions is or contains the same representation given by the third. A molecule on a real surface will always belong to a point group of lower symmetry than the same molecule in the gas phase, mainly due to the loss of a (horizontal) symmetry plane. The exact definition of that symmetry will also depend on the adsorption site and on the crystallographic nature of the

surface. Depending on the strength of the interaction of the
molecule with its environment (both in the initial and final
states), these factors will determine a so-called site group as
opposed to the molecular point group [77]. In photoelectron spectroscopy of adsorbate valence levels we can use a particular property of the final state: for emission along the proper axis of
the system (which is always the surface normal) the final state
wave function must be symmetric with respect to the symmetry elements contained in that axis (i.e., always totally symmetric).
Similarly for emission in a symmetry plane the final state wave
function must be symmetric with respect to reflection in that plane.
Knowing the final state symmetry we can vary the symmetry of the
dipole operator (by varying $\vec{E}$) to distinguish between initial
states of differing symmetry. The potential of these dipole
selection rules in valence level spectroscopy has recently been
discussed in several papers [55,58,67,68,78-80]. A more detailed
discussion of these ideas, illustrated by reference to the C_{2v}
point group, has been published by one of us elsewhere [79]. An
essential prerequisite for the method is a model configuration for
the adsorbate, thus enabling, in certain circumstances, structural
information to be obtained. In the case of CO adsorbed on Ni(100)
in a configuration of C_{4v} symmetry, the 4σ and 5σ orbitals are
totally symmetric (A_1) whereas the 1π orbitals belong to the
representation E. Under conditions of normal emission the dipole
operator must also be totally symmetric to observe emission from
the 4σ and 5σ orbitals. This condition is only fulfilled for A_z,
the component of the vector potential normal to the surface. The
components A_x and A_y in the surface plane belong to the E representation and can only produce emission from the 1π orbital. The
dipole selection rule for the normal emission case is thus particularly simple: the initial state wavefunction must have the same
symmetry as a component of the vector potential. It also provides
an effective method of separating the 5σ and 1π contributions in
the P_2 peak of adsorbed CO: p-polarised light (A_z+A_x) will give
rise to both π- and σ-levels in the spectrum, whereas s-polarised
light (A_y) only to the π-levels. Examination of published spectra
for the CO|Ni(100) system [81] shows that the 5σ level lies at
∿ 0.3 eV higher binding energy than the 1π level with a 4σ - 1π
separation of ∿ 3 eV.

It is not necessary that polarised light be used in such experiments as long as the angle of incidence can be varied [67,79].
The variation of the angle of incidence will vary the relative
intensities of A_z, A_x and A_y, giving rise to characteristic curves
for the photoionisation cross-sections for orbitals of a particular
symmetry (Figure 15). It is important to realise that the <u>form</u> of
such a curve is independent of the detailed character of the
initial state wave function, depending only on its symmetry, as well
as on the strength of the photon field. The application of the

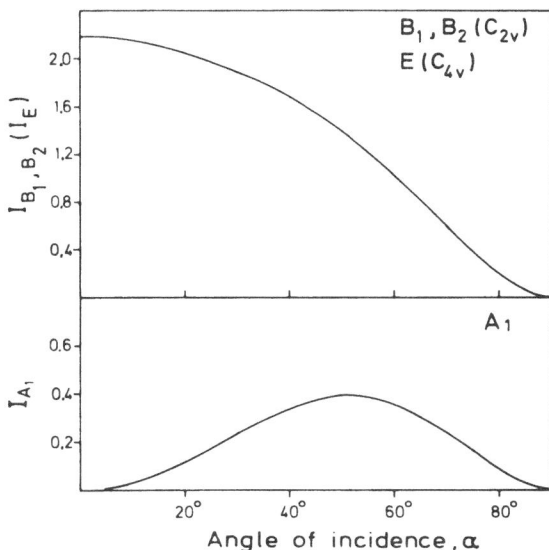

Figure 15. Intensity vs. angle of incidence relationships for normal exit photoemission from initial states of A_1 and B_1, B_2 symmetry in the C_{2v} point group (A_1 and E in C_{4v}). Nickel, $\hbar\omega$ = 21.2 and unpolarised light. After Horn et al. [79] and Scheffler et al. [67], with permission.

method to CO adsorbed in the $(\sqrt{2} \times \sqrt{2})R45°$ structure on Ni(100) is shown in Figure 16 [55]. The P_1 peak (4σ) clearly belongs to A_1. The P_2 peak was decomposed by taking intensities on the low (1π) and high (5σ) binding energy sides. The good agreement for the 1π orbital may be fortuitous: there appears to be a bulk feature of identical symmetry at this energy, the intensity of which may be enhanced by adsorption. The adsorption of more complicated molecules, ethylene and acetylene, has also been investigated with the unpolarised light method [79]. Under such circumstances non-bonding molecular orbitals (non-bonding with respect to the chemisorption **bond**) can belong to representations other than those allowed by the normal emission dipole selection rule. From such states normal exit photoemission is forbidden for all components of the vector potential. A simple example is seen in the case of ethylene adsorbed on Ni(100) in the $(\sqrt{2} \times \sqrt{2})$ R45° structure, suggesting a C_{2v} point group if the molecule is π-bonded as in inorganic complexes. The highest energy σ_{C-H} level in the gas phase ($1b_{1g}$) correlates with a state belonging to A_2 in the C_{2v} point group [77]. In fact normal exit photoemission is observed from this state, indicating either severe distortion of the molecule or very poor ordering in the adlayer [79].

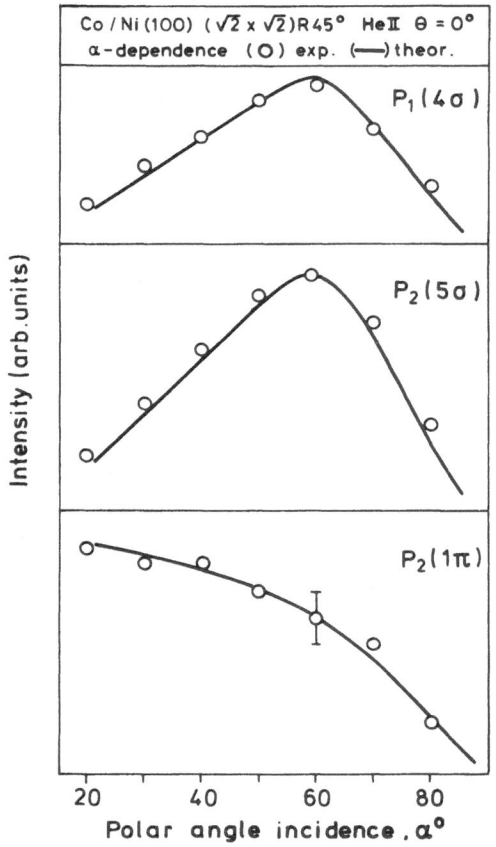

Figure 16. Intensity vs. angle of incidence relationships for CO|Ni(100) under normal exit emission conditions. After Horn et al. [55].

Having discussed various methods of assignment with particular reference to molecular CO adsorption, it is now appropriate to derive the chemical information on bonding in the adsorbate complex. The usual description [82] of the CO standing upright and bonding through the carbon atom as in the metal carbonyls [74-76] is confirmed by the results [83]. The favoured bonding mechanism is then donation of the 5σ lone pair into the metal accompanied by synergic back-donation of metal electrons into the 2π orbital of the CO moiety. Both are expected to be bonding with respect to the chemisorption bond but anti-bonding with respect to the C-O bond.

The photoelectron results show that the 4σ and 1π separation is about 3 eV as in the gas phase. The 5σ is shifted down relatively in energy by 3 eV due to the bonding interaction [84]. Using equation (3) the difference in E_B^V (gas) and E_B^V (ads.) for the 4σ and 1π orbitals is $\sim$ 3 eV, which is composed of the change in final state effects as well as the influence of the electrostatic potential of the solid in the surface region. The observation of back-donation effects, which appear to influence strongly the position of the C-O stretch frequency in the IR spectrum [85], is hindered by the complexity of the changes taking place in the d-band region. Agreement of the empirical orbital energy diagram with CNDO [86], SWX_α [87] and HF [88] cluster calculations is good.

The simplest use of valence level spectroscopy is in character-isation of molecular species with non-interacting orbitals, thus allowing comparison with gas phase spectra. Particular use of this method has been made in the investigation of adsorbed hydro-carbons. There are already many examples to be found in the literature, see, e.g., Refs. [54, 89-91]. The adsorption of acetylene and ethylene on Ni(100) and (111) is particularly illus-trative. The difference spectrum for ethylene adsorbed on Ni(111) at 100 K is shown in Figure 17. Comparison with the gas phase

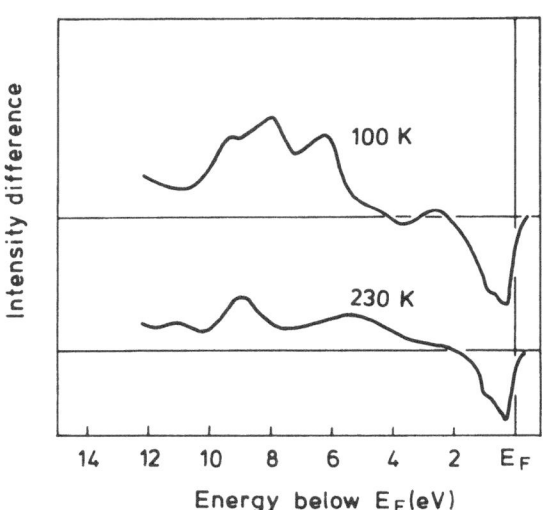

Figure 17. The thermally activated conversion of ethylene to acetylene on Ni(111). $\hbar\omega$ = 21.2 eV. After Eastman and Demuth [54] with permission.

spectrum indicates essentially undisturbed σ-orbitals in the range
6-10 eV below E_F, whereas the π-orbital at ∿ 4.5 eV suffers a
relative bonding shift of ∿ 0.9 eV [54]. This indicates a π-bonded
ethylenic complex similar to that encountered in inorganic [92] and
matrix isolation [93] chemistry. Upon heating to 230 K thermally
activated dehydrogenation to chemisorbed acetylene takes place.
The latter is characterised by the π-orbitals at ∿ 5.5 eV (presumably no longer degenerate) and the σ_{CC} and σ_{CH} orbitals at 9
and 11 eV respectively. Adsorption of acetylene at low temperatures gives a result almost identical with the 230 K spectrum of
Figure 17. No changes occur on heating to room temperature; only
at ≳ 470 K does dehydrogenation take place with the formation of a
carbon-rich layer. The results on the Ni(100) surface are quite
different, indicating a high degree of surface specificity [79].
A gradual loss of intensity in all peaks of the C_2H_4|Ni(100)
spectrum occurs between 100 and 200 K. At 273 K a broad weak feature between 6 and 8.5 eV remains with some evidence for a peak
at ∿ 11 eV. Dehydrogenation occurs at ∿ 370 K leaving a sharply
defined (2 x 2) carbonaceous overlayer. The corresponding

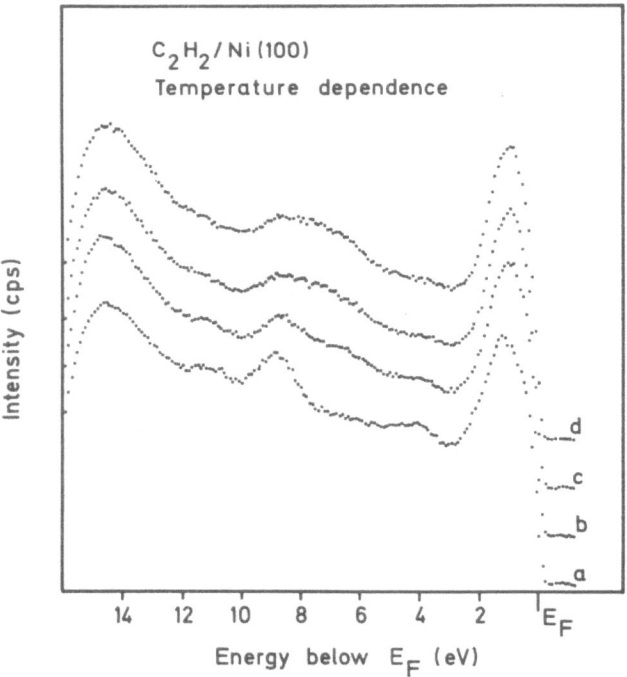

Figure 18. Changes in the valence level spectrum of acetylene
on Ni(100) as a function of increasing temperature. $\hbar\omega$ = 21.2 eV.
a) 115 K. b) 210 K. c) 273 K. d) 300 K. After Horn et al. [79].

sequence for acetylene as a function of increasing temperature is shown in Figure 18. By 300 K a broad structureless feature between 6 and 9 eV remains, which appears to be due to a conjugated polymeric species [94]. Similar results have been obtained on Pd(111) and Pt(111) surfaces [95].

CONCLUSIONS AND FURTHER DEVELOPMENTS

We have attempted to show in this short review article how the photoelectron technique has already contributed to the study of adsorption phenomena. Among the main attributes of core level spectroscopy is its ability to differentiate non-destructively between multiple adsorption states and to follow their relative concentrations quantitatively. Despite the difficulties associated with the exact description of adsorption-induced binding energy shifts, such measurements will continue to be a source of real chemical information. The use of satellite structure on core level peaks and X-ray -induced Auger spectroscopy are also useful, but as yet little exploited, fingerprinting techniques. The full impact of valence level spectroscopy on our understanding of the chemisorption bond can only be really felt in the context of parallel developments in chemisorption theory. The experimental possibilities of angular resolved measurements and of the polarised continuum source offered by synchrotron radiation are just beginning to be realised in this field. The use of photoelectron angular distributions for precise structural determinations has not as yet lived up to its early promise. A necessary prerequisite here is a series of careful measurements on systems already characterised by LEED structural analysis.

ACKNOWLEDGEMENTS

The authors would like to thank their past and present colleagues who have contributed to many of the ideas and experiments used as examples in this paper.

REFERENCES AND NOTES

1)   For an example of this philosophy, see T. Edmonds and J. J. McCarroll, these Symposia Proceedings.

2)   The term "photoionisation" is usually applied to the process of photoelectron emission in molecules and emphasises quite correctly the ion which is produced in the final state. In surface and bulk studies the term "photoemission" is in general use, but this should not detract from the importance of the hole state.

3)   A. M. Bradshaw, L. S. Cederbaum and W. Domcke, in Structure
      and Bonding, vol. 24, p. 133 (1975), Springer-Verlag,
      Heidelberg.

4)   E. W. Plummer in The Physical Basis for Heterogeneous
      Catalysis (ed. E. Drauglis and R. I. Jaffee), 1975, Plenum
      Press, New York.

5)   D. Menzel, in Photoemission and the Electronic Properties of
      Surfaces (ed. B. Feuerbacher, B. Fitton and R. F. Willis),
      Wiley, London (in press).

6)   D. Menzel in Critical Reviews in Solid State Science, CRC
      Press, in press.

7)   E. W. Plummer in Photoemission and the Electronic Properties
      of Surfaces (ed. B. Feuerbacher, B. Fitton and R. F. Willis),
      Wiley, London (in press).

8)   J. A. R. Samson, Techniques of Vacuum UV Spectroscopy (1967),
      Wiley, New York.

9)   C. Kunz in Photoemission and the Electronic Properties of
      Surfaces (ed. B. Feuerbacher, B. Fitton and R. F. Willis),
      Wiley, London (in press).

10)   D. Roy and J. D. Cavette, in Topics in Current Physics (4):
      Electron Spectroscopy for Surface Analysis (ed. H. Ibach),
      p. 13 (1977), Springer-Verlag, Heidelberg.

11)   P. A. Redhead, J. P. Hobson and E. V. Kornelson, The Physical
      Basis of Ultrahigh Vacuum (1968), Chapman and Hall, London.

12)   L. S. Cederbaum and W. Domcke, Adv. Chem. Phys. 36:205 (1977).

13)   L. Kleinmann, Phys. Rev. B3:2982 (1971).

14)   J. C. Ashley and R. H. Ritchie, Phys. Stat. Sol. (b) 62:253
      (1974).

15)   H. Kanter, Phys. Rev. B1:2357 (1970).

16)   J. C. Tracy, cited by C. J. Powell, Surface Sci. 44:29 (1974).

17)   Extrinsic coupling is the creation of the plasmon by the
      excited photoelectron, whereas intrinsic coupling corresponds
      to the creation of the plasmon during the photoionisation
      event by image charge screening. See, e.g., D. C. Langreth,

Phys. Rev. B1:471 (1970) and, for a useful review article, J. W. Gadzuk in Photoemission and the Electronic Properties of Surfaces (ed. B. Feuerbacher, B. Fitton and R. F. Willis), Wiley, London (in press).

18) S. A. Flodström, L.-G. Petersson and S. B. M. Hagström, J. Vac. Sci. Tech. 13:280 (1976).

19) U. Gelius, Physica Scripta 9:133 (1974).

20) P. H. Citrin and D. R. Hamann, Phys. Rev. B10:4948 (1974).

21) J. Bardeen, Phys. Rev. 49:653 (1936); C. Herring and M. H. Nichols, Rev. Mod. Phys. 21:185 (1949).

22) J. W. Gadzuk, Phys. Rev. B14:2267 (1976).

23) We assume that the difference in correlation energy can be neglected.

24) J. J. Chang and D. C. Langreth, Phys. Rev. B5: 3512 (1972).

25) A. M. Bradshaw, W. Domcke and L. S. Cederbaum, Phys. Rev. to be published.

26) R. J. Baird, M. Mehta and C. S. Fadley, Proceedings of the Third International Conference on Solid Surfaces, Vienna, 1977, p. 2221.

27) N. D. Lang and A. R. Williams, to be published.

28) K. Schönhammer and O. Gunnarsson, Proceedings of the Third International Conference on Solid Surfaces, Vienna, 1977, p. 795; K. Schönhammer and O. Gunnarsson, Solid State Comm., in press.

29) C. R. Brundle and K. Wandelt, Proceedings of the Third International Conference on Solid Surfaces, Vienna, 1977, p. 1171; C. R. Bundle, private communication.

30) J. C. Fuggle and D. Menzel, Proceedings of the Third International Conference on Solid Surfaces, Vienna, 1977, p. 1003.

31) T. E. Madey, J. T. Yates and N. E. Erickson, Chem. Phys. Lett. 19:487 (1973).

32) E. Umbach, J. C. Fuggle and D. Menzel, J. Elect. Spect. 10:15 (1977).

33) See, for example, R. Gomer, Jap. J. Appl. Phys. Suppl. 2, Pt. 2, p. 213 (1974).

34) S. D. Worley, N. E. Erickson, T. E. Madey and J. T. Yates, J. Elect. Spect. 9:355 (1976).

35) J. C. Fuggle, T. E. Madey, M. Steinkilberg and D. Menzel, Surface Sci. 52:521 (1975).

36) C. R. Brundle, J. Vac. Sci. Tech. 13:301 (1976).

37) R. W. Joyner and M. W. Roberts, Proc. Roy. Soc. (London) A350:107 (1976).

38) P. R. Norton, R. L. Tapping and J. W. Goodale, Surface Sci. 65:13 (1977).

39) A. M. Bradshaw, P. Hofmann and W. Wyrobisch, Surface Sci., in press.

40) J. T. Yates, T. E. Madey and M. J. Dresser, J. Catalysis 30:260 (1973).

41) E. Umbach and D. Menzel, to be published.

42) A. M. Bradshaw, D. Menzel and M. Steinkilberg, Jap. J. Appl. Phys., Suppl. 2, Pt. 2, p. 841 (1974).

43) As mentioned in the first section, angular effects associated with the incoming photons and the outgoing photoelectrons also play an important role in core level spectroscopy. Such effects are discussed by C. S. Fadley in Prog. Solid State Chem. vol. 11 (1976). The measurements discussed in this paper were invariably performed under angle integrating conditions.

44) M. Barber, J. A. Conner and I. H. Hillier, Chem. Phys. Lett. 9:570 (1971).

45) For the energetically accessible core levels studied in surface chemistry the most important decay mechanism for the hole state is the Auger process. Thus it is clear that Auger electrons should be observed in the kinetic energy spectrum. In electron excitation of Auger spectra, which is most frequently used for surface analysis, the data is usually collected in the first derivative of the energy distribution, $N'(E)$, because of the high sloping background.

46) M. A. Chesters, B. J. Hopkins and R. I. Winton, Surface Sci. 59:46 (1976).

47) T. Kawai, K. Kunimori, T. Kondow, T. Onishi and K. Tamaru, Phys. Rev. Lett. 33:533 (1974).

48) J. C. Fuggle, E. Umbach, P. Feulner and D. Menzel, Surface Sci. 64:69 (1977).

49) T. A. Carlson, M. O, Krause and W. E. Moddeman, J. Phys. (Paris) C4:76 (1971).

50) J. C. Fuggle, E. Umbach and D. Menzel, Solid State Commun. 20:89 (1976).

51) D. W. Turner et al., Molecular Photoelectron Spectroscopy (1970), Wiley-Interscience, New York.

52) E.g., D. E. Eastman in Techniques of Metals Research Part VI (ed. E. Passaglia), 1972, Wiley-Interscience, New York.

53) D. E. Eastman and J. K. Cashion, Phys. Rev. Lett. 27:1520 (1971).

54) See, e.g., D. E. Eastman and J. E. Demuth, Jap. J. Appl. Phys. Pt. 2, Suppl. 2, p. 827 (1974).

55) K. Horn, A. M. Bradshaw and K. Jacobi, Surface Sci., in press.

56) E. W. Plummer, T. Gustafsson, W. Gudat and D. E. Eastman, Phys. Rev. A15:2339 (1977).

57) E. Dietz, H. Becker and U. Gerhardt, Phys. Rev. B12:2084 (1975); Phys. Rev. Lett. 36:1397 (1976).

58) J. Hermanson, Solid State Commun. 22:9 (1977).

59) J. K. Sass and H. Gerischer, in Photoemission from Surfaces (ed. B. Feuerbacher, B. Fitton and R. Willis), Wiley, London (in press).

60) J. W. Gadzuk, Phys. Rev. B10:5030 (1974).

61) I. G. Kaplan and A. P. Markin, Opt. Spectrosc. 24:475 (1968).

62) J. C. Fuggle, M. Steinkilberg and D. Menzel, Chem. Physics 11:307 (1965).

63) J. W. Davenport, Phys. Rev. Lett. 36:945 (1976).

64) C. L. Allyn, T. Gustafsson and E. W. Plummer, Chem. Phys. Lett. 47:127 (1977).

65) Using the dipole selection rule the final state resonance with Σ^+ symmetry can only be excited with $A_\parallel$

66) A. Liebsch, Phys. Rev. Lett. 32:1203 (1974) and Phys. Rev. B13:544 (1976).

67) M. Scheffler, K. Kambe and F. Forstmann, Solid State Commun. 23:789 (1977).

68) M. Scheffler, K. Kambe, F. Forstmann and K. Jacobi, Proceedings of the Third International Conference on Solid Surfaces, Vienna, 1977, p. 2227.

69) J. B. Pendry, Low Energy Electron Diffraction (1974), Butterworths, London.

70) A. Liebsch, Phys. Rev. Lett. 38:248 (1977).

71) T. Gustafsson, E. W. Plummer, D. E. Eastman and J. L. Freeouf, Solid State Commun. 17:391 (1975).

72) J. C. Fuggle, T. E. Madey, M. Steinkilberg and D. Menzel, Phys. Lett. 51A:163 (1975).

73) K. Siegbahn et al. in ESCA Applied to Free Molecules (1969), North-Holland, Amsterdam.

74) D. R. Lloyd, J. C. S. Far. Disc. 58:136 (1974).

75) R. F. Fenske, Prog. Inorg. Chem. 21:179 (1976).

76) H. Conrad, G. Ertl, H. Knözinger, J. Küppers and E. E. Latta, Chem. Phys. Lett. 42:115 (1976).

77) This analysis has been rigourously developed by Halford [J. Chem. Phys. 14:8 (1946)] for three-dimensional crystals. The situation in a two-dimensional lattice is analagous.

78) K. Jacobi, M. Scheffler, K. Kambe and F. Forstmann, Solid State Commun. 22:17 (1977).

79) K. Horn, A. M. Bradshaw and K. Jacobi, J. Vac. Sci. Tech., in press.

80) J. K. Sass, K. Horn and A. M. Bradshaw, to be published.

81) J. Anderson and G. J. Lapeyre, Phys. Rev. Lett. 36:376 (1976). This is not a specific conclusion of this paper, but is clear from the published results.

82) See, e.g., R. R. Ford, Adv. Catalysis 21:51 (1970).

83) It should perhaps be mentioned that IR spectroscopists came to an identical conclusion over twenty years ago!

84) This assumes that other adsorption-induced shifts are identical for all three orbitals.

85) See, e.g., F. M. Hoffmann and A. M. Bradshaw, <u>Proceedings of the Third International Conference on Solid Surfaces</u>, Vienna, 1977, p. 1167.

86) G. Blyholder, J. Vac. Sci. Tech. 11:865 (1974).

87) I. Batra and P. Bagus, Solid State Commun. 16:1097 (1975).

88) L. S. Cederbaum, W. Domcke, W. von Niessen, and W. Brenig, Z. Physik B21:381 (1975).

89) E. W. Plummer, B. J. Waclawski and T. V. Vorburger, Chem. Phys. Lett. 28:510 (1974).

90) J. E. Demuth and D. E. Eastman, Phys. Rev. B13:1523 (1976).

91) G. Brodén, T. Rhodin and W. Capehart, Surface Sci. 61:143 (1976).

92) L. Manojlović-Muir, K. W. Muir and J. A. Ibers, J. C. S. Faraday Disc. 47:84 (1969).

93) G. A. Ozin and W. J. Power, Inorg. Chem., in press.

94) J. C. Bertolini, G. Dalmai-Imelik and J. Rousseau, Le Vide, in press.

95) J. E. Demuth, Chem. Phys. Lett. 45:12 (1977).

IMPACT OF SURFACE PHYSICS ON CATALYSIS

T. Edmonds

J.J. McCarroll

BP Research Center
Middlesex, England

ABSTRACT

No real catalyst is a single component system. It is a multi-component cocktail designed to meet three requirements - activity, selectivity and stability. The reality is at least one major component usually dispersed finely over a high surface area refractory support, which may or may not perform an active catalytic role itself, plus one or more modifiers often called promoters. We have used surface physics to define the role of the modifiers and will be discussing our conclusions in respect to nickel and platinum catalysts.

INTRODUCTION

Establishing a link between surface science and heterogeneous catalysis can be attempted at various levels. One can, for instance, attempt to establish a conceptual link and show how the techniques of surface physics should be capable of filling the gaps left by the historical methods of study of surface chemistry and catalysis. At a second level one may itemise those results in which improvements in the knowledge of the chemical behaviour of species at surfaces can be detected. Such cases are now numerous and noted in many reviews. The relevance of such results and theories have been discussed with

imagination and persuasion. No one now doubts that, ultimately, surface physics will be of considerable benefit
to catalysis.

What has been lacking is any discussion of the extent to which attempts are being made to improve and de-
sign industrial catalysts by the use of the methods and
theories of the new surface physics, and of the results
of these attempts.

That this has not been attempted seriously should not
be surprising. Modern surface physics is still a young
science; industrial catalysis is a secretive art, possibly
becoming a secretive science. Only a very few scientists,
almost invariably working within industry, are actively
engaged in both surface physics and industrial catalysis.
We make no apology for emphasising the industrial uses
of catalysis. Of all the reasons that can be deployed
for looking at matters from that side only one need be
stated - that the usefulness of surface physics in catalysis will ultimately be judged from the catalytic
side.

What we will attempt in this review is to examine
those cases that we know of (and are in a position to
discuss) where catalytic processes have been developed
or improved with the aid of surface physics or where
considerable improvements in understanding have been
achieved and to pass some sort of judgement as to
whether surface physics has been of use. We hope that
the review will be decipherable by both sides involved.

Some non-contentious explanations are necessary to
describe the framework in which we are setting out our
chosen examples. To be of practical use, industrial catalysts must fulfil a wide range of criteria, all of which
have the economic basis that the value to society of the
products must exceed the value of the feedstock by an
amount greater than the cost of the process employing
the catalyst.

The primary criteria are twofold:

1. Catalysts should have good activity and selectivity
- they should convert feedstock at a high enough rate to
desired product(s) without producing unwanted product(s).

2.   Catalysts should have good lifetimes, implying,
inter alia, resistance to poisons normally present in
feedstocks and physical durability under frequently
severe conditions.

In addition any new catalyst should obviously represent improvements in existing catalysts, allowing an overall more efficient process.

Physically, heterogeneous catalysts are almost invariably porous solids, most frequently oxides, with surface areas ranging from a few square metres per gram to, several hundred square metres per gram. The oxides serve as the base for an 'active' metal component distributed across the base as crystallites ranging in size from 10 - 20 Å at one end to several hundred or several thousand Å at the other.

Virtually all catalyst systems used in industry employ additional components which, by themselves, play no catalytic role but confer on the catalyst package considerable advantages of activity, selectivity or stability. These 'promoter' substances are of crucial importance and their almost universal use in industrial catalysis is a feature virtually ignored and little studied in surface chemistry and physics, except by accident. How crucial their role is can be simply stated by listing a variety of major industrial processes and the catalysts used - Table 1.

That this review is heavily involved with promoters is not by deliberate intent, but as a consequence of its stress in industrial catalysts.

TABLE 1

Process	Catalyst	Promoters
NH_3 synthesis	Fe	K, Aℓ oxides
Steam reforming	Ni	K
Olefin oxidation	Ag	Ba, K, Ca, Cℓ
Hydrodesulphurisation	$MoS_2/A\ell_2O_3$	Co
Hydrocarbon reforming	$Pt/A\ell_2O_3$	Re, Sn, Ge etc. in 'bimetallics'. Use of Cℓuniversal
Hydrogenation	$WS_2/A\ell_2O_3$	Ni

The surface science results quoted concentrate on two techniques - Low Energy Electron Diffraction (LEED) and X-ray Photoelectron Spectroscopy (XPS) alias ESCA (Electron Spectroscopy for Chemical Analysis). The former can be simply regarded as a method for investigating the structure of surfaces, albeit only of large single crys-

tals in high or ultra high vacuum conditions. The latter
is a method for interrogating the electronic state and
environment of surface species again in vacuo, but widely
used to study real catalysts.

1.   METAL CATALYST SYSTEMS

                         (a) Ni/S

     The first metal catalyst systems to be extensively
studied by surface science techniques were Ni/S selective hydrogenation catalysts. In these catalysts S is
added to the system, either in careful presulphiding
steps (1,2), or by sulphiding direct from low sulphur
containing feedstocks. Concentrations of sulphur on the
catalysts are sub-monolayer and the promoting effect is
seen in the changed selectivity of the catalyst in hydrogenation. By suppressing the natural exuberance of
the nickel, mono-olefins in gasoline streams are hydrogenated at much lower rates than conjugated di-olefins.
This enables the latter, which would otherwise cause
gum formation, to be selectively hydrogenated while re-
taining the octane number advantages of the former.

     The explanation, given by the LEED work of Edmonds,
McCarroll and Pitkethly (3,4,5) showed the active role
of a Ni (100)(2 x 2) - S structure with the active site
almost certainly the central, uncovered Ni atom, though
possibly not in a straightforward way (5). This explanation improved the picture suggested by Bourne, Holmes
and Pitkethly (1) by identifying ordered rather than dis-
ordered structures of S on Ni. What was unexpected, and
what conferred a degree of uniqueness and certainty on
the answer, was that not only was the (100)(2 x 2)-S
structure found on the (100) face in both of the sulphiding routes studied but it was also found on the (111)
face. The catalyst system was, in effect, making use of
the face change or 'reorientation' available in the Ni/S
surface system to add to its selectivity and activity.

                         (b) Pt

     A second catalyst system studied in a similar way,
but which we developed in the opposite sense, was the Pt
reforming catalyst system. Here the start point was the
known reorientation of the square Pt (100) face (6). That
this was caused or at the very least facilitated by the
presence of alkali/alkaline earth metals was shown by a

combination of LEED and soft X-ray appearance potential
spectroscopy (SXAPS). The catalytic consequences of this
were shown by the promoter effects of alkali metals in
Pt/graphite dehydrocylisation catalysts, converting n-
hexane to benzene (5).

The mirror imaging, in a chemical sense, of the surface chemistry of the Ni/S and Pt/Ca systems is significant. Both are fcc metals, one with an electro-negative
element (S) producing a (111) to (100) face change, the
other with electro positive element addition (Ca, Na)
reversing the direction and giving a (100) to (111)
face change. In this there was sufficient basis to con-
template a general involvement of reorientation in catalysis.

In the Pt case considerable further work was in-
vested in attempting to relate the information to practical situations. The conversion of n-paraffins (such
as n-hexane) to aromatic ring compounds is the principal
reaction (though not the only one of significance) in
the operation of Pt hydrocarbon reforming catalysts.
These are arguably the most important catalysts in
current technology, upgrading low octane naphtha to high
octane motor spirit and, less importantly, aromatic
streams for the petrochemical industry.

Attempts to produce activity comparable to those
of commercial Pt/Al_2O_3 catalysts came up against the
chemical difficulty of combining Pt with Na on a support
material. Use of oxide supports would have meant saturat-
ing the support with Na before Na could be effectively
combined with Pt. The development of high surface area
graphites (7) of controlled surface structure enabled
efficient doping to be achieved (typically 0.7 per cent
weight Pt plus 300 - 700 ppm by weight of Na). These
catalysts had activities considerably higher than current
commercial Pt/Al_2O_3 catalysts for the conversion of n-
hexane to benzene, and greater selectivity (8). Typical
figures for conversion of n hexane in hydrogen at atmospheric pressure were 85 per cent benzene at 87 per cent
selectivity (ie 85 per cent of nC_6 converted to benzene
and 87 per cent of all products being benzene) for a 0.7
per cent Pt/700 ppm Na/graphite catalyst. By contrast a
current commercial catalyst gave 18.5 per cent benzene
at 33 per cent selectivity - and this level was typical
of all the modern catalysts under the conditions of test
used. Without Na results were also good - probably showing the influence of the hexagonal graphite structure
in 'templating' to some extent the desired (111) struc-

ture of Pt - ie encouraging the epitaxial growth of (111)
faces. The differences in benzene production between undoped and doped Pt/graphites was 20 - 40 per cent - a
range simply consistent with conversion of (100) to (111)
faces and thus only allowing geometric effects to be
contemplated.

    Investigation of the Pt/Na/graphite system in pilot
plants as a potential reforming catalyst showed it to
have some flaws (9). Firstly, it had very low resistance
to S poisoning. Figure 1 shows the typical behaviour of a

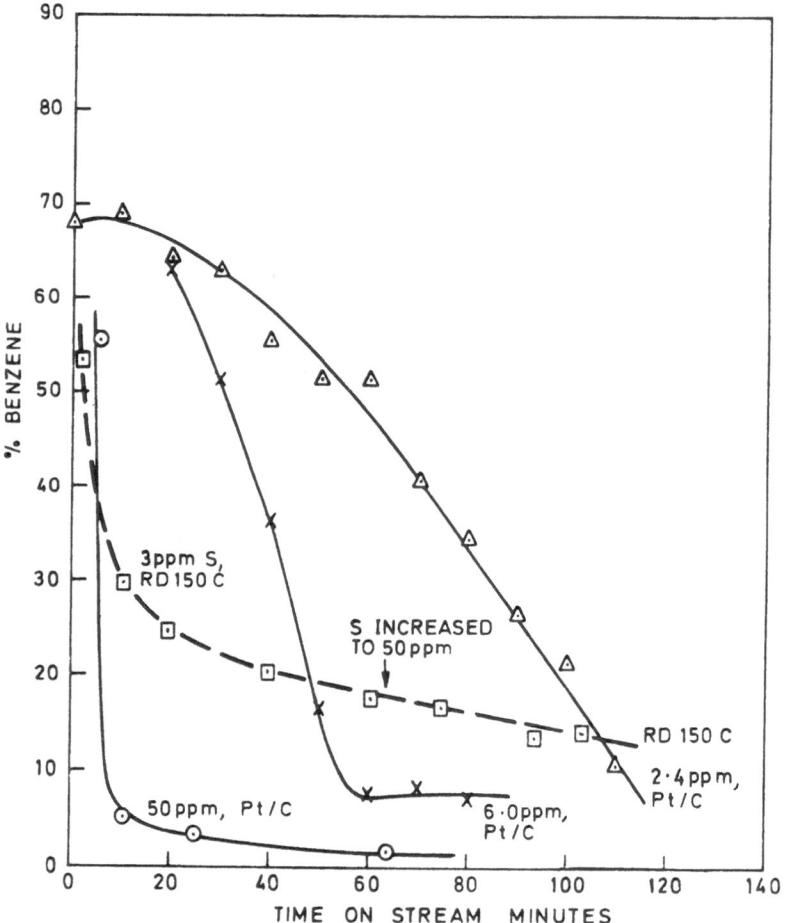

FIG 1   EFFECT OF SULPHUR CONCENTRATION ON BENZENE YIELD
        OVER PLATINUM CARBON AND RD 150 C

Pt/Na/graphite catalyst (or a Pt/graphite catalyst) versus a PtAℓ_2O$_3$ catalyst in terms of producing benzene from sulphur-containing n-hexane feeds. The activity of the graphite catalyst dropped steadily to a low level as it accumulated S. The Pt/Aℓ_2O$_3$ catalyst, starting from a lower initial rate, at first behaved similarly with benzene production dropping rapidly but levelled out at a much higher conversion than the graphite based material.

Secondly, the graphite catalysts would not convert other components of a typical naphtha feed - a mixture of hydrocarbon types - to benzene to the same extent as Aℓ_2O$_3$ based catalysts, though in their presence they would continue to convert n-paraffins at high rates.

The obvious and most simple conclusion, since the Pt/graphite catalysts could only be employing Pt or Pt/Na metal clusters, is that there exists on Pt/Aℓ_2O$_3$ catalysts an active phase which is neither metallic Pt (since this is rapidly poisoned by S) nor alumina (which is inactive).

FIG 2   DIFFRACTION PATTERNS FROM Pt (100)
AFTER OXIDATION FOLLOWING EXTENSIVE
CLEANING.   AES SPECTRUM FROM THIS
SURFACE SHOWN IN FIG 3

Recent LEED/Auger work (10) using a Pt (100) crystal has offered a possible, even likely, answer. The crystal was the same as that on which proof of calcium involvement in the (5 x 1) structure had been previously demonstrated. Further ion bombardment and heat treatment in vacuo had removed any traces of Ca as detected by SXAPS or AES, yet the crystal displayed a clear, sharp (5 x 1) pattern. The more sensitive AES gave a spectrum identical with that taken as the standard (11).

However, high temperature (1000 - 1400 $^{\circ}$K) treatment in oxygen gave patterns of obvious hexagonal symmetry (Figure 2). The AES spectra from this system are shown in Figures 3a and 3b. Apart from the obvious Pt and O Auger peaks, the only additional features lie at 54 and 1397 eV and are obviously attributable to considerable surface concentrations of aluminium. The patterns could not be removed by prolonged heating in vacuo at temperatures at and above those quoted as used in its formation and had similar stability in low pressure of H_2. No AES has been carried out on the corresponding pattern based on Pt (111). Scanning Auger Microscopy (SAM) study of the clean crystal showed what appeared to be inclusions in the crystal surface with dimensions between approximately 20 and 80 micron. These remained unaltered after prolonged bombardment and selected area AES gave spectra showing these to be predominantly Aℓ and O with varying amounts of Ca and Pt. We assume that these were originally alumina particles embedded in the crystal at its growth stage. It would appear unlikely that they were introduced at the polishing stage since all later stages of mechanical polishing were carried out with diamond abrasives from 15 micron to <1 micron finish, and substantial amounts of material were removed.

From these findings two lines of argument can be followed:-

1.   An ordered Pt/Aℓ/O phase could be prepared by heat treatment in low pressures of O_2 of Pt and Aℓ_2O_3 in intimate contact. The temperatures used were somewhat higher than those typical in the calcination stage of Pt/Aℓ_2O_3 catalyst preparation in which the alumina, impregnated with a Pt salt is dried and then roasted in air at atmospheric pressure at 973°K. However O_2 pressures were much lower in the single crystal case. Times were of the same order of magnitude. One may simply propose

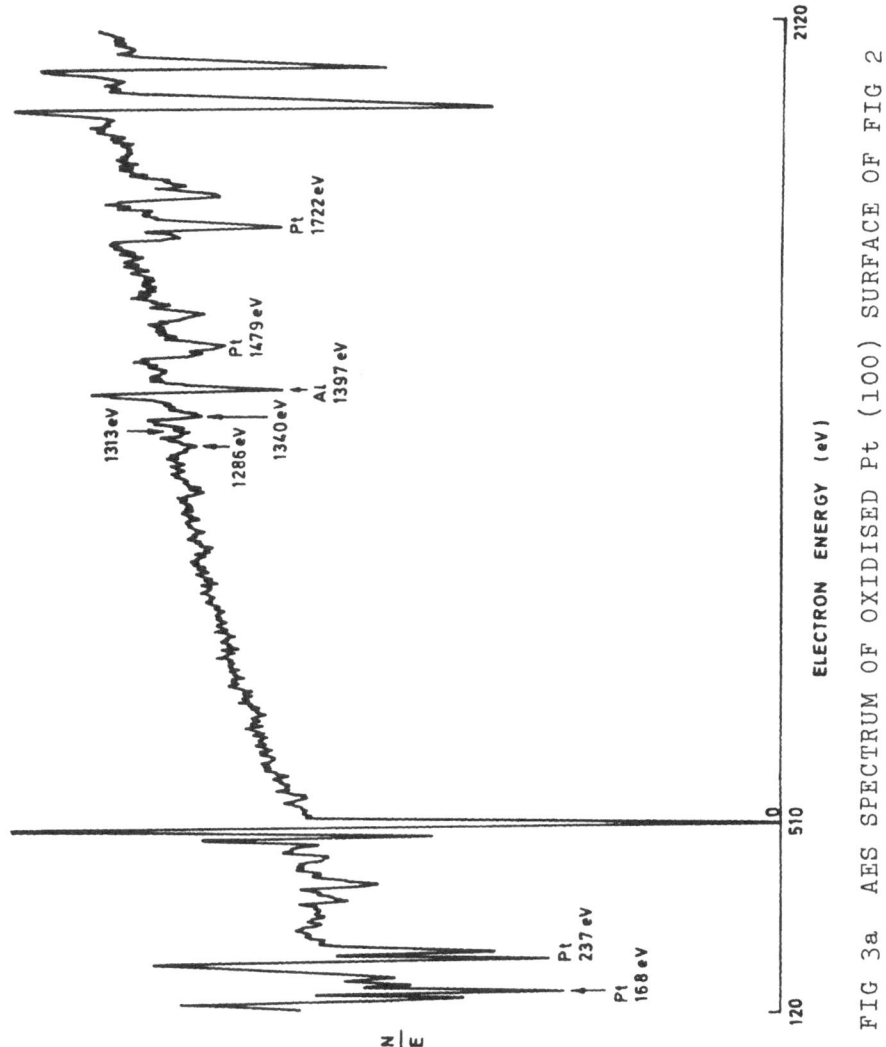

FIG 3a AES SPECTRUM OF OXIDISED Pt (100) SURFACE OF FIG 2

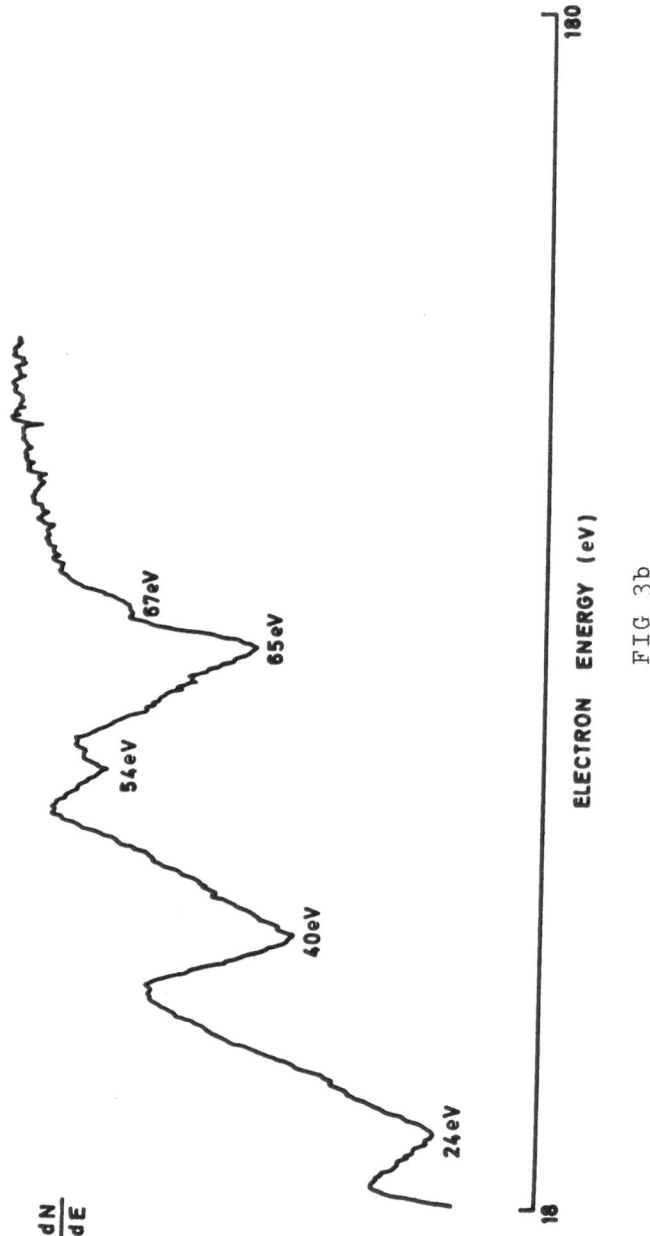

ELECTRON ENERGY (eV)

FIG 3b

that the phase on the $Pt/A\ell_2O_3$ catalyst which shows much
higher sulphur resistance than metallic platinum is similar to that seen on the crystal and composed of Pt, $A\ell$
and O. A not dissimilar postulate of a $Pt/A\ell$ phase has
been invoked by Dautzenberg (12) to explain features of
H_2 chemisorption on $Pt/A\ell_2O_3$ catalysts.

2.   It should be noted that the 'clean' AES spectrum of
Pt displays small features at 53 eV and 1397 eV which
may, or may not, be representative of Pt. If they are
not, it would suggest that the particular cause of the
Pt (100) (5 x 1) structure at this stage of this crystal's
history was $A\ell$, diffusing from the inclusions in the crystal. Such a conclusion would represent a natural extension
of the explanation af alkali and alkaline earth metals
offered previously (5), since all are electropositive in
nature.

(c) Nickel for Steam Reforming

(i) Classical Background

Steam reforming is the generic name associated with
the production of hydrogen containing gases by reacting
hydrocarbons with steam. The feedstock can be a light
naphtha or petroleum gases including methane and the product is either hydrogen or synthetic natural gas (mainly
methane). The primary reaction which occurs is:-

(a) Gasification

$$C_n H_m + nH_2O \longrightarrow nCO + (n + \frac{m}{2})\, H_2$$

The hydrogen can then react with hydrocarbon to produce
methane.

(b) Hydrogenolysis

$$C_n H_m + (2n - \frac{m}{2})\, H_2 \longrightarrow nCH_4$$

For hydrogen production carbon monoxide is recovered
as CO_2.

(c) Water Gas Shift

$$CO + H_2O \rightleftharpoons CO_2 + H_2$$

For synthetic natural gas production CO is converted by

(d) <u>Methanation</u>

$$CO + 3H_2 \rightleftharpoons CH_4 + H_2O$$

Normal operating conditions for the production of hydrogen are an inlet temperature of 725 K, an outlet temperature from the primary gasification reactor of 1025 - 1125 K, a pressure of approximately 450 psig and a steam hydrocarbon ratio of 3 (13).

The formulation of catalysts for hydrogen and SNG production has been described by ICI (13) Haldor Topsoe (14) British Gas (15) and the Japan Gasoline Company (16). All are based on nickel which is present as large particles (typically one micron diameter) on a low surface area support, to minimise support sintering at the high operating temperatures. The complete formulation which always contains an alkali, usually potassium, aims to maintain the activity of the nickel throughout the life of the catalyst. This activity is lost by a reduction of the effective surface area of the nickel either directly by sintering or indirectly by poisoning.

These catalysts have been extensively studied by classical techniques and the major conclusions are listed below:-

1.   Reaction proceeds by the following sequence of reaction steps:-

$$C_nH_{2n+2} \longrightarrow S_1$$

$$S_1 \longrightarrow S_2$$

$$S_2 + H_2O \longrightarrow \text{Products}$$

$$S_2 + C_nH_{2n+2} \longrightarrow \text{Poison or poison precursor}$$

S_1 and S_2 are different cracked products, C_yH_x, at the nickel surface. S_1 is dependent on the starting hydrocarbon but S_2 is essentially independent ($y = 1, x \geq 1$).

2.   The rate of gasification is a function of the nature of the hydrocarbon. The generally agreed trend at 775 K is cyclo-paraffin > branched chain hydrocarbon > n-paraffins (decreasing with carbon number) > aromatics. This is consistent with the hypothesis that carbon compounds with

the exception of aromatics, undergo multiple fission of
C-C bonds over the nickel surface to form $CH_x(S_2)$ radicals.

3. The specific activity for gasification does not
depend on surface area or particle size so C-C bond fission must be independent of site geometry.

4. The major problem in maintaining the life of the
catalyst is poisoning via coking. It is least severe in
methane reforming where carbon deposition is readily
controlled by the steam ratio. With naphthas the severity
increases with the temperature of the cut point.

5. The coking rate which maximises between 825 K and
850 K bears an inverse relationship to gasification activity as a function of the type of hydrocarbon falling
as :-

C_2H_4 > > benzene > n-heptane   =   cyclohexane

Since CH_x radicals at the surface are precursors for both
the formation of products and coke it is clear why there
is this inverse relationship and why the reaction products can be controlled by the steam hydrocarbon ratio.

6. Carbon is deposited in a variety of forms. This is
indicated by the deviation of carbon-forming reactions
from equilibria based on graphite.

7. For naphtha feedstocks the problem of coking is
controlled by incorporating mobile alkali in the catalyst.
This promotes the removal of carbon by reaction with
steam. In this context the choice of support for its
ability to adsorb steam may also be important.

8. However, the addition of alkali causes a major
decline in the specific activity for gasification and
methanation. This arises because there is a lower degree
of dissociation of hydrocarbon molecules relative to
the undoped catalyst.

9. The catalyst is also poisoned by sulphur but this
poisoning is reversible.

10. The reduction in activity caused by alkali is considerably greater than that caused by sulphur poisoning
yet there is no evidence for a fall in the nickel surface area.

(ii) Impact of Surface Physics Techniques

In viewing the surface physics literature on this topic the key question to which we have sought an answer is 'what is the role of alkali in this system'? This involved asking the subsidiary questions:-

a) Does it affect the structure of the nickel surface?
b) Does it affect the interaction of hydrocarbons and steam?
c) Why does it reduce the rate of gasification so much more than sulphur?

There is enough information available from our own work and from the literature to build up an hypothesis for the role of the promoter in this system.

Does the presence of alkali affect the structure of the nickel surface?

At the outset of this discussion we must be quite clear that we are considering metal particles of about one micron diameter reacting at temperatures of at least 800 K. Thermodynamically such large particles should display a marked preponderance of (111) faces (17) and the scope for any marked effect of conversion of (100) to (111) faces must be small, if it occurs at all. The evidence, however, is that no face changes were observed when the interaction of alkali metals with Ni (100) and (111) faces was studied (18, 19, 20). Only simple ordered structures were observed, these exhibiting order-disorder transitions at temperatures as low as 500 - 600 K. Temperature-programmed desorption results support this conclusion (18). The major desorption peak from sodium atoms adsorbed on both (111) and (100) faces of nickel occurs around 600 K. However, in each case a small component, θ_{Na} = 0.06 at (111) and θ_{Na} = 0.12 at (100), was not desorbed until 900 K.

No observations have been made of Ni/alkali metal/ oxygen systems. Pronounced structural effects were noted in the Pt/alkaline earth or aluminium/oxygen systems as noted above. One must bear this in mind but at this time we must conclude that the presence of alkali is not influencing the structure of nickel particles and that its concentration at the surface will be relatively small at steam reforming temperatures.

Does it affect the interactions of hydrocarbons and steam?

Structures formed during the interaction of hydrocarbons with nickel surfaces fall into three regimes. At
low temperatures adsorption is non-dissociative (21).
At intermediate temperatures a single structure is
formed by the dissociative interaction of all hydrocarbon
molecules. This temperature range extends from around room
temperature to approximately 725 - 775 K depending on the
adsorbate. Above 775 K, with readily dissociated molecules, graphite co-exists with the structure from the
intermediate range (22).

Only the structure formed in the intermediate
temperature range may be intrinsically significant to
steam reforming, since those at low temperature are not
due to the dissociation of the hydrocarbons and the
high temperature structure is graphite (probably at
higher pressures co-existing with amorphous carbon). The
unit mesh of the structure is Ni (100) (2 x 2) - C
at both (100) and (111) faces ie the latter reorientates.
The clue to its significance is that the temperature at
which Ni (100) (2 x 2) - C forms with various hydrocarbons
increases in the order (22):-

acetylene $<$ ethylene $<$ benzene $<$ cyclohexane $<$
paraffins

ie it is directly related to the incidence of coking and
inversely related to the incidence of gasification in
classical studies.

Two hypothetical structures, shown as Figures 4
and 5 (22), can be drawn to explain the two-fold symmetry
of the diffraction patterns and the non-appearance of
Ni (100) c (2 x 2) - C structures vis à vis sulphur.
Both are carbon polymers. The linear, Figure 4, has the
stoichiometry C_2H and the cross-linked, Figure 5, the
stoichiometry CH_0. The latter is already coke. The former
could be S_2 although the C:H ratio is already < 1. Unfortunately its reactivity with steam has never been investigated.

Since there is a direct link between the temperature
at which this structure forms with various hydrocarbons
and the onset of coking in steam reforming, it seems unlikely that the linear polymer is a reaction intermediate
to hydrogen or SNG. On the other hand, if the rate limiting step is the formation of S_2, assuming this is the
linear polymer, the balance of the products between
H_2 and coke will then depend on the rate of transformation
between the linear polymer and cross-lined polymer if it
is assumed the latter cannot form gaseous products. The

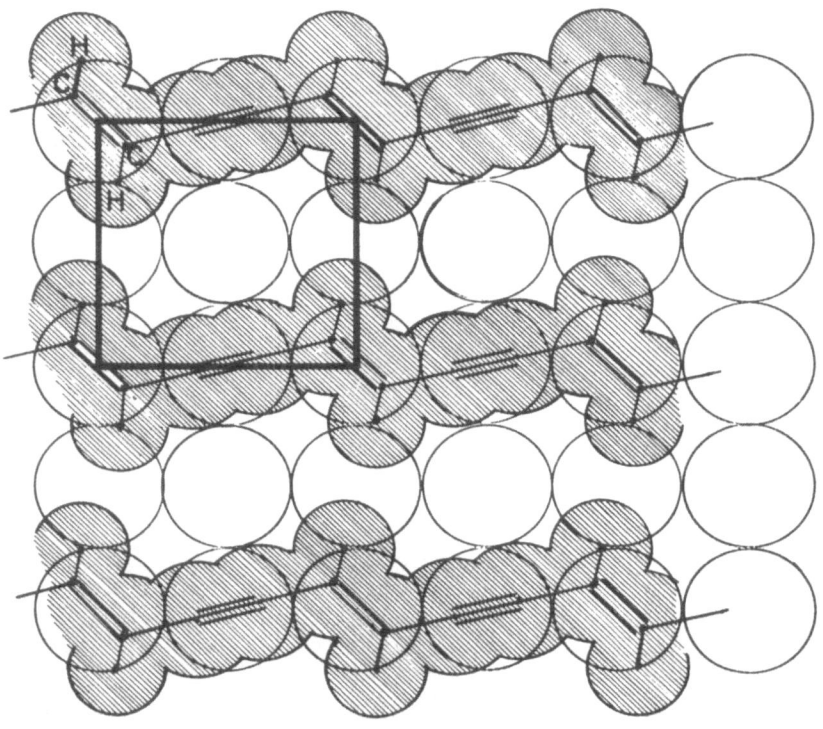

FIG 4   LINEAR POLYMER TO SATISFY Ni (100) (2x2)-C

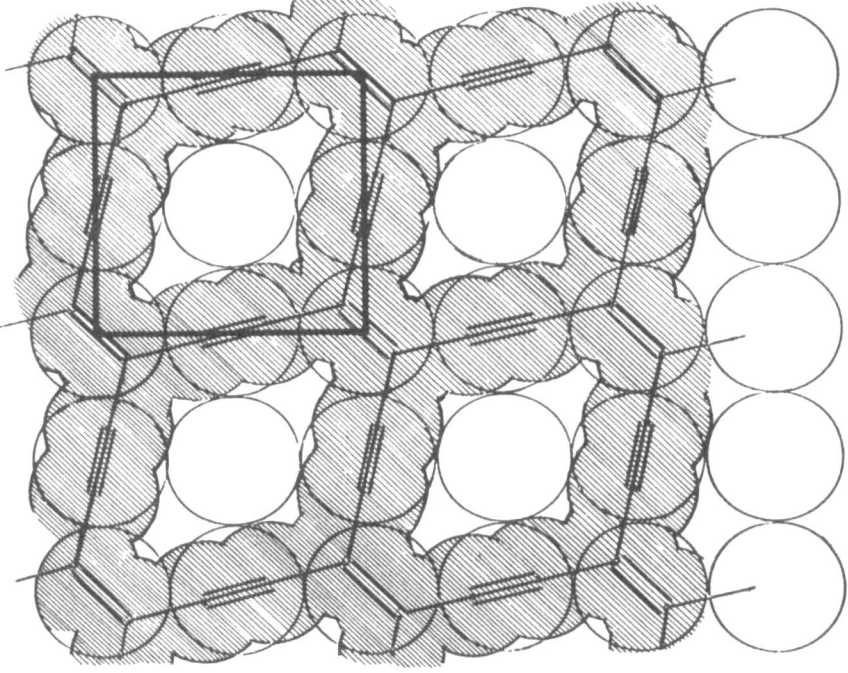

FIG 5   CROSS-LINKED POLYMER TO SATISFY Ni (100) (2x2)-C

rate of this transformation will increase with the ease of formation of Ni (100) (2 x 2) - C while the rate of formation of gaseous products will show the opposite behaviour.

We conclude that the difficulty of formation of Ni (100) (2 x 2) - C could monitor the ease of formation of gaseous products in steam reforming.

Although the adsorption of oxygen at nickel surfaces has been studied extensively, the adsorption of water has not. Norton, Tapping and Goodall (23) using XPS find that both clean and oxidised nickel surfaces show a low reactivity towards H_2O at 295 K. In particular clean nickel surfaces are ∿ 10^3 times less reactive towards H_2O than to oxygen. The O (1s) binding energy was 531.3 eV, compared with 529.7 eV for oxygen in nickel oxide. The authors proposed that an NiO (H) species similar to Ni $(OH)_2$ has been formed. With LEED we have found that at 625 and 725 K on the (111) face a simple Ni (111) (2 x 2) structure is formed but the oxide structure does not develop as with oxygen itself. In addition the intensity-voltage distribution was not the same as for the (2 x 2) - O pattern which suggests that hydrogen is still associated with the oxygen.

The overall conclusion is that the interaction of water with clean nickel surfaces is weakly dissociative.

Having described the background information on the interaction of hydrocarbons and steam with nickel we can now turn to the limited information on the influence of alkali. No surface physics study of the influence of alkali on the interaction of hydrocarbons with nickel exists. However, a classical study of the effect of metallic potassium on the hydrogenation of ethylene over nickel has been performed by Shigehara and Ozaki (24). They found for a nickel powder with a surface area of 2.1 $m^2 g^{-1}$ onto which metallic potassium is evaporated to a coverage of one potassium atom per 5 $Å^2$.

1.   Neither self-hydrogenation nor polymerisation was detected after contacting with ethylene at room temperature.

2.   The rate of isotopic mixing in ethylene was reduced at 250 K ie the dissociation of ethylene on nickel is suppressed by the addition of potassium.

3.   The inhibition of hydrogen chemisorption is somewhat weaker.

The major conclusion is that the addition of potassium reduces the extent of C-H dissociations of ethylene on nickel (and thereby the subsequent polymerisation steps). The hypothesis for the action of potassium is that it causes a reduction of the unsaturation of the nickel surface via electron donation and not because the potassium layer restricts the access of the ethylene molecules to the nickel surface. This seems reasonable because APS (5) showed that the effective valency of calcium at a Pt (100) surface is + 1.5 and, as noted above, a small fraction of the sodium at nickel surfaces is not desorbed even at 900 K (18). It must be more strongly bound and hence ionic.

The influence of pre-adsorbed alkali on the interaction of water must be developed from the adsorption of oxygen. This has been studied with pre-adsorbed cesium at the (100) face (25). Papageorgopoulus and Chen found that 0.14 monolayers enhanced both the sticking probability of oxygen and the oxidation rate of nickel which was some thirteen times greater than in the absence of alkali. Similarly, enhanced dissociation of water should be induced by the presence of alkali.

Therefore the answer to the question whether alkali affects the interaction of hydrocarbons and steam is yes. The dissociation of the hydrocarbons is reduced while that of the steam is enhanced.

Why does alkali reduce the rate of gasification so much more than sulphur?

The loss of gasification activity caused by alkali is a net consequence of reduced hydrocarbon decomposition i.e. reduced CH_x formation in its presence which we have noted in the previous section. As CH_x is the reactive intermediate for gaseous product formation the net activity is reduced.

On the other hand, as we have seen in selective hydrogenation (5) the loss of activity caused by sulphur is structural and arises because sulphur atoms are adsorbed on the nickel surface and block the access of the hydrocarbon molecules to the nickel sites where they can dissociate to CH_x fragments. If the effect with sulphur was electronic sulphur in extracting electrons from the surface would increase its unsaturation and hence pro-

mote the rate of gasification. Structural poisoning by
sulphur under reaction conditions must be reduced by
reaction with oxygen dissociated from the steam. This
will always maintain a low number of free nickel sites
even with sulphur compounds in the feedstock.

Since alkali limits the formation of CH_x it thereby
reduces coke formation. It also promotes the removal of
carbon and even graphite (26). Presumably it aids the
latter by the dissociation of steam.

Thus the answer to the question is that the effect
of alkali is electronic/chemical while the effect of
sulphur is structural. Structural poisoning should reduce the rate of gasification to zero except that the
reactants cause its removal by oxidation.

(d) Summary of Metal Catalyst System Results

In all of the three cases, Ni/S, Pt, and Ni steam
reforming catalysts, we have shown how, in varying degrees, the use of surface science information can lead
to considerable improvements in the understanding of
catalysts. In one case - Pt - we have shown how the abstraction and use of precise information on an active
structure and its mode of formation led to a new catalyst,
spectacularly active for the reaction it was designed to
assist.

In the case of the Ni/S catalyst the amount of detail that has been acquired by LEED is such that we can
be confident that we have identified the actual site on
which reaction occurs. In the case of platinum two active
'sites' can now be postulated - the (111) face of Pt
itself and a $Pt/A\ell/O$ structure identified as yet only
by an hexagonal diffraction pattern. In all cases common
structures are found on different faces - ie reorientation has been shown to occur.

The Ni steam reforming catalyst was selected to
demonstrate how results obtained in experiments not
specifically aimed at a catalytic system could still be
used to attempt solution. Although it cannot be claimed
that the explanations attempted in this paper are unique
it clearly casts considerable fresh light on the system.

In all three cases the understanding is based on
comprehending the role of the promoter. The Ni/S and
Pt cases encourage us to think of promoters as carrying
out the structural modification of reorientation. The

steam reforming case encourages contemplation of the electronic role of promoters in affecting the catalytic reaction. The two are not necessarily separate, as has already been pointed out. The role of electronegative S
causing (111) to (100) face changes of Ni contrasts with
the reverse transition of Pt surfaces promoted by alkali,
alkaline earth and possibly aluminium ions - all electropositive in nature.

2.   NON-METALLIC CATALYST SYSTEMS

   Our approach to metallic catalyst systems has relied
extensively on structural information supported by spectral data. With non-metallic systems which are frequently
insulators this approach is no longer apt because of the
severity of sample charging problems with LEED. In addition, with non-metallic systems the chemical state during
reaction is less well defined than with metals. Thus a
major use of surface physics techniques in this area is
to define the state of the catalyst after activation and
by implication during reaction and a review of these developments forms the first part of this section.

   In no system has the role of the promoter been investigated. In NiW and CoMo systems the chemical state
has been discussed and this will form the second topic
of this section.

   Perhaps the area where surface science can make the
biggest impact in oxide catalysis lies in acid-base
(electron donor - electron acceptor) catalysed reactions.
The binding energies of the same element in different
compounds determined by ESCA is a direct determination
of the ability of that element to donate or accept electrons. Consequently, the binding energy can be used to
predict the reaction mechanism in, for example, elimination reactions.

1.   Definition of the State of the Catalyst after
                       Activation

                 (a) CoMo Catalysts

   CoMo catalysts are activated by treatment with a
sulphur compound, usually H_2S or thiophen, under reducing
conditions at a temperature near 700 K. The most unambiguous ESCA study of this sulphiding, in which reaction
occurred in a specially constructed chamber and the
sample was transferred without contact with the air, has

been described by Hercules and his coworkers (27). In
fact, these results do not differ significantly from
those in which the reactor and the spectrometer were separated, (28, 29) since the rate of reoxidation is slow.
During this activation molybdenum trioxide is converted
to molybdenum disulphide. However, sulphiding is never
complete. Hercules (27) reports that it is less than 62
per cent in 10 per cent H_2S after ten hours at 675 K and
a total pressure of one atmosphere.

Stevens and Edmonds (30) have studied the subsequent
reactivity of presulphided catalysts for both hydroge-
nolysis and hoydrogenation. In this work, more detailed
analysis of the spectrum suggested that not only MoS_2 but
also states akin to MoO_2 and MoS_3 were formed especially
at lower degrees of sulphiding. When chemical state and
reaction rate were compared, the first order rate constant for both hydrogenation, see Figure 6, and hydroge-
nolysis was proportional to the combined areas under
the two Mo^{4+} states.

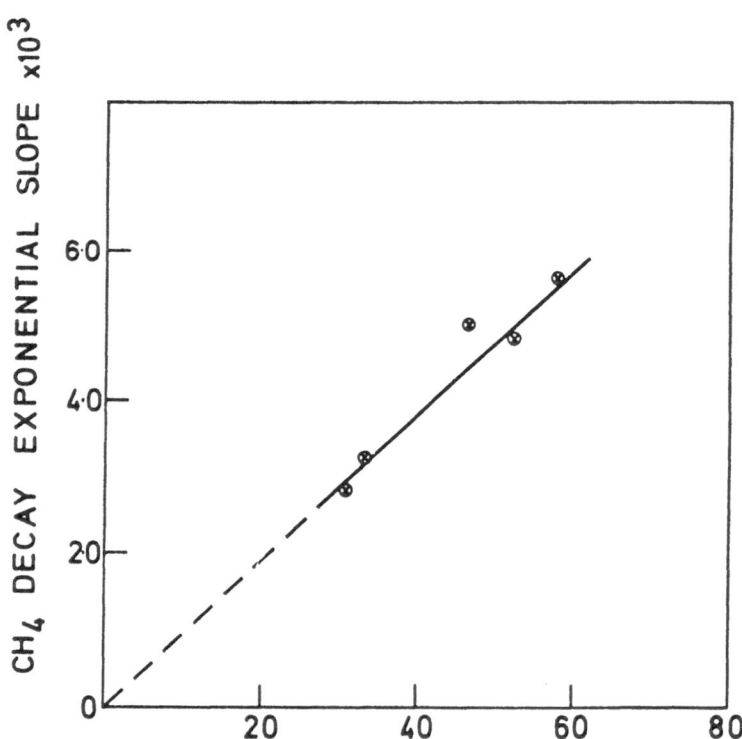

FIG 6   SURFACE COMPOSITION (% $Mo^{4+}-S+Mo^{4+}-O$) SURFACE
COMPOSITION AND RATE OF METHANE RELEASE

Prereduction does not improve the rate of sulphiding of CoMo catalysts. Patterson (27) found that activation with thiophen in hydrogen at 825 K achieved 65 per cent sulphiding in 35 minutes and 89 per cent sulphiding after 250 minutes. In contrast, following prereduction for 3 hours at 825 K, the subsequent activation with thiophen in hydrogen at the same temperature only gave 32 per cent sulphiding after 100 minutes. After 3 hours reduction at 775 K the surface molybdenum composition of the CoMo catalyst would have reached an equilibrium composition of~ 25 per cent Mo^{6+}, 35 per cent Mo^{5+} and 40 per cent Mo^{4+}. Similar conclusions with respect to the retarding effect of prereduction on sulphiding were drawn by Stevens and Edmonds (30), when sulphiding with H_2S.

In the interpretation of the Mo (3d) spectrum following sulphiding, Walton (31) has noted the relatively small spread of Mo (3d) binding energies within the series of thio-anions MoS_4^{2-}, $MoOS_3^{2-}$ and $MoO_2S_2^{2-}$ reported by Muller (32). As a consequence he has argued that the presence of mixed molybdenum oxysulphide species on the catalyst surface cannot be ruled out as an alternative interpretation of the MoS_2 peak. Muller (32) did, in fact, show that, although the spread between the thio-anions is only 1 eV, their binding energies lie at a minimum of 1 eV and at a maximum 2 eV below MoO_3. Thus there can be no ambiguity in the assignment of MoS_2 which lies at least 3 eV below MoO_3. Nevertheless, it is possible that these compounds are formed during intermediate stages of sulphiding. The more detailed analysis of Stevens and Edmonds (30) did deconvolute a state between MoO_3 and MoS_2 which was tentatively assigned to MoO_2. This could be a thio-anion.

(b) NiW Catalysts

NiW catalysts are activated either by reduction or sulphiding, prior to use as hydrogenation catalysts. In another study with direct sample transfer, Hercules (33) has investigated both activation procedures. After 6 hours reduction in an atmosphere of hydrogen at 725 K the tungsten trioxide remained unchanged while approximately 80 per cent of the nickel oxide was converted to the metal. Since the unsupported catalytic components are more readily reduced these results support the hypothesis that the WO_3 forms a monolayer surface interaction complex with the alumina and that at least some of the nickel oxide is incorporated in a spinel.

The sulphiding study with NiW was more detailed than with CoMo. Tungsten trioxide was converted to WS_2 using

both H_2S and thiophen and complete sulphiding was observed with 9.2 per cent H_2S in hydrogen at 725 K after
15 hours. The rate of sulphiding increases with reaction
temperature and concentration of H_2S. It was always higher
with H_2S than with thiophen, at the same concentration
and temperature. This is consistent with the prior conversion of thiophen to H_2S before reaction.

If, as with CoMo catalysts, the active state for
hydrogenation is W^{4+}, these results indicate why sulphiding NiW catalysts is the preferred mode of activation.
Reduction does not lead to the formation of W^{4+} in marked
contrast to the CoMo system.

Hercules (33) believes that an intermediate WO_xS_y
state is also formed with NiW catalysts during sulphiding
of previous section. His evidence is that the ratio of
intensities for S (2p): $W(4f_{7/2})$ (ex WS_2) is always less
than the value for WS_2 itself.

(c) Phillips Catalysts

The normal Phillips catalyst contains ~ 1 per cent
Cr as CrO_3 supported on silica. It is activated by
calcining above 775 K and then by reduction with either
ethylene itself or CO. Because chromium is difficult to
detect at 1 per cent weight most results are reported
at higher loadings, typically 9 per cent (34, 35). These
show that calcining at 775 K and above causes the reduction of Cr^{6+} to Cr^{3+}. Further reduction to Cr^{2+} was noted
after 3 hours reaction with CO at 575 K (35). However, in
Cimino's work (34), the Cr^{6+} state showed greater stability with calcining as the chromium loading decreased.
This trend was clearly observed from 9 per cent weight
to 3 per cent weight Cr but although Cr^{3+} could be detected at 1 per cent weight Cr the relative proportions
of Cr^{6+} and Cr^{3+} could not be quantified. It is evident
that better sensitivity is required to give an unambiguous answer at the commercial catalyst loading of the
exact chemical state of chromium following activation.
Every indication from this work suggests that it is Cr^{2+}.

2.   The Chemical State of Promoters

(a) Cobalt in CoMo Catalysts

The chemical state of cobalt in CoMo catalysts has
not yet been resolved. Both Friedman (29) and Patterson
(27) find that in the fresh catalyst the cobalt is pre-

sent as both the oxide and as the aluminate. Upon reduction and sulphiding it is only the oxide that is converted to sulphide. All the cobalt is never sulphided
and the unreacted cobalt is believed to be in the aluminate which was shown to sulphide very slowly (27). On
the other hand in a recent detailed study of the cobalt
signal Brinen (36) now concludes that cobalt is present
in a reduced state. This conclusion is based on an extensive study of CoMo catalysts with a range of cobalt
loadings and was deduced by studying the Co ($2p_{3/2}$) -
Co ($2p_{1/2}$) and satellite separations. Corroboration was
found in the observation that the Mo to S ratio remained
close to that for MoS_2 independent of cobalt loading.

The role of cobalt is equally obscure. Brinen (28)
has made the enigmatic statement that the most significant information obtainable from the measurement of
ESCA spectra of catalysts is related to the relative
concentrations of promoters on the surface. This he
exemplified with a CoMo catalyst having both cobalt and
phosphorus as promoters, the latter causing shell im-
pregnation. However, this probably is only related to
diffusion limitations in catalyst extrudates and gives
no direct information on the function of the promoter.
Similarly Friedman (29) could find no obvious correlation
between the activity of a CoMo catalyst for hydrodesul-
phurisation and the sulphided cobalt/molybdenum ratio.
Indeed a better correlation was found with cobalt aluminate. However, later results from the same group (37)
based on more detailed studies reversed this conclusion.

(b) Nickel in NiW Catalysts

The chemical state of nickel in NiW catalysts following sulphiding is more clearly resolved (33). Nickel
sulphides, Ni_3S_2 and NiS, are formed fairly quickly and
their intensities suggest that approximately 70 per cent
of the nickel has been sulphided. The proportion of NiS
depends on the concentration of H_2S. Thus with thiophen
only Ni_3S_2 is formed but as the concentration of H_2S
increase so does NiS. Sulphiding changes the ratio of
Ni:Mo since Ni^{2+} ions diffuse to the surface. This was
not noted with CoMo.

(c) Conclusions

Surface spectroscopic techniques have not yet revealed the role of the promoter in sulphide catalysis.
Various models have been proposed - the intercalation
model (38) the monolayer model (39) contact synergy (40),

vacancies (41,42) and a dual site model (43), consisting
of anion vacancies for heterolytic hydrocarbon adsorption
and surface oxide-sulphide sites for hydrogen adsorption
and transfer. None is universally accepted. Progress has
been made but further detailed studies are required and
must include the conduction and valance bands.

3.   OXIDE CATALYSIS

     In an elegant paper, Noller (44) and his coworkers
have recently progressed the potential of ESCA as a tool
for predicting activity and selectivity in acid-base
catalysis. They have argued that, provided the structure
of the transition state is known or reasonable assumptions
are possible, the catalytic properties of the solid
should depend upon two characteristics of the surface
atoms involved in the bonds - the bond forming capacity
and the accessibility. Ignoring the second factor they
developed at detailed picture based on the bond forming
capacity using elimination reactions as the main example.

     The simple transition state adsorption structure
they have drawn is:

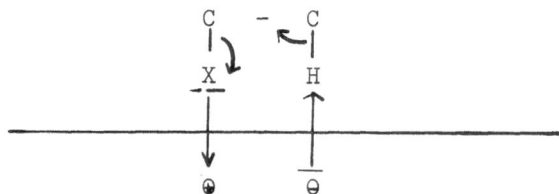

In bond rupture, X and H are abstracted as ion X^- and H^+
and the binding energy electron pair is taken over by
one of the atoms to be separated. Both X^- and H^+ interact with the catalyst surface in either an electron pair
acceptor (EPA) or an electron pair donor (EPD) interaction. The EPD X has to interact with an EPA site at
the surface which is usually a cation whereas the EPA
H interacts with an EPD site, an anion. Thus every cation-
anion pair in the surface is capable of being an active
centre provided it is accessible.

     The activity as well as the mechanism, which is
closely related to the selectivity should depend on the
EPA strength and EPD strength of these centres. Three
possible mechanisms are usually distinguished which are

termed E1, E2 and E1 CB and differ in the timing of the
bond rupture.

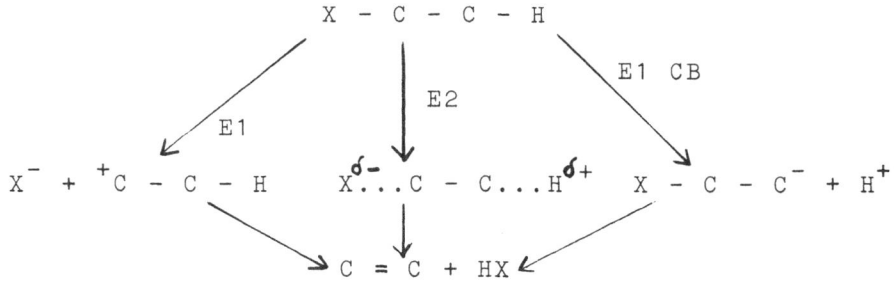

The following factors affect the mechanisms above as
indicated:

1. Increasing EPA strength of cation $\longleftarrow$
2. Increasing EPD strength of anions $\longrightarrow$
3. Increasing C-X bond strength $\longrightarrow$
4. Increasing C-H bond strength $\longleftarrow$

It was noted that the E1 mechanism involves a carbonium
ion intermediate which can undergo isomerisation. E2 is
a one-step mechanism without an intermediate. While in
E1 CB the most acidic proton is abstracted preferentially.
Therefore but-1-ene is the preferred product from 2-X-
butane and dehydrogenation can also be observed.

     The problem is to determine the EPA and EPD strength
of the surface sites particularly in compounds of the
same element. The binding energy and the bond forming
capacity of various cations and anions are related in the
following manner. Decrease of the binding energy of a
cation indicates a decrease in its EPA strength and vice
versa, while decreasing the binding energy of an anion
indicates higher EPD strength.

     An example is the reaction of butan-2-ol over three
magnesium compounds, MgO, MgSO$_4$ and MgHPO$_4$. Their binding energies are given below. Arrows indicate increasing
EPA for the cation and EPD for the anion.

Compound	Mg(2p)	O(1S)	EPA	EPD
MgO	48	530.2		
MgSO$_4$	49.1	531.1		
MgHPO$_4$	49.8	531.8		

Over MgO, an E1 CB mechanism is anticipated because of
the high EPD of the anion. Butan-2-ol was almost exclusively dehydrogenated over MgO (96 per cent dehydrogenation at 625 K) in agreement with the above principles.
Over MgSO$_4$, and MgHPO$_4$, only dehydration took place according to an E2 mechanism with notable E1 character at
higher temperatures.

Noller used ESCA to determine this information and
hence understand the catalytic behaviour of magnesium
compounds towards the dehydration of alcohols, via the Mg
(2p) binding energies; the behaviour of a range of oxides
towards the same reaction via their O(1S) binding energies;
the behaviour of alumina relative to spinels via Al(2p)
binding energies and the behaviour of various zeolites
via their cation binding energies.

CONCLUSION

In this review we have shown that in metal catalyst
systems the role of the promoter can be mainly structural, or mainly electronic or indeed a combination of
the two factors depending on the system.

The study of sulphided and oxide promoted catalysts
is still in its infancy. In many systems the active
chemical state of the main component and the promoter
has not been defined. Future work will require better
controlled experiments, in which there is no possibility
of chemical change in transit between the reactor and
the spectrometer. Where possible structural and spectroscopy techniques should be combined and, because vacancy
models may be important, the valence band must be studied
with greater care. The recent study with unpromoted inorganic systems is a most encouraging sign that surface
science techniques will surely lead to a greater understanding of catalysts even in this difficult area.

ACKNOWLEDGEMENT

The authors thank the Directors of BP for permission
to publish, and their colleagues Dr. A.I. Foster, Dr.
G.C. Stevens, Mr. S.R. Tennison and Miss S. Jain for
their parts in the development of many of the items
noted in this paper.

REFERENCES

(1) Bourne, K.H., Pitkethly, R.C. and Holmes, P.D., Proc 3rd Int Conf on Catalysis, 1965, 2, 1400.

(2) Duyverman, C.J., Vlugter, J.C. and van der Weerdt, W.J., Proc 3rd Int Conf on Catalysis, 1965, 2, 1416.

(3) Edmonds, T., McCarroll, J.J. and Pitkethly, R.C., Ned Tijd voor Vacuumtechniek, 1970, 8, 162.

(4) Edmonds, T., McCarroll, J.J. and Pitkethly, R.C., J Vac Sci Tech, 1971, 8, 68.

(5) McCarroll, J.J., Surface Science, 1975, 53, 297.

(6) Fedak, D.G. and Gjostein, N.A., Acta Metallurgia, 1967, 16, 827.

(7) British Patent 1 468 441, Assigned to BP.

(8) British Patent 1 471 233, Assigned to BP.

(9) Foster, A.I., McCarroll, J.J. and Tennison, S.R., Unpublished Data.

(10) Jain, S. and McCarroll, J.J. Unpublished Data.

(11) PHI Handbook of Auger Electron Spectroscopy, Published by Physical Electronics Industries Inc, 6509 Flying Cloud Drive, Eden Prairie, Minnesota 55 343, 1976.

(12) Dautzenberg, F.M., den Otter, G.J. and Walters, H.B.M., Rideal Conf, London, April 1977.

(13) Bridger, G.W. and Chinchen, G.C., Catalyst Handbook, Wolf Scientific Books, 1970.

(14) Topsoe, H., IGE Journal, 1966, 6, 401. Rostrup Nielsen, J.R., J Catal, 1973, 3, 173.

(15) US Patent 3 511 624. Phillips, T.R., Mulhall, J. and Turner, G.E., J Catal, 1969, 15, 233.

(16) US Patent 3 429 680. Okagani, A., Uemoto, K. and Morikawa, K., ACS 165th National Meeting Dallas, April 1973, Div Petr Chem, 1973, 18 (2), 401.

(17) Romanowski, W., Surface Science, 1969, 18, 373.

(18) Gerlach, R.L. and Rhodin, T.N., Surface Sci, 1969, 17, 32.

(19) Andersson, S. and Kasemo, B., Surface Sci, 1972, 32, 78.

(20) Papageorgopoulos, C.A. and Chen, J.M., Surface Sci, 1975, 52, 40.

(21) Bertolini, J.C. and Dalmai-Imelik, G., Structures and Propriété des Surfaces des Solid, 1970, p 139.

(22) Edmonds, T., McCarroll, J.J. and Pitkethly, R.C., Meeting on Low Temperature Carbon Deposition, University of Glasgow, 1972.

(23) Norton, P.R., Tapping, R.L. and Goodall, J.W., Surface Sci, 1977, 65, 13.

(24) Shigehara, Y. and Ozaki, A., J Catal, 1973, 31, 309.
(25) Papageorgopoulos, C.A. and Chen, J.M., Surface Sci, 1975, 52, 53.
(26) Cairns, J.A., Keep, C.W., Bishop, H.E. and Terry, S., J Catal, 1977, 46, 120.
(27) Patterson, T.A., Carver, J.C., Leyden, D.E. and Hercules, D.M., J Phys Chem, 1976, 80, 1700.
(28) Brinen, J.S., J Electron Spect, 1974, 5, 377.
(29) Friedman, R.M., Declerck-Grimée, R.I., and Fripiat, J.J., J Electron Spectr, 1974, 5, 437.
(30) Stevens, G.C. and Edmonds, T., J Catal, 1976, 44, 488.
(31) Walton, R.A., J Catal, 1976, 44, 335.
(32) Muller, A., Jorgensen, C.K. and Dieman, E., Z Anorg Allg Chem, 1972, 391, 38.
(33) Ng, K.T. and Hercules, D.M., J Phys Chem, 1976, 80, 2094.
(34) Cimino, A., De Angelis, B.A., Luchetti, A. and Minelli, G., J Catal, 1976, 45, 316.
(35) Best, S.A., Squires, R.G. and Walton, R.A., J Catal, 1977, 47, 292.
(36) Brinen, J.S., 174th ACS Meeting, Chicago, September 1977.
(37) Gajardo, P., Declerk-Grimmée, R.I., Delvaux, G., Olodo, P., Zabala, J.M., Canesson, P., Grange, P. and Delmon, B., 2nd International Conference on the Chemistry and Uses of Molybdenum, Oxford, 1976.
(38) Farragher, A.L. and Cossee, P., Proceedings of the 5th International Congress on Catalysis, Hightower, J.W. (Editor), 1973, p 1301.
(39) Schuit, G.C.A. and Gates, B.C., AIChE Journal, 1973, 19, 417.
(40) Canesson, P., Delman, B., Delvaux, G., Grange, P. and Zabala, J.M., Proceedings of the 6th International Catalysis Congress, 1977, p 929.
(41) Aoshima, A. and Wise, H., J Catal, 1974, 34, 145.
(42) Wentrcek, P.R. and Wise, H., J Catal, 1976, 45, 349.
(43) Massoth, F.E. and Kibby, C.L., J Catal, 1977, 47, 300.
(44) Vinek, H., Noller, H., Ebel, M. and Schwarz, K., JCSFT, 1977, 73, 734.

SPECTROSCOPY OF SURFACE VIBRATIONS

Stig Andersson

Department of Physics,
Chalmers University of Technology
Fack, S-402 20 Göteborg 5, Sweden

1. INTRODUCTION

The intention of this paper is to give a brief account of recent progress in the study of vibrational spectra of atoms and molecules adsorbed on single crystal surfaces of metals. Such adsorption systems offer unique possibilities to investigate ordered arrangements of adsorbed species and the spectroscopic information achieved can be discussed in terms of structural models suggested from electron diffraction analysis (LEED, RHEED) and electron structure models derived from electron spectroscopic analysis. Vibrational spectroscopy is certainly highly sensitive to the adsorption site symmetry and there exist a vast number of important structural problems related to different adsorbate structure – substrate surface configurations. At present most such analysis work is performed empirically relying on vibrational spectroscopic data for known molecules and solids. Chemisorption on metal surfaces constitute, however, a very metal rich chemical phase and specific coordinations do occur, that are never met in other systems. For that purpose vibrational frequencies derived from adsorption energy curves calculated for high symmetry adsorption sites would resolve ambiquities. Such calculations are within the grasp of modern chemisorption theory. The analytical power of vibrational spectroscopy as applied to chemisorption is of course not only related to questions concerning the adsorption site (which may be most relevant for atoms and simple molecules) but most importantly to the capability of identifying species formed during any surface reaction /1/.

Several different spectroscopic techniques are currently being used to determine vibrational spectra of adsorbed species: infrared

absorption /2,3/ and reflection spectroscopy /4,5/, high resolution
electron energy loss spectroscopy /1/ inelastic electron tunneling
spectroscopy /6/, inelastic neutron scattering spectroscopy and
Raman spectroscopy. Of these methods, infrared reflection spectroscopy (IRS) and high-resolution electron energy loss spectroscopy
(EELS) are particularly well suited to determine vibrational spectra
of atoms and molecules adsorbed on single crystal metal surfaces.
IRS has primarily been applied to the study of CO chemisorption
/7-9/ while EELS has found a wider area of applications comprising
atomic adsorbates like hydrogen, carbon and oxygen but also various
molecular adsorbates. This paper will deal with EELS as applied to
comparatively simple atomic and molecular adsorption system: H_2, O_2
and CO chemisorption on Ni(100) and W(100). Vibrational spectroscopy of simple chemisorption systems will contribute to the pro-
found understanding of chemisorption where knowledge about structure, vibrations and electron structure are basic requisities but
such systems will also provide a useful test ground for the validity
of empirical assignments of the character of a chemisorption bond
as deduced from experimental vibrational frequencies. Such knowledge will facilitate more reliable interpretations of infrared
adsorption spectroscopic data. This technique has been the prime
analytical tool to determine vibrational spectra of adsorbed species
and is undoubtedly the practical method to study the nature of
adsorbed species on e.g. an operating catalyst.

2. ELECTRON ENERGY LOSS SPECTROSCOPY

In EELS the vibrational excitation occurs via inelastic
scattering of the interacting electron. Both emission and absorption of vibrational quanta are possible. Depending on the cross
section multiple excitations may also be observed. In a general EELS
experiment the angular and energy distribution of the scattered
electrons are analyzed as a function of energy and direction of the
incident electrons. Knowledge of angular and energy dependence is
of particular importance when one wants to explore resonant scattering phenomena which in fact may provide very detailed information
about the orientation of adsorbed molecules such as CO, N_2, NO and
O_2 /10/. Such scattering is due to short range interaction related
to the formation of a compound state involving the molecule and the
interacting electron /11/. In general one assumes that the prominent contribution to the vibrational excitation cross section derives from the long range dipole interaction between the surface
oscillators /12-16/ (the adsorbed atoms and molecules) and the
electric field induced at the surface by the incident electron. At
frequencies corresponding to vibrational excitations (i.e. <0.5 eV)
a metal substrate can reasonably well be considered as a perfect
conductor. Accordingly the electric field due to the incident

electron is given by the classical image construction

$$\mathbf{E} = \frac{1}{4\pi\varepsilon_o} \, \mathbf{z} \, \frac{2\mathbf{z}\cdot\mathbf{r}}{|\mathbf{r}|^3}$$

where $\mathbf{z}$ is a unit vector normal to the surface and $\mathbf{r}$ is the position of the electron with respect to an oscillator in the surface plane. Thus the electric field vector is normal to the surface and only vibrational modes that are associated with dipole changes, $\frac{d\mu}{dz}$, normal to the surface can couple to the field and extract energy from it. Treating the surface as a perfect electron reflector and the incident and scattered electron beams as plane waves the cross section for inelastic scattering from a point dipole oscillator in the surface plane is given by /16/

$$\frac{d\sigma}{d\Omega} \sim (\frac{d\mu}{dz})^2 \, \frac{|\mathbf{k}_1|}{|\mathbf{k}_o|} \cdot \frac{1}{\cos\alpha} \, \frac{|\mathbf{k}_1'' - \mathbf{k}_o''|^2}{|\mathbf{k}_1 - \mathbf{k}_o|^4}$$

where $\mathbf{k}_o$, $\mathbf{k}_1$ and $\mathbf{k}_o''$, $\mathbf{k}_1''$ are the wavevectors and their surface components of the incident and scattered electrons respectively and α is the angle of incidence with respect to the surface normal. This cross section is shown schematically in Fig. 1 for primary energy E_o = 1.4 eV, α = 47.7^{o} and a vibrational quantum energy $\hbar\omega$ = 0.25 eV /16/. The cross section peaks strongly in lobes around the specular direction.

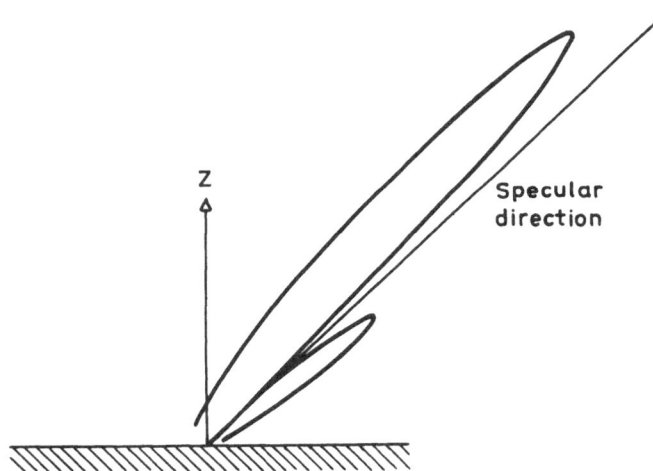

Fig. 1. Angular distribution of electrons scattered inelastically from a point dipole on a metal surface (from Ref. 16).

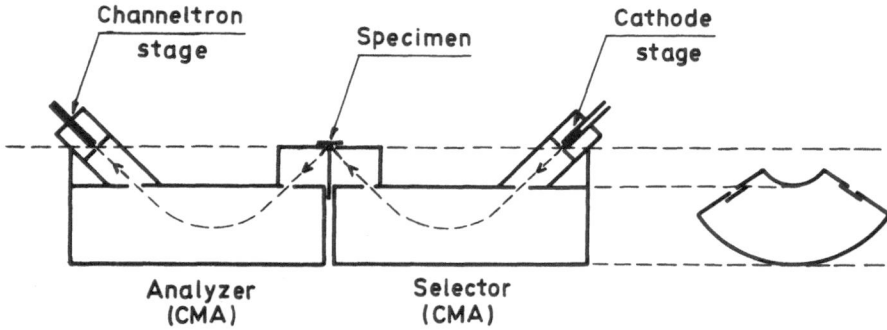

Fig. 2. Schematic drawing of the experimental EELS arrangement used by the author. The scattering angle is fixed at
95.4°.

The experimental EELS arrangement used by the author is
sketched in Fig. 2. The collimated monoenergetic electron beam
formed by the selector impinges on the specimen surface at the fixed
angle $\alpha = 47.7°$ and the electrons scattered in a narrow angular
interval around the specular direction are analyzed as concerns
their energy distribution. Such an energy loss spectrum provides
knowledge about vibrational excitations for the scattering parameters given. The selector and analyzer are of electrostatic cylindrical mirror construction. The angular half-width of the electron
trajectories that leave the selector and enter the analyzer is 3°
and the overall energy resolution is 0.7% i.e. 7 meV FWHM at a
primary energy of 1 eV.

3. CHEMISORPTION OF HYDROGEN AND OXYGEN

Hydrogen chemisorption on metal surfaces plays a key role in
surface physics and chemistry. It is simple enough to be treated in
detail by theory and is simultaneously of great practical importance
in a vast number of surface reactions. Knowledge about the state
and site of chemisorption is of course of central interest and the
way EELS can contribute will be illustrated below for hydrogen
adsorbed on Ni(100) and W(100).

Fig. 3 shows a set of EEL spectra for H_2, D_2 and HD chemisorbed
at 200 K on the clean Ni(100) surface /17/. The two spectra for H_2
correspond to two different surfaces converages 0.5 θ_m and 0.9 θ_m
where θ_m is the maximum coverage of the β-adsorption state /18,19/.
The corresponding H_2 exposures were 0.5 L and 5 L respectively.
(1 L = $1\cdot10^{-6}$ torr sec) and the relative coverages were determined
from desorption measurements. LEED shows formation of a quasi-

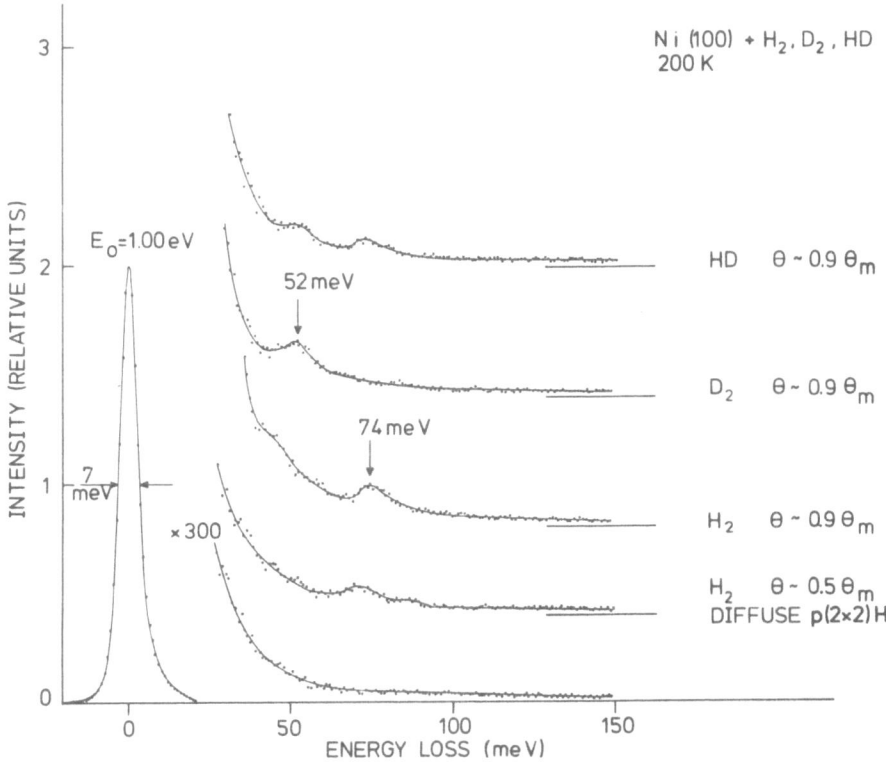

Fig. 3. EEL spectra of H_2, D_2 and HD adsorbed on Ni(100) at
200 K. Angle of incidence α = 47.7°, the analyzer accepts
the specularly reflected beam.

ordered p(2x2) structure around 0.5 θ_m which slowly falls off as
the hydrogen coverage increases.

At θ = 0.9 θ_m the EEL spectra for H_2 and D_2 reveal vibrational
losses at 74 ± 1 meV and 52 ± 1 meV, the ratio being close to $\sqrt{2}$
as predicted by the mass ratio. The corresponding spectrum for HD
shows loss lines at 73 ± 1 meV and 52 ± 1 meV i.e. within the
experimental accuracy the same energies as for H_2 and D_2 separately.
No high-frequency H-H, D-D or H-D stretching vibrational losses
have been observed. These observations provide an almost definite
spectroscopic proof that the adsorption is dissociative as evidences
from adsorption-desorption kinetics suggest /18,19/. The possibility
of an accidental degeneracy of the vibrational energies for dipole
active modes of molecularly adsorbed H_2, D_2 and HD is very small
(see below for W(100)).

TABLE I

Adsorption energies, E, and vibrational energies, ℏω, at
equilibrium positions, h_o, of hydrogen atoms adsorbed on
Ni(100) (from Ref. 20)

Adsorption site	E (eV)	h_o (Å)	ℏω (meV)
top (A)	2.6	1.50	157
bridge (B)	2.5	1.15	105
center (C)	2.3	0.75	69

In Table I is listed adsorption energies, E, equilibrium positions h_o and vibrational excitation energies, ℏω, obtained from a
semi-emperical calculation /20/ for hydrogen adsorbed at the three
high symmetry sites, top (A) bridge (B) and center (C) of a Ni(100)
surface. These sites are shown in Fig. 4 and the vibrational motion
considered is perpendicular to the surface plane. The vibrational
excitation energies were derived by the author from the adsorption
energy curves E(h) /20/ where h is the height of the H atom above
the center of the Ni surface plane:

$$\hbar\omega = \hbar\sqrt{\frac{k}{\mu}}$$

$$k = \frac{d^2E}{dh^2}\bigg|_{h_o}$$

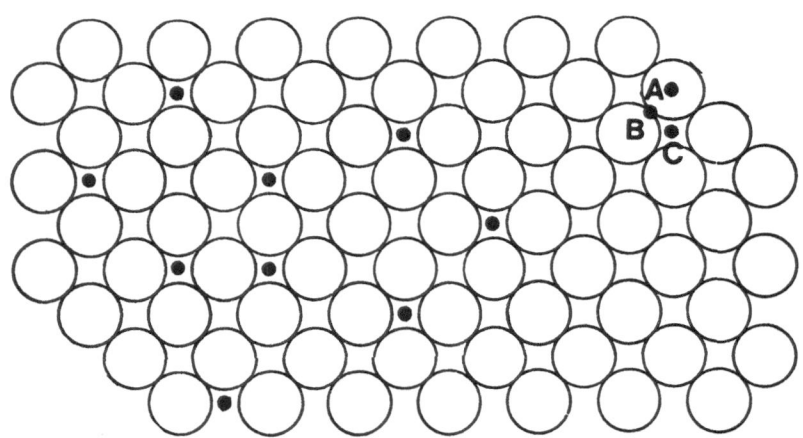

Fig. 4. Model structure of the quasiordered p(2x2)H
structure on Ni(100). The small filled circles represent
the hydrogen atoms. A, B, and C are three possible high
symmetry sites.

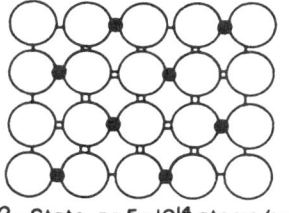
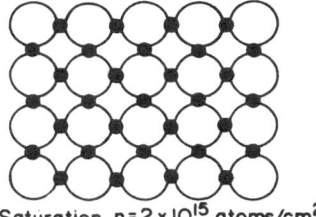

β_2 State, n = 5 x 10^{14} atoms/cm^2 Saturation, n = 2 x 10^{15} atoms/cm^2

Fig. 5. Model structures for hydrogen adsorption on W(100)
(from Ref. 21a). The EEL spectra suggest that the β_2 state
occupies the on top sites rather than the bridges (Ref. 26).

where k is the force constant and μ is the reduced mass of the
system (i.e. ~ the mass of the hydrogen atom). The experimental
figure agrees remarkably well with the value calculated for the
centered site. The quasiordered p(2x2)H structure produced a loss
line at 72 ± 1 meV i.e. close to the value for the higher coverage
and the model structure in Fig. 4 accordingly adopts the center
site. Thus for hydrogen adsorption on Ni(100) one kind of adsorption site characterises the β-adsorption state.

     It is interesting to compare the observations for Ni(100) with
those for W(100). Hydrogen chemisorption on this surface has been
extensively investigated using a number of different experimental
techniques and these observations are summarized in several recent
reviews /21/. Two adsorption states β_2 and β_1 are recognized
The saturation coverage has been found to be θ = 2 i.e.2 hydrogen
atoms per W surface atom /22/. LEED shows the formation of a c(2x2)
structure anticipated to correspond to θ = 0.50 at the completion of the β_2 adsorption state /23/. The adsorbate induced LEED
reflections then split, and the pattern transforms to (1x1) at
saturation. The proposed structure models are shown in Fig. 5.
Vibrational excitations have been observed in inelastic field
emission /24/ and by EELS /25-27/. The two absorption states have
been distinctly observed in recent high-resolution EELS work
/26,27/. Fig. 6 shows a series of spectra for H$_2$ and HD adsorption
obtained by Adnot and Carette /27/. The H$_2$ spectra show a progressive
transition from a spectrum with a loss line at 159 meV at low coverage to a loss at 132 meV at saturation coverage as was observed
previously by Froitzheim, Ibach and Lehwald /26/. The vibrational
loss lines for HD just coincides with those for H$_2$ and D$_2$ (as for
Ni(100)). Different molecular configurations were investigated /27/
for a range of reasonable parameters and it was found that an
accidental degeneracy was not possible i.e. the spectra correspond
to dissociative chemisorption. The 132 meV vibrational loss for
H at θ = 2 was interpreted /26/ to correspond to the bridge bond
model in Fig. 5 since that model is compatible with the requirements;

atomic adsorption at one type of site. The binding energies of the
two β states are quite similar so the 159 meV loss of the $β_2$ state
was accordingly proposed /26/ to correspond to H atoms bonded on
top of the W atoms (see the sequence in Table I) and this should be
the adsorption site of the W(100)c(2x2)H structure in Fig. 5.

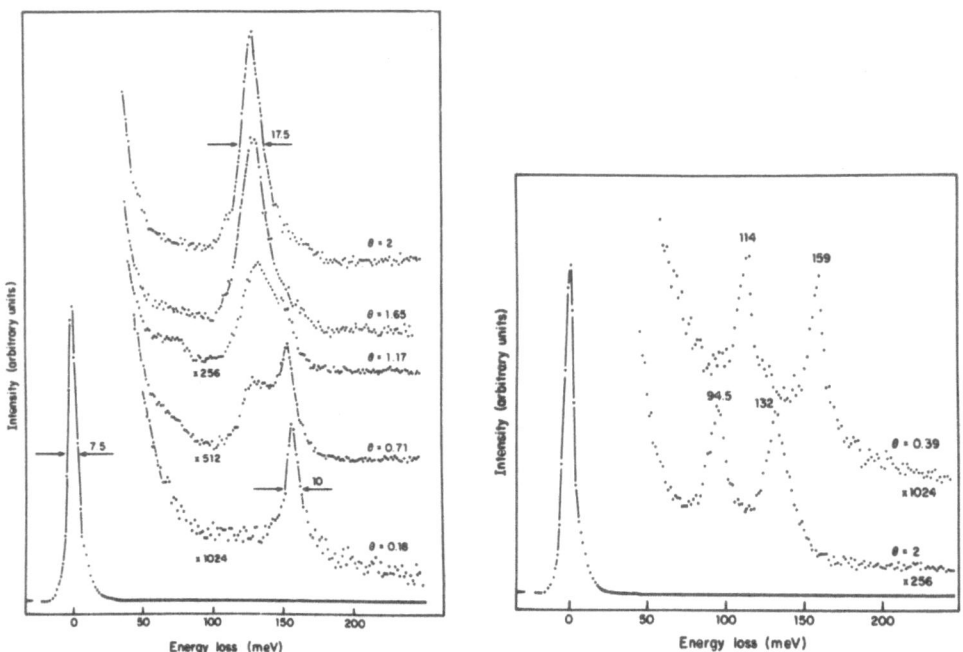

Fig. 6. EEL spectra of H_2 and HD adsorbed on W(100). The
primary electron energy is 5 eV and the angle of incidence
is 45° (from Ref. 27).

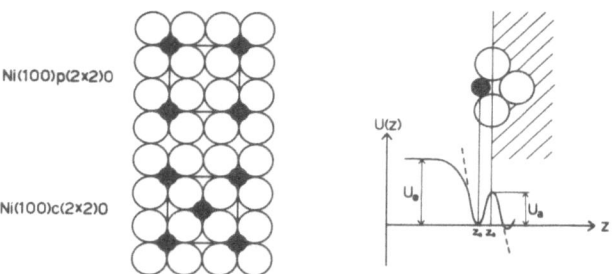

Fig. 7. Models for the Ni(100)p(2x2)O and c(2x2)O structures
and a model potential well for the adsorbed oxygen atom.

It is interesting to notice that EEL spectra determined at low coverages of hydrogen on W(110) and W(111) /28/ also reveal vibrational losses around 160 meV which suggests that the initial adsorption state on these three low index W planes occupies the top site.

The chemisorption of oxygen on single crystal surfaces of Ni and W has been the subject of comprehensive experimental work. Early LEED experiments of oxygen chemisorption on the clean Ni(100) surface revealed the formation of two consecutive surface structures denoted p(2x2)O and c(2x2)O. In more recent LEED investigations these structures have been the subject for structure determinations by means of dynamical LEED intensity analysis and several investigators have arrived at the same structural interpretation /29,30/. The structure models adopted are shown in Fig. 7. One finds for both structures a separate O layer situated 0.9 Å outside the center of the last Ni layer with the O atoms occupying the centered site.

The EEL spectra in Fig. 8 show vibrational loss lines at 53 meV and 40 meV for the p(2x2)O and c(2x2)O structures respectively /31/. These frequencies are considerably lower than that for a free NiO molecule /32/, 76 meV, and the surface mode of an epitaxial NiO(100) film /33/, 68 meV. This suggests that the chemisorbed O is multiply bonded in the Ni(100) surface. Since only one mode is observed it is attributed to the excitation of a motion of the O atoms perpendicular to the surface plane. The large difference in vibrational excitation energy between the p(2x2)O and c(2x2)O structures is intriguing. For a simple harmonic oscillator the difference would imply a decrease in the force constant by about 80%. Unfortunately there are no reliable binding energy measurements for these systems. LEED and electron spectroscopic measurements indicate, however, no

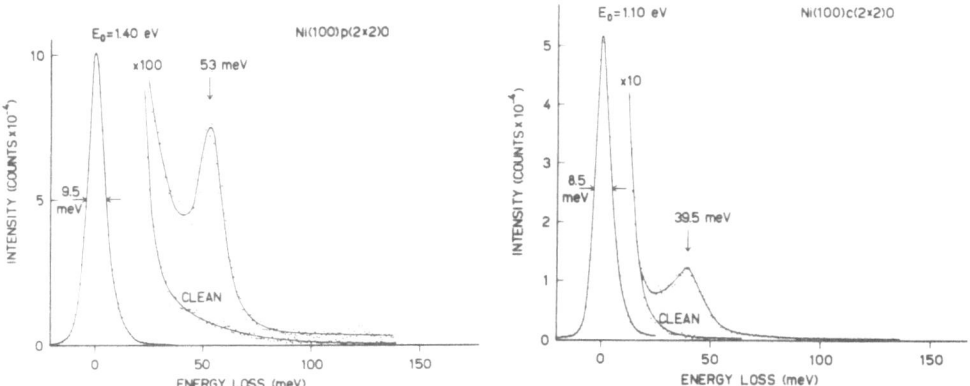

Fig. 8. EEL spectra from the Ni(100) p(2x2)O and c(2x2)O structure (conditions; as Fig. 3.).

substantial change in structure or electron structure. It is known
that c(2x2)O transforms to p(2x2)O at elevated temperatures primarily because of oxygen dissolution into the bulk of the Ni. Thus the
potential barrier towards the Ni surface may be rather low as
sketched in the model potential of Fig. 7. If the barrier height,
U_a, is about 2 eV for the p(2x2)O structure and the binding energy,
U_e, is lowered by only 1 eV in going to c(2x2)O thus reducing U_a
to about 1 eV, the observed change in vibrational excitation energy
is in fact accounted for.

Oxygen chemisorption on W(100) is considerably more complicated.
Desorption and electron spectroscopic measurements suggests formation of at least two different binding states at room-temperature.
At the coverage θ = 0.5 a two-domain (4x1) LEED structure is observed
/34,35/, which has been proposed to correspond to double rows of
oxygen atoms (to account for θ = 0.5 and the (4x1) symmetry. Fig. 9a
show a set of EEL spectra due to Froitzheim, Ibach and Lehwald /36/.
The spectra were recorded for increasing oxygen exposures at 300 K
substrate temperature and substantiate in a dramatic fashion the
complexity of this adsorption system. The spectral observations have
been discussed in terms of the structure models in Fig. 9b /36/. At
the low coverage, θ = 0.17, i.e. well below the formation of the
(4x1) LEED pattern just one vibrational loss line at 75 meV is observed. This line is tentatively attributed to the vibrational ex-

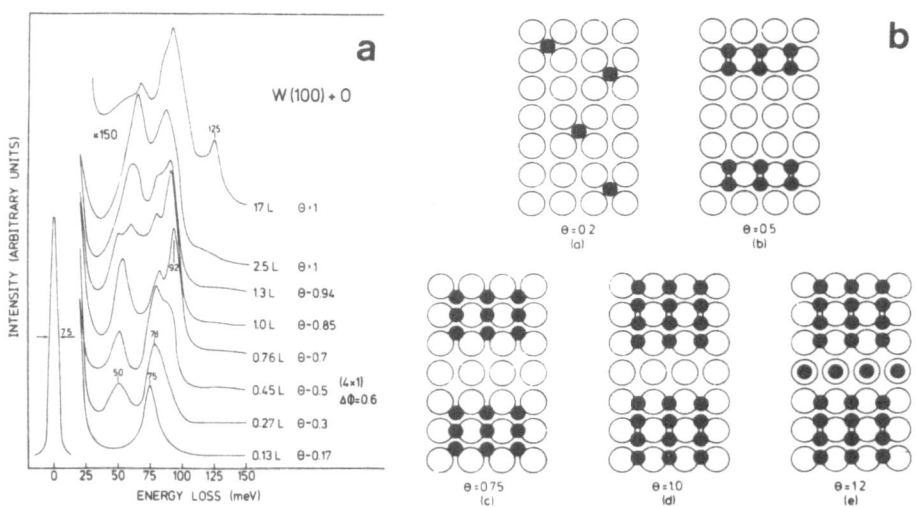

Fig. 9a. EEL spectra of oxygen adsorbed on W(100) at 300 K.
Primary electron energy 5 eV, angle of incidence 70°.

9b. Model structures for oxygen adsorbed on W(100) at
300 K (from Ref. 36).

citation of O atoms adsorbed in the fourfold center site (Fig. 9b,
θ = 0.2). When the coverage increases to 0.3 a new loss line appears
at 50 meV and the 75 meV loss shifts to 78 meV. This spectral
change is related to the formation of the (4x1) structure and the
spectrum is consistent with the double row model (Fig. 9b, θ = 0.5)
for this structure. The (4x1) unit mesh contains two O atoms in C_{2v}
symmetry which is compatible with the observation of two modes. The
appearance of a loss line at 92 meV is proposed /36/ to be due to
O atoms adsorbed in pure bridge sites (Fig. 9b, θ = 0.75) within
the (4x1) unit mest and the 125 meV loss for θ > 1 was proposed to
correspond to the on top positions (Fig. 9b, θ = 1.2). These new
sites are occupied during the adsorption stage when the (4x1) LEED
pattern becomes "streaky". In summary this EELS investigation suggests the structural issue of Fig. 9b /36/ i.e. a subsequent formation of double, triple and quadrupole rows of O atoms and the
occupation of different adsorption sites.

4. CHEMISORPTION OF CARBON MONOXIDE

On transition metal surfaces, CO may chemisorb dissociatively or non-dissociatively. The molecular chemisorption bond
is considered to be closely related to that of metal carbonyl compounds, i.e. the carbon atom is linked to the metal substrate via
the 5σ orbital and the metal d-electrons back-donate into the
otherwise empty $2\pi^*$ antibonding CO orbital. This process results
in a weakening of the C-O bond which results in a lowering of the
C-O stretch frequency from the free molecular value 266 meV. From
the carbonyls one knows that this effect is related to the coordination, for a terminally bonded CO molecule the frequency is in
the range 250-260 meV while a bridge bonded CO group produces
frequencies around 220-230 meV. This knowledge has been extensively
used in infrared absorption spectroscopy to propose structures for
CO adsorbed on small metal particles and thin metal films /2,3/.
From the discussion above concerning the vibrational frequency of
atoms adsorbed at sites of different coordination one expects
qualitatively that the metal-carbon (M-C) frequency for adsorbed
CO should decrease with increasing coordination of the adsorption
site i.e. the C-O and M-C frequencies should behave in a similar
manner.

On the clean Ni(100) surface CO chemisorption produces a
c(2x2)CO structure around 2L exposure (θ = 0.5) which transforms
to a compressed quasihexagonal CO structure at θ = 0.61. The LEED
patterns and model structures are shown in Fig. 10 (bond configurations are inferred from the EEL spectroscopic observations) /38,39/.

The EEL (Fig. 11) spectra reflects the nucleation of the
c(2x2)CO structure /38/. At low CO exposures where no ordering is
observed by LEED (< 0.8 L; CO lattice gas /37/) the spectra show

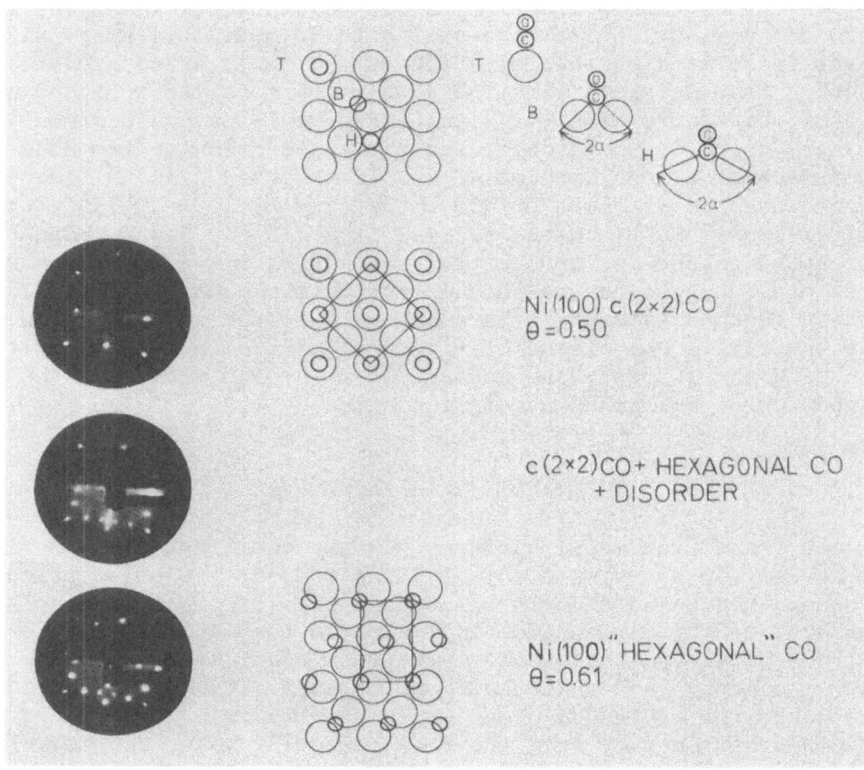

Fig. 10. LEED patterns observed at 175 K and model structures for CO adsorption on Ni(100).

two C-O stretching vibrational loss lines at 239 and 256 meV. Nucleation of the c(2x2)CO structure causes the 256 meV loss peak to increase in intensity and become the dominating loss line. The average C-O stretch frequency in $NiCO_4$ is 256 meV and this close correspondence to the observed loss line strongly suggests that CO is linearly bonded to the Ni atoms in the c(2x2)CO structure as in the model of Fig. 10. The 59 meV loss is interpreted as the Ni-C stretching vibrational excitation of this linear Ni-CO arrangement.

The EEL-spectra /39/ in Fig. 12 correspond to the LEED pattern sequence in Fig. 10. At 175 K the c(2x2)CO LEED pattern is very sharp indicating high structural perfection and the 240 meV shoulder is barely observable in the EEL spectrum. The structural transition; c(2x2)CO-quasihexagonal CO is most striking. The intermediate mixed

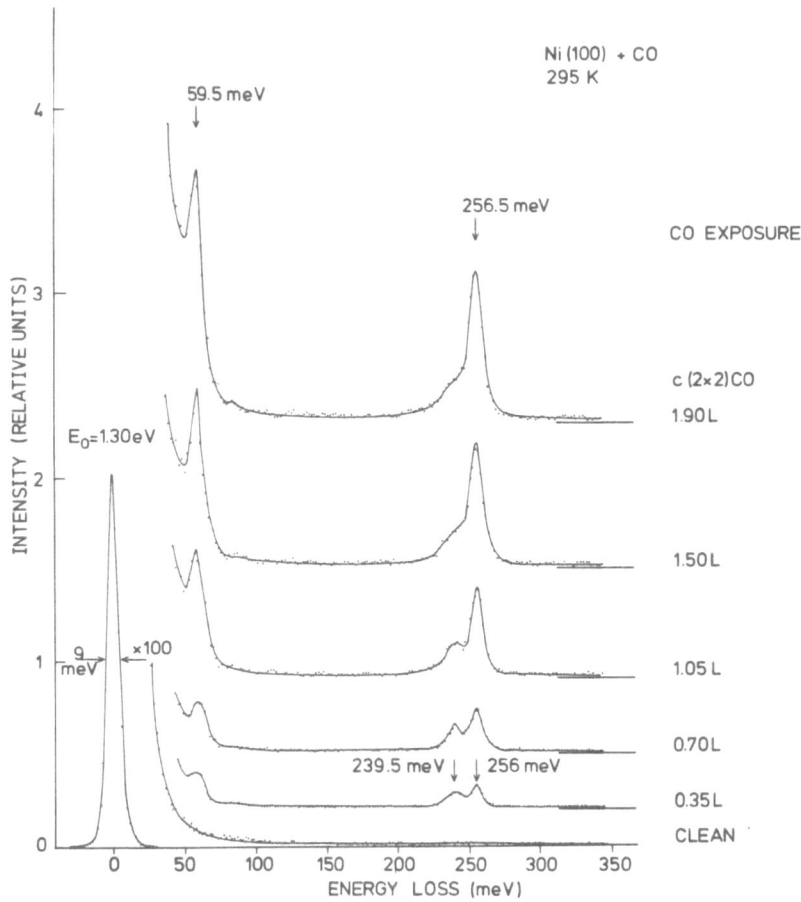

Fig. 11. EEL spectra of CO adsorbed at 295 K on Ni(100) (conditions as in Fig. 3).

CO structure produces an EEL spectrum where the 239 meV and 256 meV loss lines are of comparable intensity. This shows that the sites corresponding to the 239 meV loss are as abundant as the linear ones. Thus they are intrinsic sites of the clean flat Ni(100) surface. At this stage no more CO molecules can be incorporated in the linear configuration and the new sites are accordingly suggested to correspond to bridge sites (presumably to two n.n. Ni atoms). For the well-ordered quasi-hexagonal CO structure ($\theta = 0.61$) the C-O vibrational loss peak has almost turned into an evelope of the spectrum for the preceeding mixed structure. This can be understood

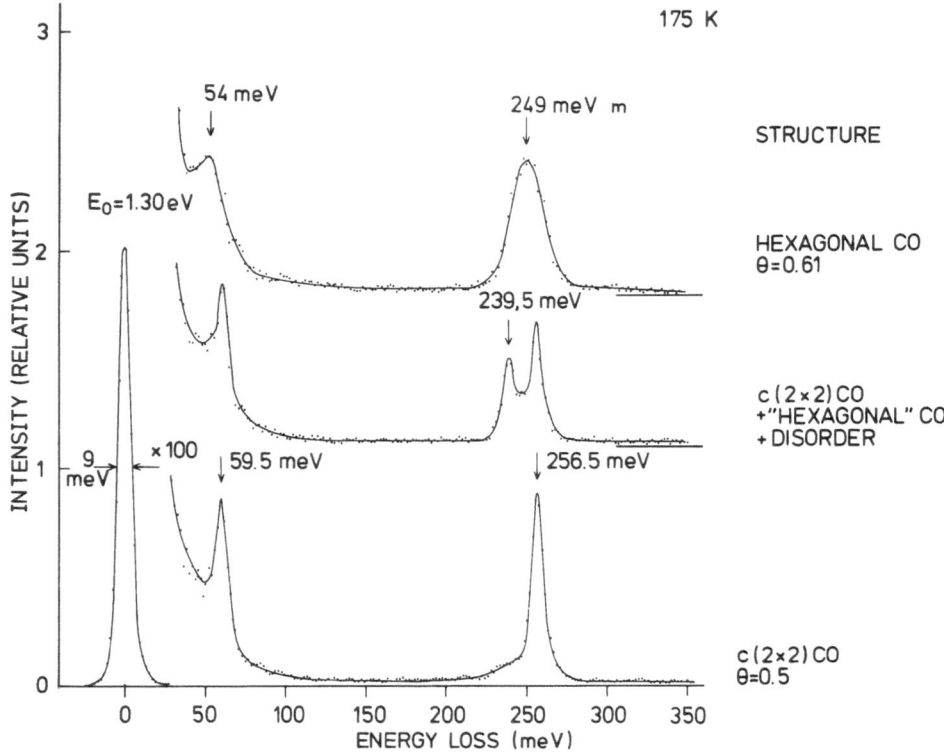

Fig. 12. EEL spectra of CO adsorbed at 175 K on Ni(100).
The spectra correspond to the structural sequence of the
LEED patterns in Fig. 10 (conditions as in Fig. 3).

in terms of the incoherency of this structure with respect to the
Ni(100) net which means that the CO molecules adopt adsorption
sites ranging from approximately linear to bridged as sketched in
Fig. 10.

CO chemisorption on W(100) shows a complex behaviour that
depends on the adsorption conditions /21/. At 300 K substrate
temperature a β-state adsorbs firstly. This state is thought to be
dissociated. It produces a (1x1) LEED pattern at saturation.
Further CO adsorption results in a molecular α-CO state. This behaviour is recognized in the EEL spectroscopic investigation by

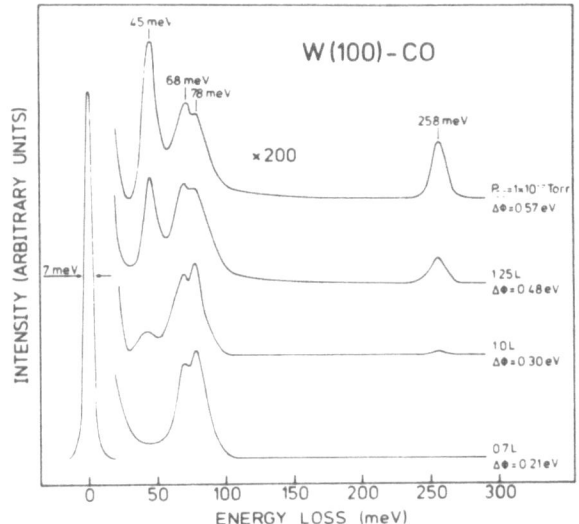

Fig. 13. EEL spectra of CO adsorbed on W(100) at
300 K. Primary electron energy 5 eV, angle at
incidence 75° (from Ref. 40).

Froitzheim, Ibach and Lehwald /40/. Fig. 13 shows a series of their
spectra for different CO exposures at 300 K substrate temperature.
For small CO exposures (< 1 L) corresponding to the β-CO state,
the spectrum shows two losses at 68 meV and 78 meV which are identified as the vibrational excitations of C and O atoms adsorbed in
the fourfold center site of the W(100) surface (see Fig. 9b).
Adsorption of α-CO produces a distinct molecular CO spectrum with
C-O and W-C stretching vibrational losses at 258 and 45 meV
respectively. These figures are close to those of W(CO)$_6$ and α-CO
is thus identified as CO linearly bonded to the W surface atoms.

REFERENCES

1. H. Ibach, Proc. 7th Intern. Vac. Congr. and 3rd Intern. Conf.
    Solid Surfaces, Vienna, p. 743 (1977).

2. L.H. Little: "Infrared spectra of adsorbed species", Academic
    Press, London and New York (1966).

3. M.L. Hair: "Infrared spectroscopy in surface chemistry", Marcel
    Dekker Inc., New York (1967).

4. R.G. Greenler, J. Chem. Phys. 44, 310 (1966).

5. J. Pritchard and T. Chatterick in:"Experimental methods in
    Catalytic Research" Vol. III (edited by R.B. Andersson and
    P.T. Dawson) Academic Press, New York (1976).

6. P.K. Hansma, Phys. Rep. 30, 145 (1977).

7. J. Pritchard, J. Vac. Sci. Technol. 9, 895 (1972).

8. R.A. Shigeishi and D.A. King, Surf. Sci. 58, 379 (1976).

9. F.M. Hoffman and A.M. Bradshaw, Proc. 7th Intern. Vac. Congr.
    and 3rd Intern. Conf. Solid Surfaces, Vienna p. 1167 (1977).

10. J.W. Davenport, W. Ho, and J.R. Schrieffer, to be published.

11. G.J. Schulz, Rev. Mod. Phys. 45, 423 (1973).

12. M. Sunjic and A.A. Lucas, Progr. Surf. Sci. 2, 2 (1972).

13. E. Evans and D.L. Mills, Phys. Rev. B5, 4126 (1972).

14. D.M. Newns, Phys. Lett. 60A, 461 (1977).

15. H. Ibach, Surf. Sci. 66, 56 (1977).

16. B. Persson, Solid State Commun., in press.

17. S. Andersson, to be published.

18. J. Lapujoulade and K.S. Neil, Surf. Sci. 35, 288 (1973).

19. K. Christmann, O. Schober, G. Ertl and M. Neumann, J. Chem.
    Phys. 60, 4528 (1974).

20. D.J.M. Fassaert and A. van der Avoird, Surf. Sci. 55, 291 (1976).

21. See

    a) L.D. Schmidt in: "Interactions on Metal Surfaces" (editor R.
    Gomer) Springer-Verlag, Berlin, Heidelberg, New York (1975).

    b) E.W. Plummer in: "Interactions on Metal Surfaces" (editor
    R. Gomer) Springer-Verlag, Berlin, Heidelberg, New York (1975).

22. T.E. Madey, Surf. Sci. 36, 281 (1973).

23. P.J. Estrup and J. Anderson, J. Chem. Phys. 45, 2254 (1966).

24. E.W. Plummer and A.E. Bell, J. Vac. Sci. Technol. 9, 583 (1972).

25. F.M. Propst and T.C. Piper, J. Vac. Sci. Technol. 4, 53 (1967).

26. H. Froitzheim, H. Ibach and S. Lehwald, Phys. Rev. Lett. 36,
    1549 (1976).

27. A. Adnot and J.-D. Carette, Phys. Rev. Lett. 39, 209 (1977).

28. C. Backx, B. Feuerbacher, B. Fitton and R.F. Willis, Phys. Lett.
    A60, 145 (1977).

29. J.E. Demuth, D.J. Jepsen and P.M. Marcus, Phys. Rev. Lett. 31, 540 (1973).

30. M. van Hove and S.Y. Tong, J. Vac. Sci. Technol. 12, 230 (1975).

31. S. Andersson, Solid State Commun. 20, 229 (1976).

32. G. Herzberg: "Spectra of Diatomic Molecules" 2nd edition van Nostrand, New York (1959).

33. G.Dalmai-Imelik, J.C. Bartolini and J. Rousseau, Surf. Sci. 63, 67 (1977).

34. P.J. Estrup and J. Andersson, Proceedings of the 27th Physical Electronic Conference, 47 (1967).

35. C.A. Papageorgopoulos and J.M. Chen, Surf. Sci. 39, 313 (1973).

36. H. Froitzheim, H. Ibach and S. Lehwald, Phys. Rev. B14, 1362 (1976).

37. J.C. Tracy, J. Chem. Phys. 56, 2736 (1972).

38. S. Andersson, Solid State Commun. 21, 75 (1977).

39. S. Andersson, Proc. 7th Intern. Vac. Congr. and 3rd Intern. Conf. Solid Surfaces, Vienna, p. 1019 (1977).

40. H. Froitzheim, H. Ibach and S. Lehwald, Surf. Sci. 63, 56 (1977).

SURFACE PHOTOVOLTAGE SPECTROSCOPY

Hans Lüth

2. Physikalisches Institut der Rheinisch-Westfälischen Technischen Hochschule Aachen, 5100 Aachen, F.R. Germany

ABSTRACT

Surface photovoltage (SPV) spectroscopy is discussed with respect to its application for the investigation of semiconductor surfaces. Principles of the method and experimental techniques are presented. As examples, recent experimental results on the electronic structure of ultrahigh vacuum (UHV) cleaved surfaces of Ge, Si and GaAs are briefly reviewed. The application to the investigation of organic layers is also described.

1. INTRODUCTION

The electronic properties of a semiconductor surface are primarily determined by the energetic distribution of electronic surface states. A clean well defined surface exhibits intrinsic surface states; gas adsorption usually changes their energetic distribution and produces new extrinsic surface states which are related to the chemisorption bond and the molecular or atomic orbitals of the adsorbate. Changes in the energetic distribution and occupation of surface states are directly correlated with band bending changes in the space charge layer and therefore with the electric properties of the semiconductor surface. Studies of the coverage dependence of extrinsic surface states can further yield information about the adsorption process itself.

The knowledge about the electronic structure of semiconductor surfaces has greatly been improved in recent years by the application of a variety of electron scattering and optical techniques. The fundamental problem in optical surface spectroscopy is the inherently low surface sensitivity. For semiconductors, however, where the density of free carriers in the space charge layer can be controlled very sensitively, the electrical response to light irradiation can be measured instead of the reflectance or absorption. This is done in surface photoconductivity (SPC) and surface photovoltage (SPV) spectroscopy where the change of surface conductivity and the change of the surface potential of a semiconductor surface are measured upon irradiation with light. In SPC spectroscopy, the photocurrent through the whole crystal is measured so that the spectra contain information from more than one surface if light with a photon energy of below band gap is used. Furthermore the measurement becomes difficult for highly conducting material. These difficulties are avoided in SPV spectroscopy where the change of the work function due to light irradiation is measured on only one surface. Reviews about optical spectroscopy of electronic surface states in general have been given recently [1,2]. The present paper is concentrated on SPV spectroscopy and the main emphasis lies on results which have been obtained recently on ultrahigh vacuum (UHV) cleaved surfaces of some important semiconductors. Furthermore, it is shown that SPV spectroscopy can also be used for the investigation of thin films of organic semiconductors.

2.   Method and Experimental Equipment

SPV is related to a charge transfer normal to the surface of a semiconductor. The most common mechanism of photovoltage generation with light of energy above the band gap produces free electron-hole pairs which are separated in the electric field of the space charge layer or due to different mobilities. Additional contributions to the effect are due to charge transfer normal to the surface during recombination and trapping processes of the nonequilibrium carriers. Electron-hole pair production near the photon energy of the band gap usually shows up as a sharp flank in the SPV spectra with a broad plateau extending to higher photon energies. For semiconducting layers of organic molecules sharper structures are expected because of the energetically sharper optical response of these molecules.

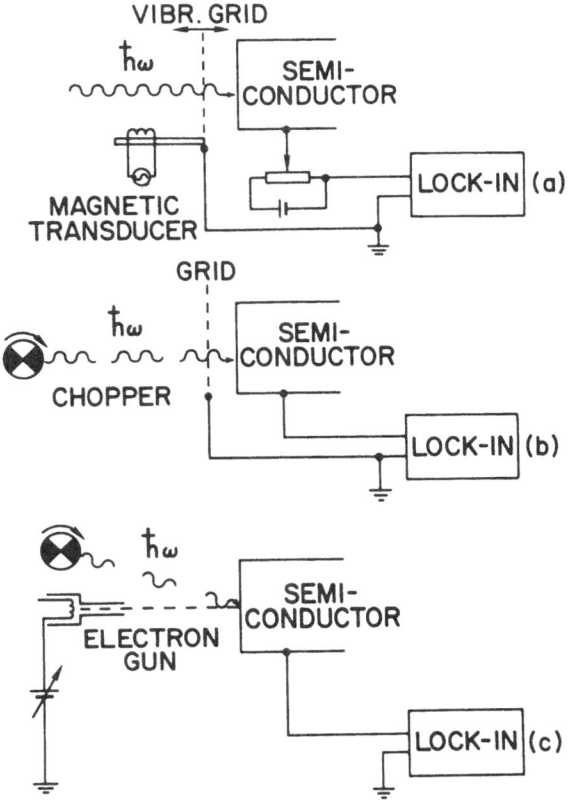

Fig. 1 Scheme of experimental set-ups for SPV spectros-
       copy: (a) Kelvin probe with semi-transparent
       vibrating electrode (grid) and steady illumina-
       tion, (b) chopped illumination modulates the con-
       tact potential difference between semiconductor
       surface and grid. (c) chopped illumination modu-
       lates the electron beam current between gun and
       semiconductor surface; a convenient bias allows
       most sensitive performance near the threshold of
       the current-voltage characteristics.

     If electrons from occupied surface states in the
energy range of the forbidden band are excited by light
into the bulk conduction band (space charge layer) they
give rise to an increase of the density of free electrons
(SPC) which is accompanied by a change of the band bending and thus of the work function (SPV). A similar effect
is caused by the excitation of electrons from the bulk
valence band into empty surface states (hole conductivity).

A distinction between transitions involving surface
states and the bulk conduction or valence band, respectively, can usually be made by means of the knowledge of
the Fermi level position at the surface.

Besides excitations involving surface states, transitions from and into bulk impurities may also be involved for energies below the band gap. In the presence
of underlying bulk impurity effects, surface excitations
can only be detected if the spectral structure is sensitive to surface treatment in a way which can not be
explained assuming merely a change in the condition of
charge.

The sign of the SPV effect is determined by the
generation process as well as by trapping and recombination mechanisms. Since in many cases only few details
are known about recombination and trapping, the analysis
of experiments is often limited to spectral changes due
to surface treatment. But this can be sufficient to
evaluate energies of transitions involving surface states.
Three currently used experimental techniques for SPV
spectroscopy are shown schematically in Fig. 1. All three
involve a modulation technique with the optically induced SPV being detected by a parallel plate capacitor
arrangement with a grid (methods (a) and (b))or by measuring the change in surface potential with a reflected
electron beam (method (c)).

3.   Results and Discussion

3.1   Cleaved Ge(111) and Si(111) Surfaces

The SPV spectra measured on UHV cleaved Ge(111)
surfaces [3] as shown in Fig. 2 are obtained by method
(b) (Fig. 1). After cleavage at low temperature the
spectrum exhibits the characteristic sharp flank at 0.7
eV (band gap at 100 K: 0.7 eV) which is due to the onset
of bulk electron-hole excitations. The shoulder below
the band gap energy reaching down to about 0.55 eV is
not sensitive to surface treatment and, therefore, does
not allow conclusions to be made concerning intrinsic
surface states of the clean cleaved Ge(111) surface with
(2 x 1) superstructure, even though dangling bond states
exist on this surface [4]. The shoulder is most probably due to bulk impurity excitations as is also suggested
by simultaneous SPC measurements[3]. Surface effects
may be hidden behind these bulk transitions. After an-

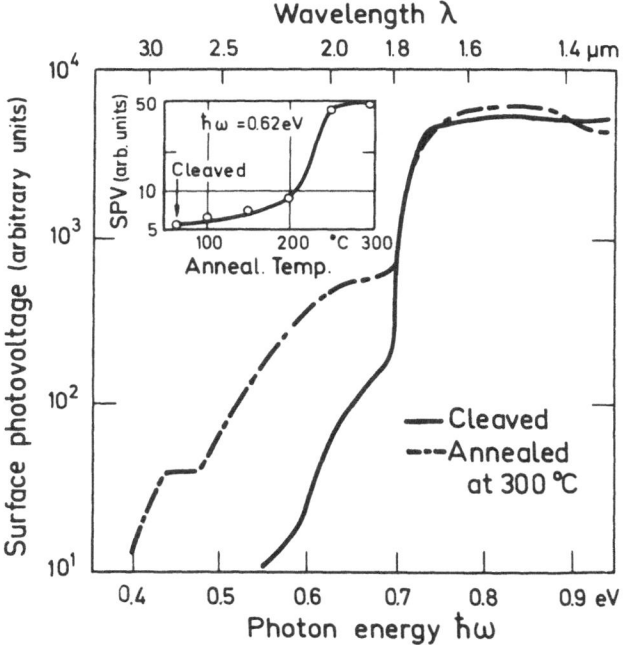

Fig. 2 Surface photovoltage (SPV) spectra of the clean cleaved Ge(111) surface normalized to constant photon flux; chopped light 13 Hz. Insert: Dependence of the SPV signal at 0.62 eV on the annealing temperature. Temperature of measurement 100 K. After Büchel and Lüth [3].

nealing to temperatures higher than about 200°C in UHV, new structures with thresholds near 0.4 eV and 0.45 eV appear (Fig. 2). The dependence of this shoulder on the annealing temperature (insert of Fig. 2) indicates that there is a critical temperature between 200 and 250°C for the appearance of the structure. At these temperatures the (2 x 1) superstructure of the clean Ge(111) surface is known to convert into a superstructure which is described as an (8)-structure with a (2 x 8) or (8 x 8) unit mesh (see e.g. [5]). Both excitations with onsets near 0.4 and 0.45 eV, therefore, are due to sur-

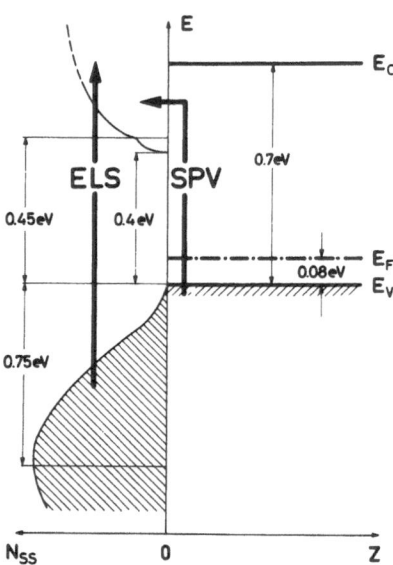

Fig. 3 Band scheme for the annealed Ge(111) surface with
(8)-superstructure; z distance from the surface
N_{ss} surface state density, E_F Fermi energy, E_c
and E_v edges of conduction and valence band.
The occupied surface states hatched are known
for the (2 x 1) superstructure and assumed to
be present also for the (8)-superstructure. After
Büchel and Lüth[3].

face states which belong intrinsically to the (8)-super-
structure. Fig. 3 shows the interpretation of the observed effects in terms of a band scheme. Excitation
from the valence band into empty surface states must
occur because of the Fermi level position. If the occupied surface states cross-hatched in Fig. 3 are assumed to be present on the (8)-structure, agreement
with data from electron energy loss (ELS) spectroscopy
[6] is obtained. Details of the surface state density in
the lower half of the forbidden band are not known, but
the present measurements suggest that there is at most
a low state density in that energy region.

Extrinsic surface states due to adsorbed oxygen on
the cleaved (2 x 1)Ge(111) surface have also been detected (Fig. 4) in SPV spectroscopy. An analysis similar
to that for the (8)-superstructure leads to the conclusion that adsorbed oxygen produces empty surface states
about 0.08 eV below the bottom of the conduction band[3].
These states were already predicted by Frankl [7] from
conductivity and field effect data.

SPV spectroscopy on the clean cleaved Si(111) surfaces with (2 x 1) superstructure of p- and n-type material has been performed by Clabes and Henzler [8] by
means of the electron beam technique (Fig. 1c). On p-
type material (200 K) they find transitions with an onset near 0.57 eV which are interpreted as due to excitations from the bulk valence band into empty dangling
bond surface states, whereas on n-type material only
transitions from the occupied dangling bond surface
state band into the bulk conduction band (onset near
0.86 eV) are observed. At present it is not clear which
mechanism is responsible for suppressing dangling bond
to conduction band excitations on p-type and valence
band to dangling bond state transitions on n-type material, respectively. The gap of about 0.29 eV between
occupied and empty dangling bond surface states as well
as their position with respect to the bulk electronic
bands as derived from those measurements agrees very
well with results from other methods (see e.g.[9]).

3.2   Cleaved GaAs(110) Surfaces

The cleaved GaAs(110) surface has attracted considerable attention recently [10-15] since different experimental methods seemed to yield different results
concerning the existence of empty and/or occupied surface states within the forbidden band. Results from SPV
spectroscopy obtained on two cleaves with mirror-like
surface and unmeasurable step density are shown in Fig.
5. The steep flank near 1.45 eV is due to electron-hole
pair production (band gap~1.45 eV). On the clean cleaved
surfaces, no SPV signals are detected below 1.35 eV
photon energy, i.e. within the limit of detection there
is no indication for empty or occupied surface states
near midgap on the clean cleaved GaAs(110) surface. This
is in agreement with recent results from surface potential measurements [10] and UPS [12,13] but in contra-
diction to earlier results by Dinan et al. [11] who
suggested the existence of empty surface states near
midgap.

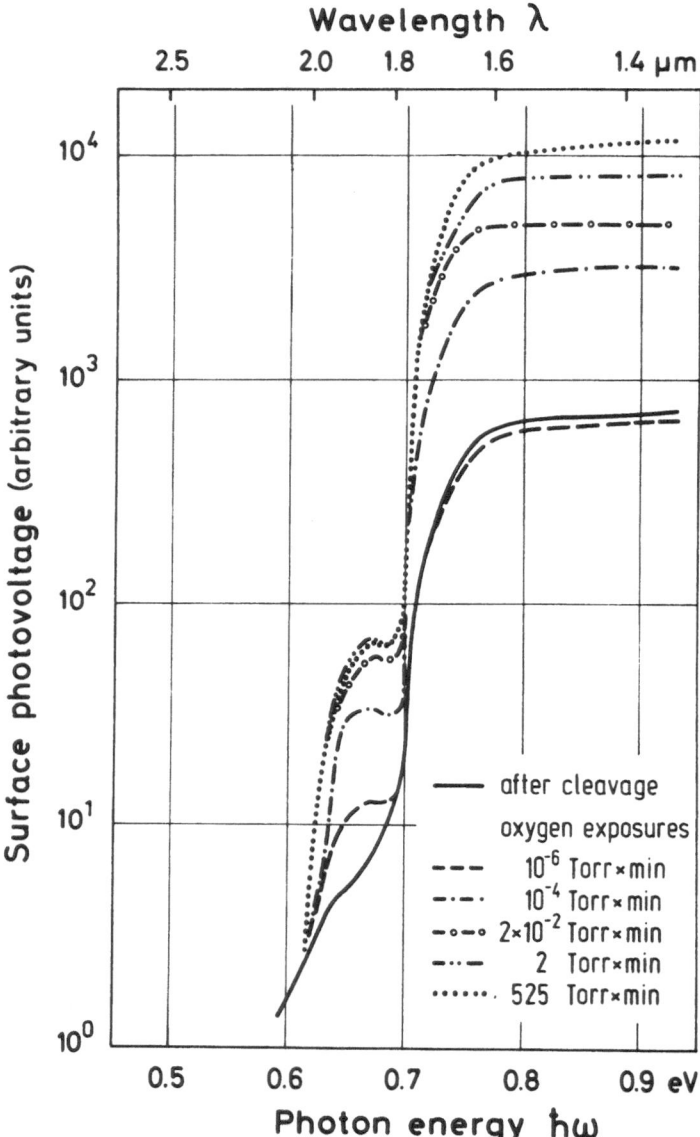

Fig. 4 Effect of oxygen adsorption on the SPV spectrum of
the cleaved Ge(111) surface; chopped light 13 Hz.
The spectra are normalized to constant photon
flux. After Büchel and Lüth[3].

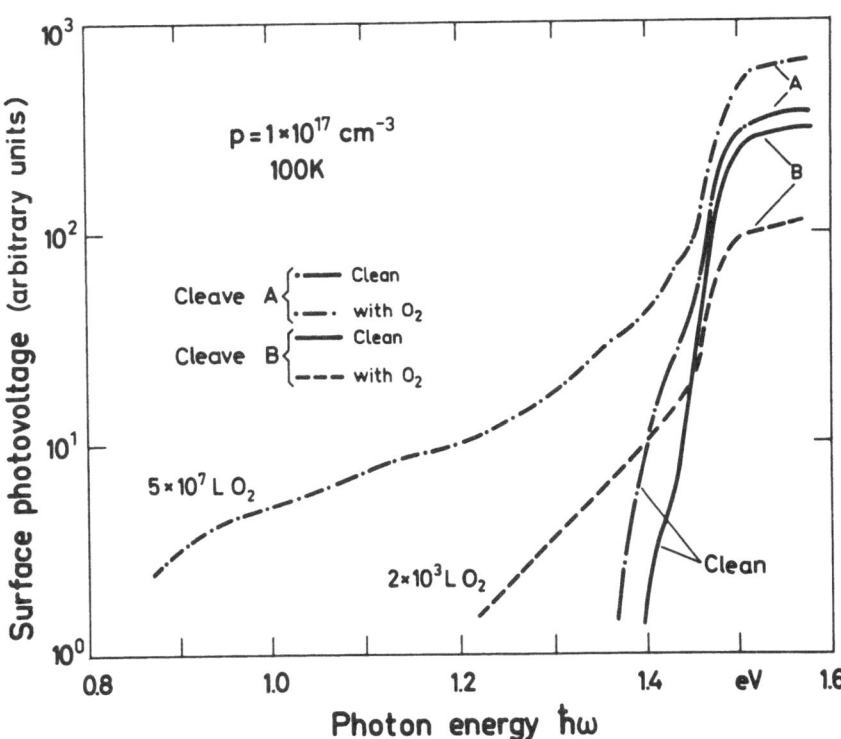

Fig. 5 Effect of oxygen adsorption on the SPV spectrum
of the cleaved GaAs(110) surface (Cd doped crystal). The cleaves A and B were optically flat
without any measurable density of atomic steps.
The spectra are normalized to constant photon
flux; chopped light 13 Hz. 1 L = 10^{-6} Torr x sec.
After Lüth et al. [14].

Oxygen adsorption up to coverages in the 0.5 monolayer range (10 L) produces a broad shoulder in the
SPV spectra for energies in the forbidden band (Fig. 5).
A threshold between 0.8 and 0.9 eV can be evaluated after
saturation with adsorbed oxygen. Since after saturated
oxygen adsorption for both n- and p-type material the
Fermi level position at the surface is near midgap
[12-14], one cannot decide if the shoulder in Fig. 5 is
due to transitions from occupied surface states (~0.8
eV below the conduction band edge) into the conduction
band or from the valence band into empty surface states
(~0.8 eV above the valence band).

A similar effect like that due to adsorbed oxygen
is found on clean surfaces of "bad" cleaves which show
visible crystallographic imperfections (incl. atomic
steps). SPV signals with a threshold below 1.0 eV suggest
surface states near midgap in that case, too. In view
of theoretical calculations of surface state densities
for various surface atom configurations [16,17] it is
suggested that the surface states near midgap which are
observed in SPV spectroscopy after oxygen adsorption
and/or due to crystallographic imperfections are produced
when the clean cleaved GaAs(110) surface with (1 x 1)
superstructure reconstructs into an ideal surface with
the atomic configuration of bulk GaAs. Namely the (1 x 1)
reconstructed surface (As surface atoms moved out of the
surface, Ga atoms moved into the surface [16,17]) being
given after cleavage does not exhibit surface states in
the forbidden band whereas on the ideal surface the
empty Ga dangling bond states have been shifted into
the band gap. These conclusions are in agreement with
LEED data [18,19] and with reflectivity changes due to
oxygen adsorption [20] which have been observed in ellipsometry [21].

3.3   Adsorption of H_2S on GaAs(110) Surfaces

Extrinsic surface states due to adsorbed gases can
be used to monitor the adsorption process itself and to
get information about different adsorption stages. The
adsorption of H_2S is used for sulfur doping of epitaxial
layers: The concentration of built-in sulfur impurities
under certain conditions may be influenced by the existence of electronic surface states in the band gap of
GaAs during the H_2S treatment [22].

Information of this kind can be derived from the
results [23] shown in Fig. 6: The SPV spectra of clean

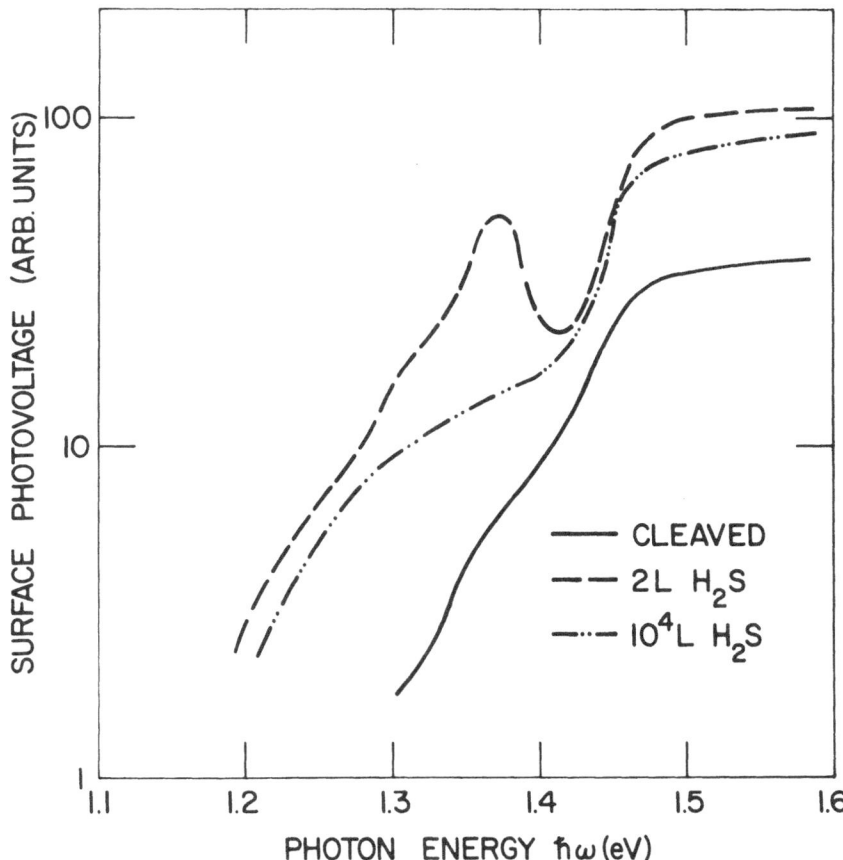

Fig. 6 Effect of H_2S adsorption on the SPV spectrum of a
cleaved $GaAs(1\bar{1}0)$ surface of n-type material
($n = 2\times 10^{17}cm^{-3}$). The spectra are normalized to
constant flux; chopped light 13 Hz. Both the
exposure and the measurements have been performed
at 170 K. After Liehr and Lüth[23].

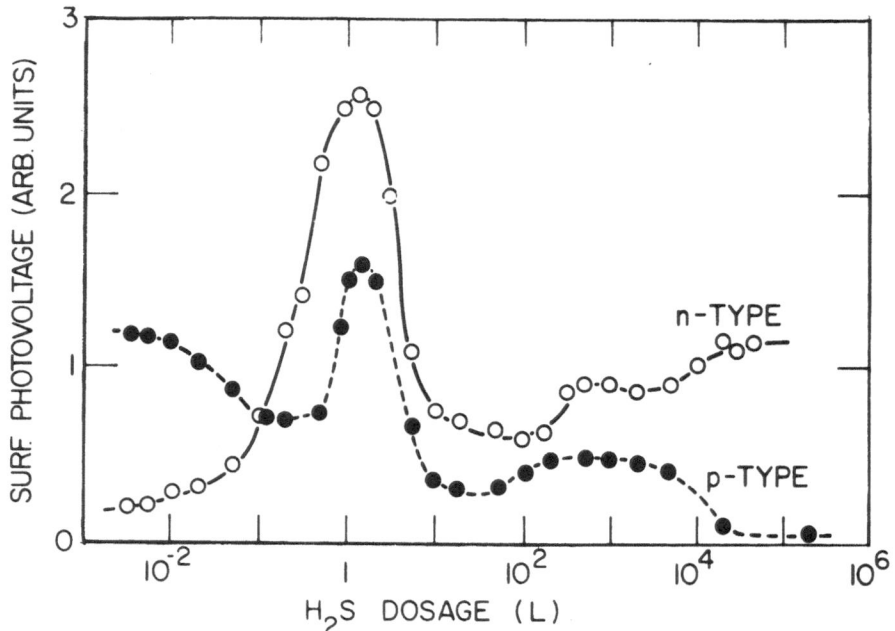

Fig. 7 Dependence of the SPV signal at the photon energy
$\hbar\omega$ = 1.58 eV on H_2S exposure for cleaved GaAs(110)
surfaces of n- and p-type material (carrier concentrations in the 10^{17} cm^{-3} range). The sign of
the SPV is different for n- and p-type material.
Chopped light 13 Hz. After Liehr and Lüth [23].

cleaved (110) surfaces of n-type material always showed
a more or less strong signal within the energy range of
the forbidden band. This signal, which extends down to
photon energies of about 1.3 eV in Fig. 6, is most probably due to bulk impurities. H_2S exposures up to the
5L range produce a broad shoulder with an onset between
1.1 and 1.2 eV and a relatively sharp structure centered
around 1.37 eV (Fig. 6). For higher exposures up to 10^4L
the threshold below 1.2 eV appears to be unchanged whereas the band near 1.37 eV vanishes for dosages above 5 to
10L. The dependence of the peak height of the structure
near 1.37 eV on exposure is correlated with the signal
intensity above band gap energy ($\hbar\omega$ = 1.58 eV) (Fig. 7):
The band at 1.37 eV develops to its maximum height at
a dosage of 1L which is the same dosage at which the
SPV signal at 1.58 eV reaches its maximum intensity
(Fig. 7). A similar effect is found on p-type crystals
only starting at somewhat higher dosages (Fig. 7). On
p-type material the low energy threshold of the SPV
spectrum is found at a somewhat different energy 1.28
± 0.02 eV [23]and the sign of the SPV signal near 1.28
eV and for energies higher than ∼1.45 eV is opposite to
that of n-type material: For p-type crystals the opposite electrode is negative whereas for n-type material
the crystal is negative and the opposite electrode positive. On p-type material the SPV signal around 1.4 eV
is quenched for dosages below 10L. This might be explained by the same excitation process being responsible
for the structure near 1.37 eV on n-type material. The
additional process near 1.37 eV therefore seems to produce SPV signals of the same sign on both p- and n-
type material [23].

     The results shown allow the following conclusions
about the adsorption of H_2S at 170K on GaAs(110):

i)   The different signs of the SPV signal for $\hbar\omega$ > 1.45
     eV and just above the thresholds (∼1.15 eV for n-
     type and 1.28 eV for p-type material) after H_2S
     adsorption might easily be explained by different
     band bendings: upwards for n-type material, and
     downwards for p-type material. Charge separation
     whithin the space charge field, then, is expected to
     be the dominant mechanism for SPV generation.

ii) The different onsets of the SPV spectra for n-
     (∼1.15 eV) and p-type (∼1.28 eV) material below band
     gap (∼1.45 eV) might be due to two different sets
     of surface states which are formed upon H_2S adsorp-

tion: The upper level (A), about 0.17 eV below the
lower conduction band edge, pins the Fermi level
on n-type crystals, whereas the lower level (B)
pins the Fermi level on p-type material about 0.3
eV above the valence band edge. For the H_2S covered
surface, if charge separation in the space charge
field rather than recombination or trapping is the
predominant SPV mechanism, then the sign of the
SPV signal near the thresholds can be explained
by an excitation of electrons from set B of totally
occupied surface states into the conduction band
for n-type, and by excitation of electrons from
the valence band into the totally empty states A
for p-type material, respectively. Sets A and B
must have different charging character because
of the different band bendings for n- and p-type
material.

Without having further information, an interpretation might also be given in terms of a broad
density distribution of surface states induced by
the adsorbed H_2S within the band gap: The thresholds,
$\sim$1.15 eV for n-type and $\sim$1.28 eV for p-type material, are then determined by the position of the
Fermi level E_F being $\sim$0.3 eV below the conduction
band (n-type) and $\sim$0.17 eV above the valence band
(p-type), respectively. On n-type material the
empty part of the broad surface state band just
above E_F yields the final states for transitions
from the valence band; on p-type material electrons are excited from the occupied part (just
below E_F) into the conduction band. For this type
of mechanism recombination and/or trapping must
be the determining factor in the SPV generation
below band gap because of the sign of the SPV
signal.

iii) The dosage dependence of the sharp structure near
1.37 eV and the results of Fig. 7 suggest a two-
stage adsorption. It is however difficult at present to give a detailed interpretation in terms
of surface state distributions. For the spectral
structure near 1.37 eV both on p- and n-type material the same mechanism seems to be responsible
because of the same sign of the SPV signal.

Additional experimental techniques, like e.g. UPS
are needed to get more insight into the nature of the
two adsorption stages. Nevertheless the information

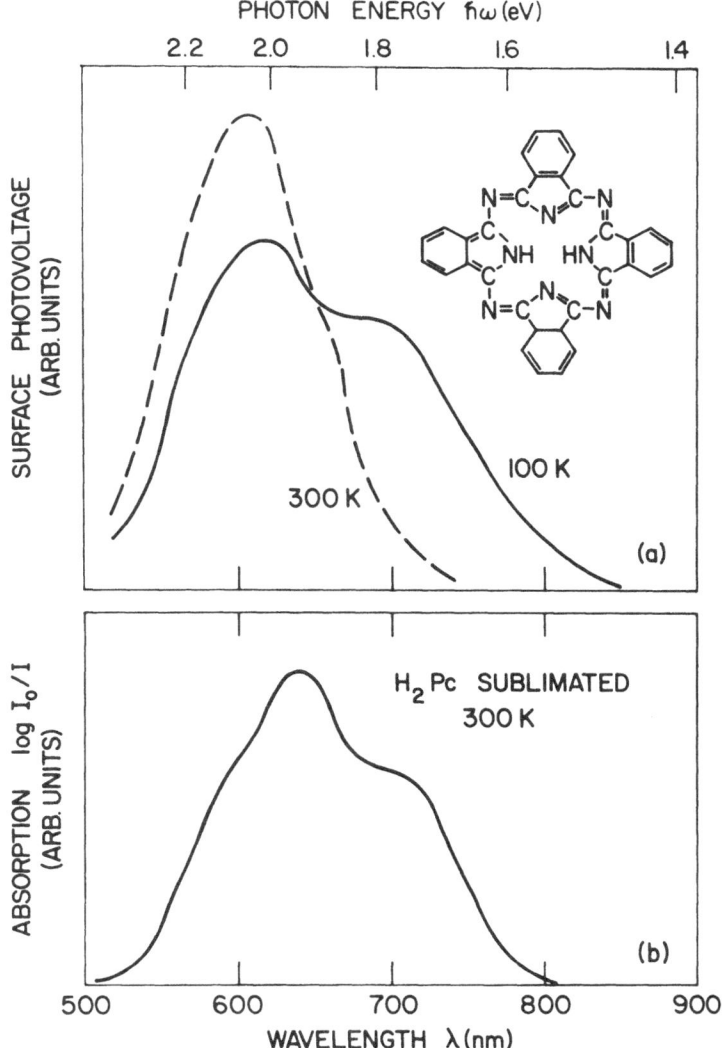

Fig. 8 (a) SPV spectra measured at different temperatures on free-base phthalocyanine (H$_2$Pc) films which are prepared by sublimation onto a stainless steel substrate (pressure during sublimation 10^{-6}Torr). Estimated layer thickness ∿500Å. The spectra are normalized to constant photon flux, chopped light 13 Hz.

Insert: Chemical structure of the H$_2$Pc molecule. (b) Optical absorption logI$_0$/I of an H$_2$Pc layer prepared like in Fig. 8(a) but on a transparent substrate. After Liehr and Lüth [23].

concerning the formation of surface states upon H_2S adsorption on GaAs might be helpful for the understanding
of sulfur doping.

3.4   Application to the Study of Organic Films

        Special problems arise in the investigation of low
conductivity organic semiconductors. Conductivity and
photoconductivity of many organic films (phthalocyanines,
merocyanines etc.) have to be studied in sandwich or in
staggered electrode arrangements. Interesting surface
properties (surface conductivity [24], surface states,
catalytic activity [25] etc.) therefore can not easily
be investigated under well defined conditions. SPV
measurements allow in these cases to probe the electric
response of one free surface under vacuum conditions
and after gas treatment.

        Fig. 8 shows some results of measurements on an
evaporated free base (acid) phthalocyanine (Pc) film. In
addition to their semiconducting properties [24] the
Pc compounds have interesting catalytic properties [25],
in particular if the hydrogen of the free base compound
H_2Pc (insert in Fig. 8a) is substituted by metal atoms.
Metal Pc's are also often studied as model substances
for the biologically active porphyrins.

        The Pc films used for the measurements in Fig. 8
were prepared by sublimation of pulverized H_2Pc from an
oven onto a stainless steel substrate under a vacuum
in the 10^{-7} Torr range. These films were then transferred into the UHV chamber, thus film surface conditions
were not clean. The film thickness was estimated to be
in the 500Å range [23]. Layers thicker than about 1000Å
can not be studied by SPV spectroscopy because of charging effects.

        The optical absorption in the visible of H_2Pc lay-
ers prepared in the above way (Fig. 8b) exhibits characteristic structure where the well known absorption bands
(610nm, 640nm, 670nm, 710nm) for H_2Pc in solution [26]
are found, but the intensity ratios are different from
those of H_2Pc in a solvent. This might be attributed to
the particular geometrical arrangement of the H_2Pc
molecules within the layer. From inelastic electron
tunneling experiments on similar H_2Pc layers it is suggested that the molecules are preferentially oriented
coplanar with the substrate [27]. The optical absorption

in the spectral region shown in Fig. 8b is due to
singlet $\pi \rightarrow \pi^*$ transitions of the inner macrocyclic ring
consisting of six N and twelve C atoms [28-31].

   As can be seen from Fig. 8a the SPV spectrum measured on the H_2Pc layer exhibits structure similar to
that seen in the optical absorption spectrum. Only the
relative intensities of the different subbands are different. The detailed shape of the SPV spectrum is furthermore strongly dependent on the temperature of measurement, which might be due to a change of recombination
and/or trapping mechanisms with temperature. A more
extensive discussion of the exact mechanism of SPV generation is not possible at present since too little is
known about the origin of the electrical conductivity
and the applicability of the band scheme model to these
organic layers.

   Nevertheless the results in Fig. 8 show that SPV
spectroscopy can be used to study bulk as well as surface properties of layers of organic semiconductors.
It should be emphasized that the results in Fig. 8 are
obtained under "dirty" conditions, i.e. the influence of
contaminations from the residual gas should be taken
into account.

4.   CONCLUSIONS

   The present paper describes several uses and limitations of SPV spectroscopy to probe the electronic
structure of semiconductor surfaces. Like SPC,SPV spectroscopy also determines transition energies with respect to the known bulk band structure and, therefore,
complements optical or electron energy loss measurements.
For organic semiconductors, SPV measurements might yield
interesting information about surface effects, but they
can also be expected to contribute to a better understanding of the electrical bulk properties. Another
interesting field of application, for which SPV is quite
sensitive namely the study of bulk defect levels [15],
[32] has not been discussed in this short review.

                    ACKNOWLEDGEMENT

   I would like to thank Dr. D.E. Eastman for critical
reading of the manuscript.

REFERENCES

/1/   H. Lüth: Appl.Phys. 8, 1 (1975).
/2/   G. Heiland and W. Mönch: Surface Science 37, 30 (1973).
/3/   M. Büchel and H. Lüth: Surface Science 50, 451 (1975).
/4/   G. Chiarotti, S. Nannarone, R. Pastore, and P. Chiaradia: Phys. Rev. (B) 4, 3398 (1971).
/5/   M. Henzler: J.Appl. Phys. 40, 3758 (1969).
/6/   H. Froitzheim: Berichte der KFA Jülich (Jülich-Report) 1179, März 1975.
/7/   D.R. Frankl: Surface Science 6, 334 (1967).
/8/   J. Clabes and M. Henzler: to be published
/9/   W. Mönch: In "Festkörperprobleme XIII" (Adv. in Solid State Physics), ed. by H.J. Queisser (Pergamon Press, Oxford 1973) p. 241.
/10/   A. Huijser and J. van Laar: Surface Science 52, 202 (1975).
/11/   J.H. Dinan, L.K. Galbraith, and T.E. Fischer: Surface Science 26, 587 (1971).
/12/   W. Gudat and D.E. Eastman: J. Vac. Science Technol. 13, 831 (1976).
/13/   W.E. Spicer, I. Lindau, P.E. Gregory, C.M. Garner, P. Pianetta, and P.W. Chye: J. Vac. Science Technol. 13, 780 (1976).
/14/   H. Lüth, M. Büchel, R. Dorn, M. Liehr, and R. Matz: Phys. Rev. (B) 15, 865 (1977).
/15/   H. Clemens and W. Mönch: CRC Critical Reviews in Sol. State Sciences 5, 273 (1975).
/16/   K.C. Pandey, J. Freeouf and D.E. Eastman: J. Vac. Sci. Technol. 14, 904 (1977).
/17/   J.R. Chelikowsky, S.G. Louie and M.L. Cohen: Phys. Rev. B 14, 4724 (1976).
/18/   A.R. Lubinsky, C.B. Duke, B.W. Lee and P. Mark: Phys. Rev. Lett. 36, 1058 (1976).
/19/   S.Y. Tong, A.R. Lubinsky, B.J. Mrstik and M.A. Van Hove: to be published in Phys. Rev.
/20/   H. Lüth: to be published in J. de Physique (Paris) 1977).
/21/   R. Dorn and H. Lüth: Phys. Rev. Lett. 33, 1024 (1974).
/22/   M. Heyen, H. Bruch, K.H. Bachem and P. Balk: to be published in J. Crystal Growth.
/23/   M. Liehr and H. Lüth: to be published.
/24/   F. Gutman and L.E. Lyons: "Organic Semiconductors", John Wiley and sons, New York 1967.
/25/   H. Kropf and F. Steinbach (Ed.): "Katalyse an Phthalocyaninen", Georg Thieme Verlag, Stuttgart 1973.

/26/ R.P. Linstead: J. Chem.Soc., p. 2873 (1953).

/27/ S. Ewert, H. Lüth and U. Roll: Proc. 7th Intern. Vac. Congr. and 3rd Intern. Conf. Solid Surfaces (Vienna 1977).

/28/ M Gouterman: J. Molecular Spectr. 6, 138 (1961).

/29/ C. Weiss, H. Kobayashi and M. Gouterman: J. Molecular Spectr. 16, 415 (1965).

/30/ M. Zerner and M. Gouterman: Theoret. chim. Acta (Berl) 4, 44 (1966).

/31/ M. Gouterman: J. Chem. Phys. 33, 1523 (1960).

/32/ P. Ulrich and G. Heiland: to be published; and H. Lüth and G. Heiland: to be published in Il Nuovo Cimento (1977).

OPTICAL SPECTROSCOPY OF SURFACE EXCITATIONS IN MOLECULAR CRYSTALS

AND MONOMOLECULAR LAYERS

Michael R. Philpott

IBM Research Laboratory

San Jose, California 95193, U.S.A.

I. INTRODUCTION

The detection of molecules at interfaces is a problem of some interest since it is the first step in following processes of more complexity. Organic molecules on solid surfaces or at the interface between media, as for example in the Helmholtz double layer region adjacent to an electrode in an electrochemical cell, can have physical and chemical properties that are quite different from their counterparts in solution and in the bulk of the solid. If these properties could be measured then one would in principle have a probe for monitoring changing surface conditions, including the possibility of following the course of chemical reactions.

In what follows, three types of somewhat interrelated experiments are described that bear on this problem of detecting and measuring the properties of molecules on surfaces and interfaces in some rather specialized systems. This work is part of an effort to understand more about the excited electronic states of organic molecules at interfaces and the optical phenomena in which they participate.

Some of the more frequently investigated interfaces containing organic molecules are summarized in Table I. There are four broad categories: organic crystal surfaces, composite systems like the interface between a metal electrode and an organic electrolyte, biological membranes, and finally systems where there are organic molecules intercalated between sheets of a host material.

Table I. Interfaces and Surfaces Containing Organic Molecules

Type	Example
Organic Crystal	Aromatic Hydrocarbons
Composite	Electrode-electrolyte Interfaces
Biological	Biological Membranes
Intercalate	Montmorillonites

We have made a fairly extensive survey of the electronic spectra of organic crystal surfaces at optical frequencies by means of reflection spectroscopy, and collaborative efforts to study selected composite systems are in progress. The study of the structure and function of biological membranes is of course an area of intense activity. It is passed over here as being outside the scope of our present interests. Likewise the properties of molecules intercalated between inorganic sheetlike structures will not be considered further, except to remark that these provide excellent examples of how the paths of chemical and physical processes can be directed by reducing the dimensionality of the system [1].

There are a number of techniques that can be utilized in the study of molecules at interfaces. The more important ones are listed in Table II. Some like ellipsometry, attenuated total reflection (ATR) and normal reflection have a long history in the investigation of surfaces and interfaces. Others like Raman, luminescence, and transmission spectroscopy have been less widely used or are applicable to only under special circumstances.

Table II. Optical Detection of Molecules on Surfaces and Interfaces

Technique	Application
Ellipsometry	Monolayers on Surfaces
Reflection	Surface Excitons
Attenuated Total Reflection	Surface Polaritons
Raman	Absorbate Vibrations
Luminescence	Absorbate Luminescence
Transmission	Absorbate Absorption

Ellipsometry [2] and differential reflectometry [3] have been widely used to study monolayer and submonolayer coverage of clean surfaces by small molecules, like CO, under ultra high vacuum conditions. Attenuated total reflection (ATR) [4] is finding increasing application to surface studies. This technique has a long history, it has been used for many years to obtain the infrared spectra of organic molecules deposited under matrix isolation conditions. In our laboratories it is being used to probe for exciton surface polaritons and to study plasmon polariton-monolayer interactions [5]. Reflection spectroscopy has long been used to measure the optical properties of opaque materials. We have been examining the exciton and polariton spectra of organic crystals in which there are strong electronic transitions. Of the last three techniques listed in Table II, Raman spectroscopy is the one most likely to find wide acceptance in the future. The increased sensitivities now available through commercially available photon counting instrumentation means that the detection of Raman scattered light from monolayers on solid substrates is now possible and will undoubtedly be carried out on a routine basis in the future.

Space and time do not permit an exhaustive review of either materials, methods or phenomena associated with organic molecules at interfaces. Therefore, as hinted at earlier, the focus will be on three topics at the core of our experimental and theoretical program. These are:

    i)   "site shift" surface excitons,
   ii)   exciton surface polaritons (ESP),
  iii)   plasmon surface polariton spectroscopy.

The connecting link between these three topics is the technique used, namely reflection spectroscopy. Some organic crystals with quasi-two dimensional structures reflect light extremely well, and within their stop-bands there are reflectivity minima associated with exciton transitions of the surface planes of the crystal. Other spatially larger types of surface excitation can occur within the frequency span of the stop-band of the exciton transition. These are surface localized electromagnetic waves which being associated with an exciton transition are called exciton surface polaritons (ESP). The first surface electromagnetic waves studied in detail were the plasmon surface polariton modes of metals [6]. These surface waves are now sufficiently well characterized on metals like silver and gold to be used to measure the optical constants of organic films and to study the excited electronic states of films one molecule thick.

II. EXCITONS, POLARITONS AND REFLECTION SPECTROSCOPY

The interpretation of the electronic transitions of organic
molecular crystals rests upon the concept of the molecular exciton
and the polariton. Many organic crystals consist of weakly
interacting (non-bonded) neutral molecular units. Within the unit
there are strong chemical bonds holding the molecule together,
and crystal forces act only as a mild perturbation on the bonds.
When one molecule absorbs light it is raised to an excited
electronic state. This excitation does not remain localized
because interactions between neighboring molecules cause the
excitation to hop from molecule to molecule. This migrating packet
of excitation energy is called an exciton, a term coined by Frenkel
in 1931 [7]. Frenkel's exciton may be thought of as an excited
electron strongly bound to the hole left by its promotion to a
higher level. The electron and its hole are never more than one
site apart in organic crystals. There is another type of
exciton, called the Wannier exciton [8], in which the electron
and hole are far apart. In the Wannier exciton the electron and
hole orbit each other at large instances and have a series of
states analogous to those of the hydrogen atom, ending in at a
limit where the conduction states of the solid begin.

In aromatic hydrocarbon crystals the electron and hole of the
Frenkel exciton are localized on one molecule because poor overlap
between π-orbitals on different molecules prevents electron
transfer without the hole. This poor overlap is a direct result
of there being no change in the principal quantum number of the
carbon atom when the molecule is excited. This is the cause of
a basic difference between Frenkel excitons in organic crystals
and those in inorganic insulators where there is a change in
principal quantum number and a concomittant increase in overlap
and delocalization of the excited electron. Tightly bound excitons
are a unique property of organic crystals and their identification
with molecular units justifies the description molecular exciton.

Each vibrational-electronic (vibronic) transition of a molecule
from its ground level gives rise to a band of exciton levels in
the crystal. For simplicity consider first one molecule with only
one excited state at energy $\Delta\omega_1$. If the molecule is trapped in
a host matrix, both the ground and excited level are altered,
generally by different amounts, with the result that the transition
is shifted by an amount D. This term is called the solvent shift,
and for aromatic hydrocarbons in n-heptane or n-octane as solvent
the shift is negative and varies from 500 cm^{-1} to 2000 cm^{-1} for the
first singlet transition. If as a final step we consider the
molecule surrounded in a regular array by identical molecules, then
the excited state, which is N fold degenerate in the absence of
intermolecular interactions, splits into a dense band with energies

$$E(k) = \Delta\omega_1 + D + I(k) .$$

(2.1)

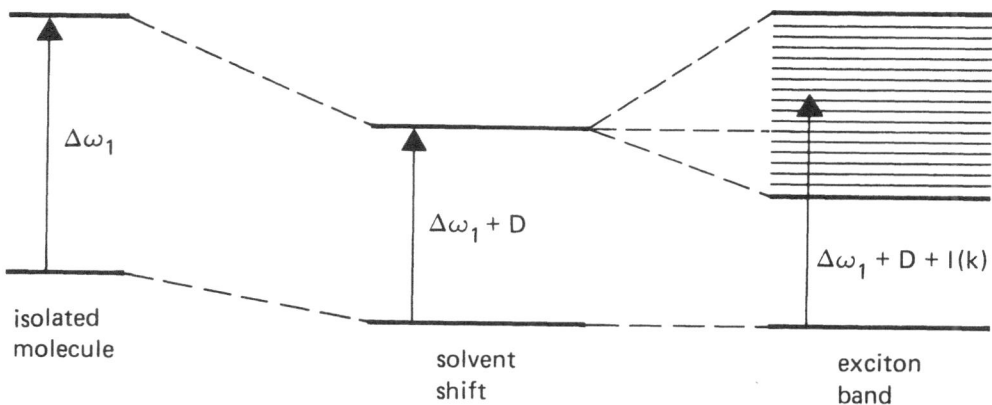

$$E(k) = \Delta\omega_1 + D + I(k)$$

Figure 1. Evolution of the band of exciton energy levels
$E(k)=\Delta\omega_1+D+I(k)$.

Invariance of the crystal Hamiltonian under translations ensures
that each of the exciton levels has a unique wavevector k. The
term I(k), which discriminates levels, is a lattice sum of exciton
transfer matrix elements modulated by a phase factor with
wavevector k. Figure 1 shows how the gas, solution and crystal
levels are related. The arrow signifies the optical transition,
which in the case of the crystal is caused by the absorption of
a photon with wavevector q. Momentum conservation requires that
k=q. In crystals with nearest neighbor exciton interactions the
optical transition occurs near or at the band edge, however where
the interactions are of longer range the optical transition can
lie well inside the exciton band as shown in Figure 1.

The full expression for I(k) is

$$I(k) = \sum_{m(\neq n)} (\phi_m^r \phi_n^o | V_{nm} | \phi_m^o \phi_n^r) \exp[ik(r_n - r_m)] \tag{2.2}$$

where V_{nm} is the Coulomb interaction operator for molecules n and
m, ϕ_m^r is the excited state wavefunction and ϕ_m^o is the ground state
wavefunction of molecule m. The matrix elements in (2.2) describe
the transfer of the rth state of excitation from site m to site
n.

The exciton levels (2.1) arise from the coulombic part of the

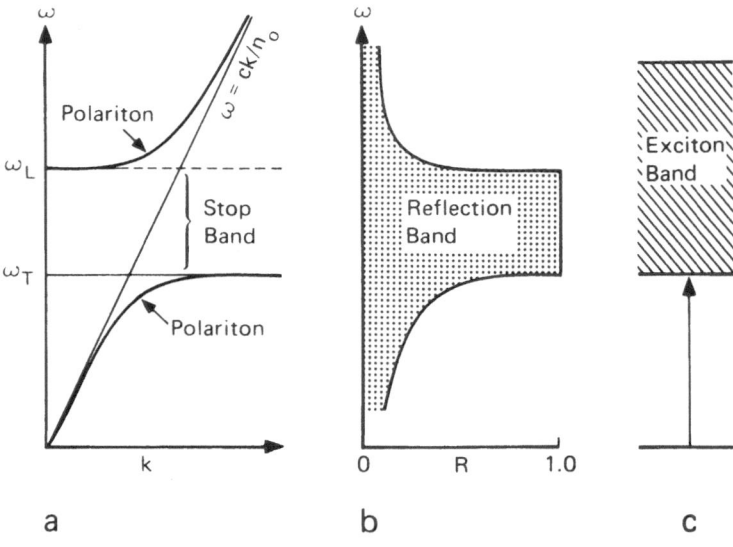

Figure 2.   Relation between the polariton branches, stop-band,
reflection band and the exciton band for a molecular crystal with
one dipole allowed transition.

total Hamiltonian of the system (crystal and radiation field) and
are therefore sometimes called coulombic excitons.   The treatment
of crystal transitions from this viewpoint neglects one essential
fact, namely that interactions are not instantaneous but retarded.
For a gauge invariant theory the radiation field must be included.
In the simplest picture a coulombic exciton with wavevector k
couples to a photon with the same wavevector to produce a new
quasi-particle excitation called a polariton [9,10].   In the
absence of a coupling interaction the unperturbed exciton ($\omega=\omega_T$)
and photon ($\omega=ck/n_0$) dispersion curves cross (thin lines Figure
2a).   In the presence of the coupling, which is proportional to
the oscillator strength per unit volume, there is an avoided
crossing (thick line Figure 2a) giving the polariton dispersion.
For a crystal with no absorption the refractive index $n(\omega)=ck/\omega$
is given by

$$n^2(\omega) = \varepsilon(\omega) = \varepsilon_0 + \frac{\omega_{po}^2 f}{\omega_T^2 - \omega^2} \ . \tag{2.3}$$

where $\omega_{po}^2 = 4\pi e^2/(m_e v_c)$ and $\varepsilon_0 = n_0^2$.   Within the stop-band region

$\omega_T < \omega < \omega_L$ where

$$\omega_L^2 = \omega_T^2 + \omega_{po}^2 f/\varepsilon_o \ , \tag{2.4}$$

the refractive index is pure imaginary and the reflection power

$$R(\omega) = \left| \frac{n(\omega)-1}{n(\omega)+1} \right|^2 \tag{2.5}$$

is unity, hence the name stop-band. Figures 2a, b and c show
schematically the inter-relations between the polariton dispersion
curve, and the associated stop-band, the reflection band and the
exciton band. Figure 2c shows schematically the band of exciton
levels without any coupling to the photon field and the allowed
transition to the bulk exciton state ω_T, which is assumed to lie
at the bottom of the band. In passing we note that the energy
spanned by the exciton band can exceed that spanned by the stopping
band. If there is no damping or relaxation process whereby light
is absorbed by the crystal then the reflectivity is unity within
the stop-band. Put another way, in the absence of damping light
incident on the crystal from the outside cannot excite a
propagating mode in the stop-band region and therefore is
completely reflected. Damping can be introduced phenomenologically
by adding $-i\omega\gamma_T$ to the denominator of Eq. (2.3) and results in a
"rounding off" of the sharp edges of the reflection band.

For non-degenerate electronic transitions such as those
occurring within the singlet manifold of a molecular crystal the
longitudinal exciton ω_L corresponds to that state of the exciton
band where the wavevector k and transition moment d are parallel.
They are to be distinguished from longitudinal excitons in
inorganic materials where s→p atomic transitions can give rise to
three excitons, two transversely and one longitudinally polarized
relative to the wavevector k.

The width of the reflection band is proportional to the
oscillator strength of the crystal transition. Thus, spin
forbidden transitions in which triplet excitons are created have
no stop-bands whereas for TCNQ° [11] and cyanine dye crystals
[12], where the oscillator strengths are greater than unity, the
stop-band may be 2 eV wide (16,000 cm^{-1}).

Real crystals absorb light within the stop-band by
multiparticle processes in which lattice phonons or vibrational
excitons are produced [13]. The edges of the reflection bands
are not sharp and do not rise to unity as shown in Figure 2b.
Structure within a reflection band yields information about how

Table III. Classification of Organic Crystals

Class	Examples
Neutral	Anthracene
Ionic	Cationic cyanine dyes
"Metallic"	TTF-TCNQ
Polymeric	Polymerized diacetylene crystals

polaritons decay into other excitations. Another complication in real crystals is anisotropy. Generally organic molecules have transitions polarized along well defined axes. This introduces the geometric factor $[1-(\hat{k}\cdot\hat{d})^2]$, where $\hat{d}$ is the direction of the transition dipole, that multiplies the oscillator strength. The width of the stop band is therefore at its maximum for transverse crystal transitions, i.e., ones polarized perpendicular to the direction $\hat{k}$ of the incident photon, and equal to zero for longitudinal crystal transitions, which are polarized parallel to $\hat{k}$. Finally it is emphasized that the band of exciton levels is always at least as wide as the stop-band. In crystals with weak transitions the stop-band may span only a small fraction of the total exciton band. This is shown schematically by Figure 2c.

III. SURVEY OF ORGANIC CRYSTAL REFLECTION SPECTRA

Organic crystals may be classified according to the physical state of the units making up the crystal. There are four major divisions and these are listed in Table III together with a representative example of each class. Subdivisions of each class are possible, for example one might classify them further using the degree of charge transfer in donor-acceptor crystals or whether the crystal packs molecules into quasi-one or two dimensional arrays. For example anthracene-pyromellitic dianhydride (A-PMDA) is a weak charge transfer mixed stack system whereas TTF-TCNQ at low temperatures is a strong charge transfer system. Likewise anthracene which has a quasi-two dimensional crystal structure can be distinguished from 9-cyano-anthracene or 9,10-dichloroanthracene which have quasi-one dimensional crystal structures.

As mentioned in the introduction the reflection band of an isolated exciton band is proportional to the oscillator strength of the transition. For weak crystal transitions the spacing of the vibronic transitions ($400-1500$ cm^{-1} for aromatic hydrocarbon molecules) is much larger than the stop-bands so that there is

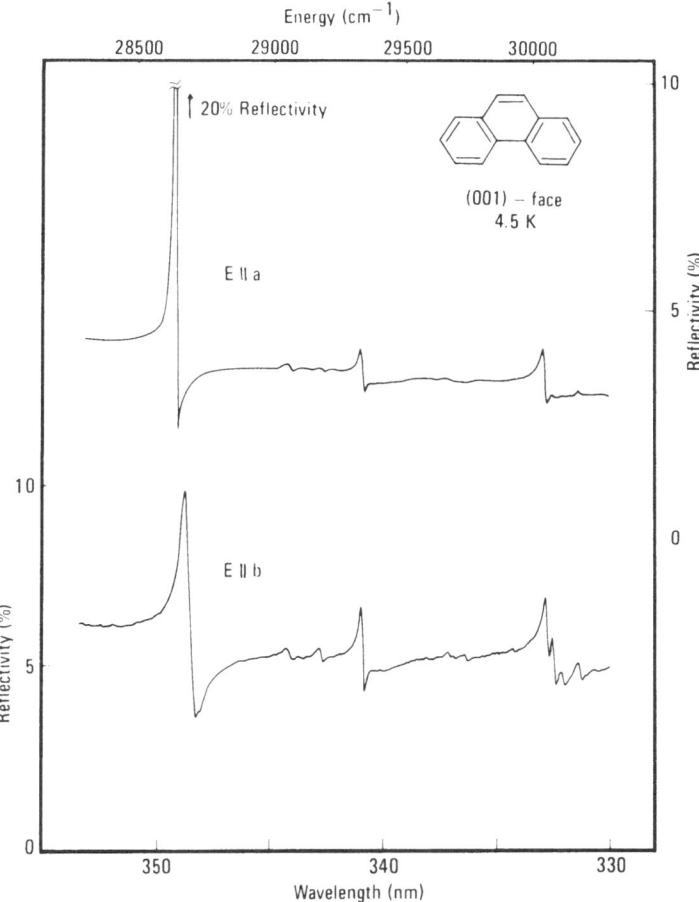

Figure 3. Polarized normal incidence reflection spectra of the first singlet transition of a phenanthrene single crystal off the (001) face at 2K.

one reflection band for every transition. Figure 3 shows an example of this type of spectrum (angle of incidence θ=0°) for a phenanthrene single crystal at 2K [14a]. The total oscillator strength of this first singlet transition of phenanthrene is 0.001 and the stop-band of the 0-0 transition for light polarized parallel to the crystallographic a axis is only 5 cm^{-1} wide. The width of the 0-0 exciton band is not known, but a lower bound is certainly the Factor Group splitting which has been measured in transmission to be 60 cm^{-1} [14b]. Phenanthrene has a quasi-two dimensional crystal structure similar to that of anthracene.

For somewhat stronger transitions, say oscillator strength f≈0.1, the reflection bands of aromatic hydrocarbon crystals merge

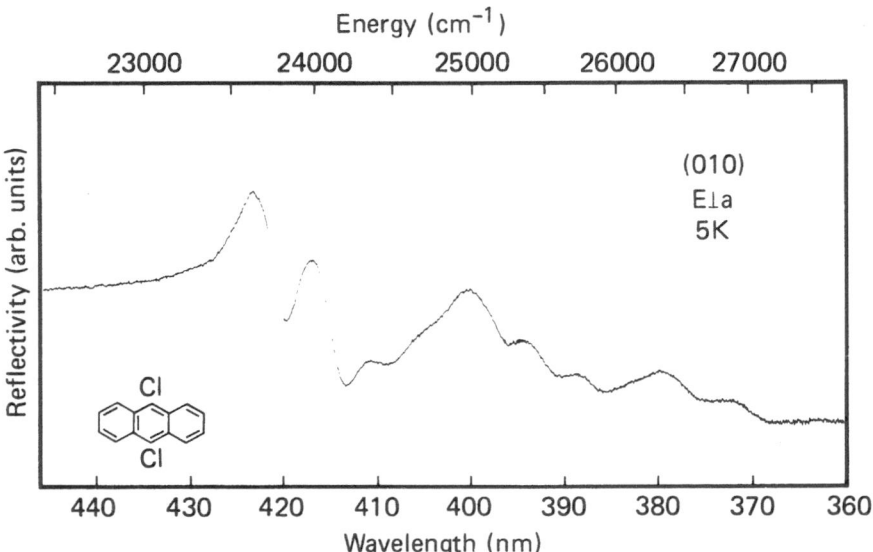

Figure 4. Normal incidence reflection spectrum of the first singlet transition of 9,10-dichloroanthracene crystal (α-form) off the (010) face at 5K. Electric vector is perpendicular to the stack axis a.

together and it sometimes becomes difficult to identify any but the most intense vibronic transitions. An example of this type of spectrum is shown in Figure 4, for 9,10-dichloroanthracene [15]. The crystal structure shows the molecules packed in chains with two molecules in the repeat unit of each chain. The reflection spectrum of another one dimensional system is shown in Figure 5. This is the (100) normal incidence spectrum of the polymerized diacetylene PTS taken at 2K with electric vector parallel to the chain axis [16]. The backbone of the polymer consists of a conjugated chain of single-triple-single-double bonds of macroscopic length. When freshly prepared the crystals are a bright gold color with reflectivities in the 70-80% range which gives them a distinct metallic appearance. Notice what appears to be a long progression in the 300-400 cm^{-1} vibration.

Organic dye stuffs typically have at least one very strong electronic transition in the visible. By a strong transition we mean one with an oscillator strength greater than unity. Reflection bands of great width, sometimes exceeding one eV, are typical of these materials. Often there are several crystal faces that have a distinct metallic lustre. We have studied this

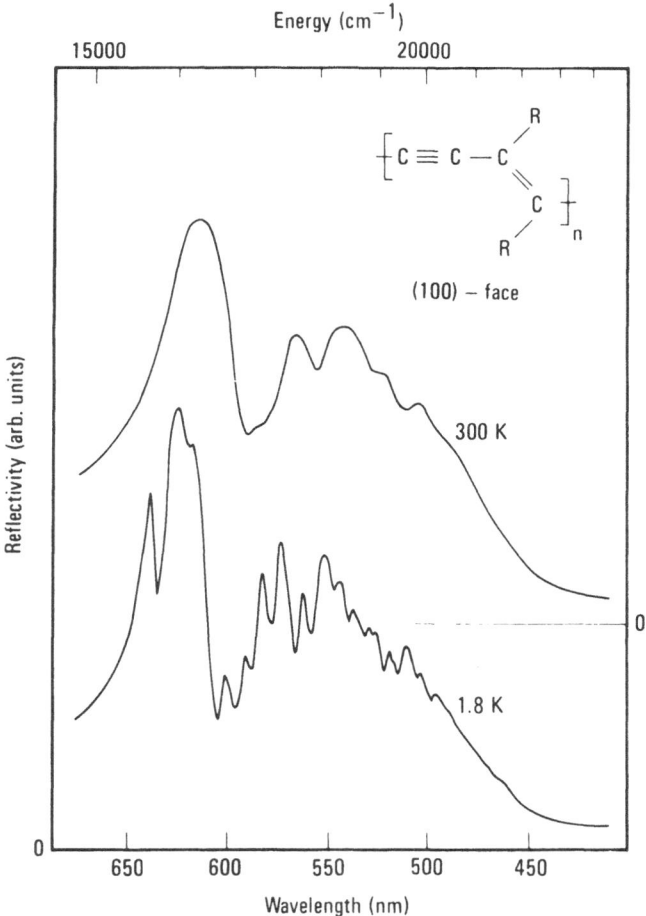

Figure 5. Normal incidence reflection spectrum of the polymerized diacetylene PTS crystal off the (100) face with E vector parallel to the b crystal axis (polymer chain axis).

"metallic reflection" from the (010) face of the TCNQ° crystal and its normal incidence spectrum is shown in Figure 6 [17]. The main reflection band is approximately 1.75 eV wide. Structure within this band arises mainly from multiparticle scattering processes in which the incident light simultaneously creates a polariton and a vibrational exciton.

IV. "SITE SHIFT" SURFACE EXCITONS

In the spectra of organic crystals the transitions of Frenkel

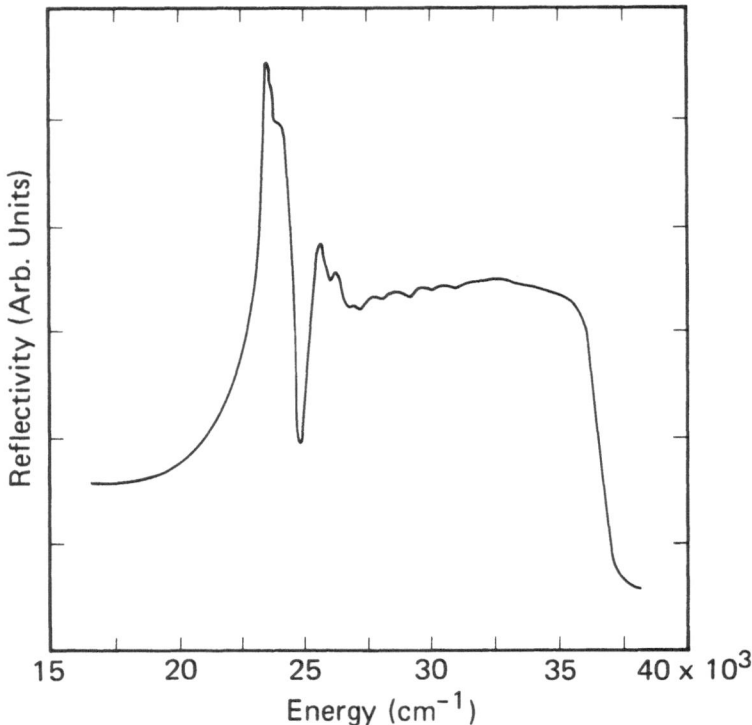

Figure 6. Near normal incidence reflection spectrum of the first
singlet transition of the TCNQ° crystal off the (010) face at 13K.
The E vector is parallel to the direction of maximum reflectivity.
This is the same direction as the projection of the long molecular
axis onto the (010) plane.

excitons associated with the surface and bulk regions are
superimposed, but since the oscillator strength of the surface
molecules is many orders of magnitude less than the bulk molecules,
surface transitions when present, generally go undetected. There
are, however, conditions which make detection of surface
transitions more favorable, the most obvious being when the crystal
is so thin that the surface to bulk ratio becomes high enough to
observe the surface transitions. A less obvious, but in some ways
more practical situation occurs when a surface transition falls
inside the energy interval spanned by the reflection band of a
bulk crystal transition, for it is then detectable as deep narrow
minimum or as a derivative-like structure consisting of an adjacent

minimum and maximum. To date, minima assigned to surface
transitions have been observed in the reflection spectra of
anthracene [18-22], tetracene (naphthacene) [23], naphthalene [24]
and some rare-gas crystals [25].

Structure attributed to surface transitions has recently been
seen in the transmission spectra of very thin (40 nm thick)
anthracene crystals [26,27] and thin crystalline films of rare
gases Xe, Kr and Ar [28]. In addition to the observations cited
there is evidence that fluorescence [29-31] occurs from the same
anthracene surface states that are seen in reflection and
transmission spectra. These fluorescence observations were
recently extended to mixed crystals of deuterated anthracenes
[32]. The fluorescence emissions are extremely weak because of
depopulation of the surface states into lower lying bulk states
by rapid radiationless decay processes. For surface levels lying
below the bulk exciton band, a high population appears feasible
since in principle the surface levels can be pumped by bulk crystal
transitions.

The surface excitons described in this section are not to be
confused with the exciton surface polariton states that are the
subject of Section V, p-polarized, well described by a set of
macroscopic Maxwell equations, and exist only for frequencies such
that some components of the dielectric function $\varepsilon(\omega) < -1$. In this
section,   the nomenclature surface exciton is used exclusively
for excitons localized on the first few crystal planes as a result
of the difference between site shift interactions (the D term in
Eq. (2.1) and Figure 1) for surface and bulk molecules. Put
another way we can say that Frenkel exciton states localized on
surface planes occur because molecules at the surface have
different transition energies than molecules lying in the bulk
region. Surface excitons can be either s- or p-polarized and can
be described by a set of microscopic Maxwell equations that take
details of the discrete nature of the crystal and its surface into
account.

The first singlet transition of crystalline anthracene consists
of a series of bands between 400 and 340 nm [33]. The transition
is M axis polarized ($^1B_{2u} \leftarrow {}^1A_{1g}$) and in solution the integrated
oscillator strength is f=0.1 [34]. In the crystal at low
temperature, narrowing of the absorption bands reduces penetration
depths of the incident light to tens of nanometers, and for the
$0-0_b$ component the oscillator strength f is sufficiently high that
optical processes can not be understood without the use of
polariton theory.

The reflection spectra with electric vector $\vec{E}$ parallel to the
a and b axes were measured for the (001) face of single anthracene
crystals at a temperatures of 2K. Spectra for other faces notably

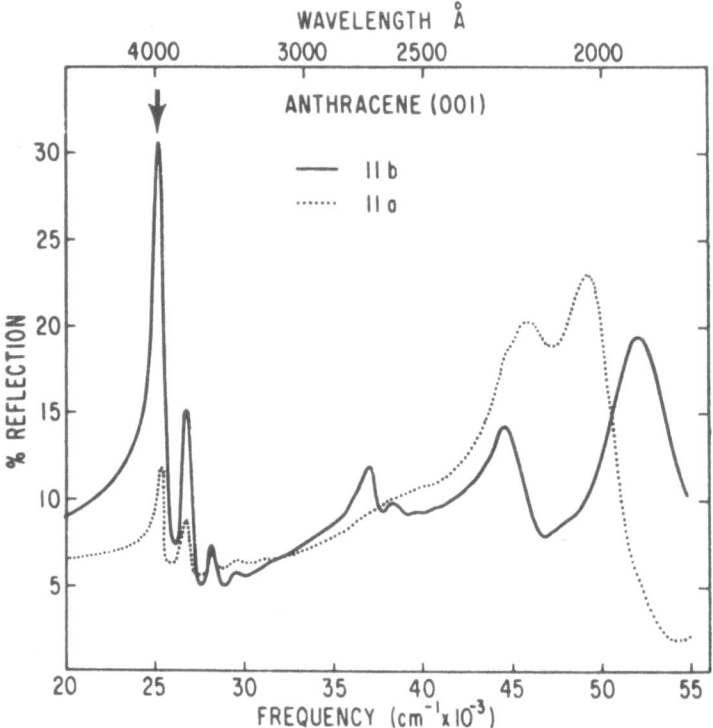

Figure 7. Polarized normal incidence reflection spectra of the
visible and UV transitions of an anthracene single crystal off
the (001) face at room temperature (adapted from Ref. 36).

(201) have also been obtained. The crystals used were either
cleaved melt grown crystals, thin sublimation flakes with (001)
developed or thick multifaceted crystals grown by vapour transport
in an evacuated cell. All samples were mounted in air and
therefore subject to some atmospheric contamination. However
since the (001) surface is known to be resistant to attack by
ozone [35] any contamination is probably due to physical absorption
rather than chemical alteration.

    The reflection spectrum of anthracene at room temperature
taken at normal incidence [36] is shown in Figure 7. In the
b-polarized spectrum we see there is a progression in what appears
to be a 1400 cm^{-1} vibration between 25,000 and 30,000 cm^{-1}, followed
by less intense transitions at 37,000 cm^{-1}, 45,000 cm^{-1} and
52,000 cm^{-1}. These latter are separate π-π* transitions. Upon
cooling to liquid He temperatures the 0-0$_b$ reflection peak (marked
by an arrow in Figure 7) undergoes the remarkable change in shape
[21] shown in Figure 8. The 2K 0-0$_b$ reflection band is shown in

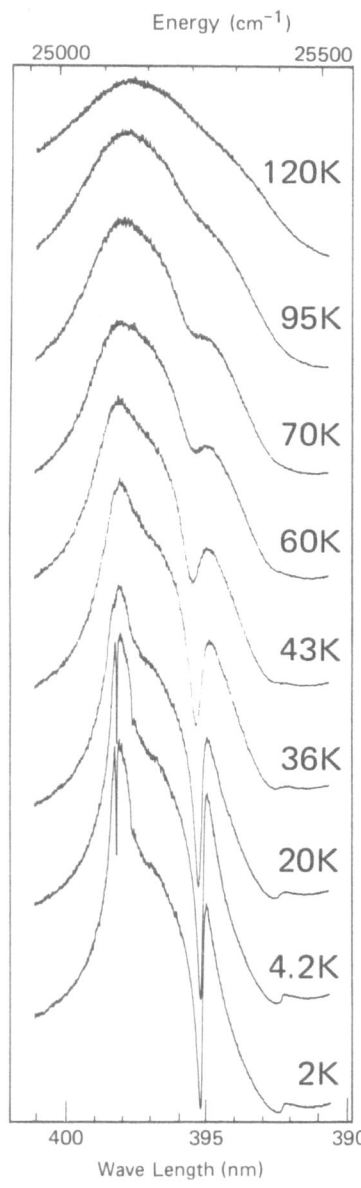

Figure 8. Effect of temperature on the $0-0_b$ reflection band of the first singlet transition of an anthracene crystal. Spectra taken off the (001) face at normal incidence.

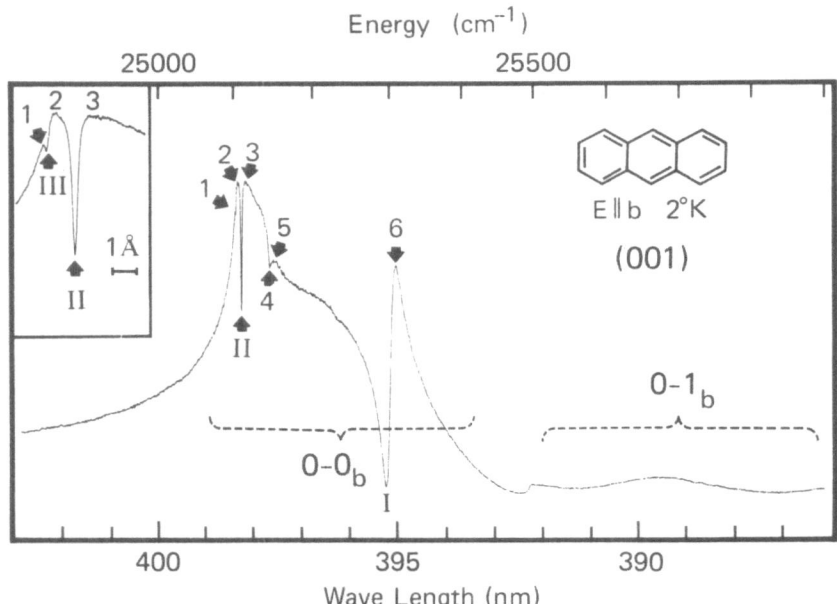

Figure 9. b-polarized normal incidence reflection spectrum off
the (001) face of an anthracene melt grown crystal that cleaved
exceptionally well. Inset is shown, in higher resolution, the
result of a slower scan through the 398 nm region. Note the small
reflectivity minimum III assigned tentatively to a third surface
transition.

greater detail in Figure 9. Also shown is the reflection band of
the $0-1_b$ vibronic transition due to the 390 cm^{-1} a_{1g} vibrational
mode of the electronically excited molecule. The reflection minima
labelled I, II and III (see inset of Figure 9 for the location of
III) have been assigned to surface and subsurface exciton
transitions, whereas the minimum labelled 4 is assigned to a
two-particle process in which a volume polariton and a 40 cm^{-1}
librational phonon are formed [23].

Transitions localized near the surface should be shifted when
the crystal is coated with a thin layer of transparent material.
We decided to try this using various gases. To deposit a frozen
layer of gas on a crystal surface the following procedure was
used. First the spectrum of the uncovered crystal surface was
recorded at temperatures between 1.6 and 2K with the crystal
completely immersed in liquid helium. Second, the flow of liquid
helium was cut off and the sample allowed to warm. After the
helium had boiled away, the sample was surrounded by helium gas
at temperatures in excess of 4.2K. By this time a modest partial
vacuum had developed in the sample chamber and could be changed

by varying the rate of pumping. The partial vacuum was used to
suck the selected gas into the sample chamber through a series of
valves that functioned as a variable leak. The amount of gas
condensing on the surface of the sample was controlled
approximately by monitoring the peak reflectivity at $\lambda=395$ nm. A
decrease in reflected intensity signalled adsorption of gas, and
in this way the exposure of the surface to the admitted gas was
limited.

Figure 10 shows the effect of an increasing amount of condensed
CH_4 gas on the $0-0_b$ reflection band. The top spectrum is that of
the cleaved melt grown crystal without any condensed gas, and the
bottom spectrum was taken after the crystal was warmed to at least
200K and the surface purged in a stream of He gas for twenty-four
hours and cooled again to 2K. All spectra are in sequence and
taken at 2K with the crystal immersed in liquid He. During the
runs more spectra than are actually shown were taken, however the
ones used are representative of the distinct stages of the whole
sequence. The shift of I was -100 cm^{-1}, this is much greater than
the shift of -70 cm^{-1} observed with air and O_2 and N_2 separately.
The fact that methane gave the greatest shift was also used to
show that the (001) surface could be saturated with a frozen air
deposit. No satellites of minimum I of the type seen with air
were observed with CH_4. The differences between spectra 2 and 3
in Figure 10 are slight and attributed to annealing of the film.

The crystal structure of anthracene, depicted in Figure 11,
shows the molecules to be packed densely in the (001) planes,
resulting in a quasi-two dimensional structure and suggesting that
exciton transfer and van der Waals interactions are weakest in the
perpendicular direction. If the interactions between a reference
molecule and all others in a particular plane are grouped together,
as in the method of plane-wise summation of dipole interactions,
then we may express the total site shift D and the transfer sum
I(k) as follows:

$$D = D_o + 2 \sum_{\ell=1}^{L-1} D_\ell \qquad (4.1)$$

$$I(k) = I_o + 2 \sum_{\ell=1}^{M-1} I_\ell \cos(ka\ell) . \qquad (4.2)$$

In arriving at (4.2) it has been assumed that the planes are
perpendicular to k. It has been shown that for the (001) planes
of anthracene I_ℓ ($\ell\geq 1$) are negligible in comparison to I_o and
therefore the energies of bulk states with wavevectors
perpendicular to (001) have no k dependence. The bulk exciton
energy is therefore given by

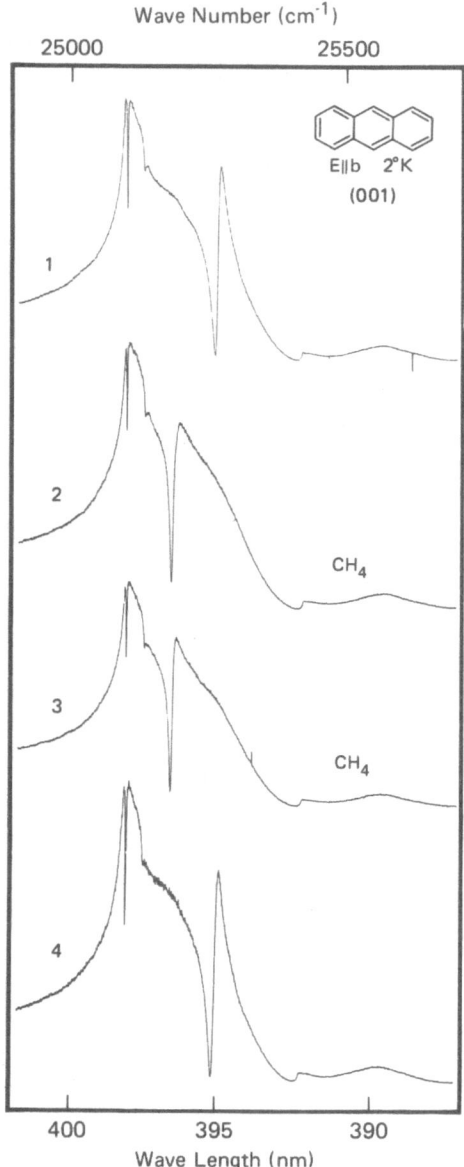

Figure 10. 2K b-polarized normal incidence reflection spectra off the (001) face of an anthracene crystal exposed to methane gas at approximately 10-20K. Spectrum 1 before exposure to CH_4 gas. Spectra 2 and 3 after exposure to CH_4 and spectrum 4 after removal of the frozen gas coating by warming to 200K in a stream of He gas.

$$E_B = \Delta\omega_1 + D + I_o . \qquad (4.3)$$

At the surface of an ideal crystal, where no reconstruction has occurred, the sum over D_ℓ in (4.1) is incomplete. If the range of the plane-sums D_ℓ is not more than twice the interlattice spacing then the first two planes have site shifts of $(D-D_1-D_2)$ and $(D-D_2)$, respectively. The missing interactions are symbolized by the arrows in Figure 12 were the dashed lines represent the planes that would supply the missing interaction. Since the I_ℓ ($\ell \geq 1$) have been assumed to be negligible, the crystal with a perfect surface has two surface states at

$$E_1 = E_B - (D_1+D_2) = E_B + \hbar\delta_1 \qquad (4.4)$$

$$E_2 = E_B - D_2 = E_B + \hbar\delta_2 . \qquad (4.5)$$

We shall refer to δ_1 and δ_2 as the barrier heights of the excitons.

According to the first order perturbation theory of coulombic exciton energies, the D term is made up of interactions of the quadrupole-quadrupole type and shorter ranged electron exchange interactions. Since the point quadrupole-quadrupole interactions fall off like $|m-n|^{-5}$, two powers _faster_ than the point

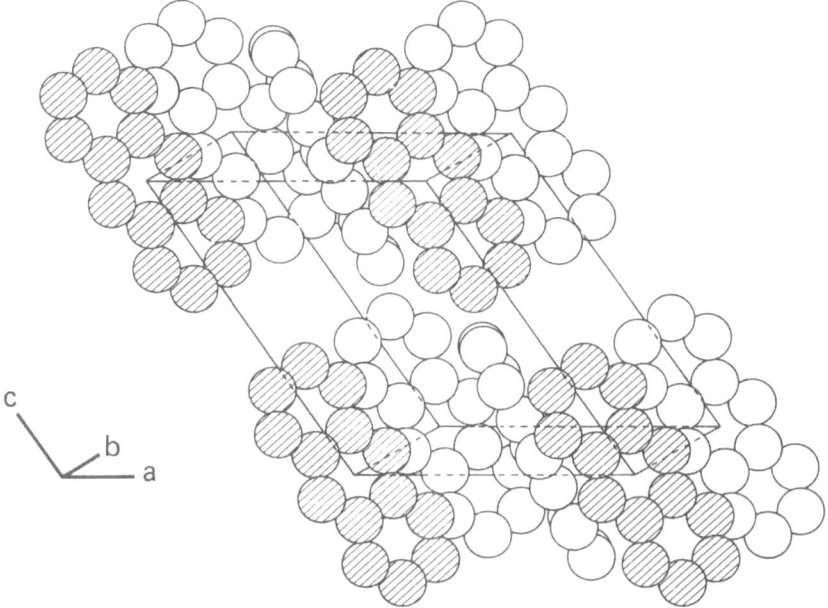

Figure 11. Crystal structure of anthracene.

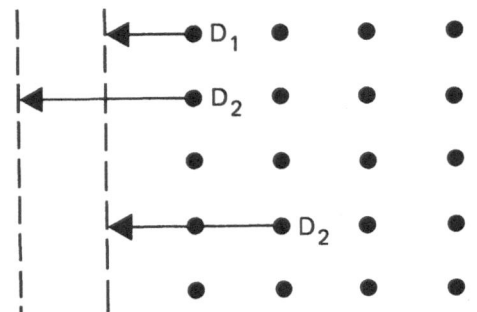

Figure 12.   Schematic diagram illustrating the origin of surface
barrier through the absence of some site shift (van de Waals)
interactions between planes at a crystal surface.

dipole-dipole interactions, it appears to be mathematically
inconsistent to assume that D_ℓ has a longer range than I_ℓ. However
in the second order theory there are contributions from higher
excited states that are primarily dipole-induced-dipole
interactions of the van der Waals type and lattice sums of these
interactions fall off more slowly than the exciton sums I_ℓ. It
is convenient therefore to replace D by D_{eff} which contains both
the first and the second order terms.

If the surface levels fall within the stop-band they give rise
to deep reflectivity minima when they occur near $E_B(=h\omega_T)$ and to
an interference-like structure at energies closer to $E_L(=h\omega_L)$.
The reflection of light from the surface and bulk regions and its
relation to the reflection spectrum and exciton band is shown
schematically in Figure 13. Light R_1 and R_2 reflected by surface
transitions 1 and 2, respectively, interfers with that from the
bulk originating from deeper lying planes of the crystal.

A microscopic theory of the effect of surface transitions on
the reflectivity of the crystal has been developed. For one
surface state the reflection power is given by [37-39]

$$R(\omega) = \left| \frac{n_{eff}(\omega)-1}{n_{eff}(\omega)+1} \right|^2 \tag{4.6}$$

where

$$n_{eff}(\omega) = n_B(\omega) - i\Omega[n_S^2(\omega)-1] \tag{4.7}$$

is given in terms of the bulk crystal refractive index $n_B(\omega)$,
$\Omega=\omega a/c$, and $n_S(\omega)$ a refractive index for a layer of surface

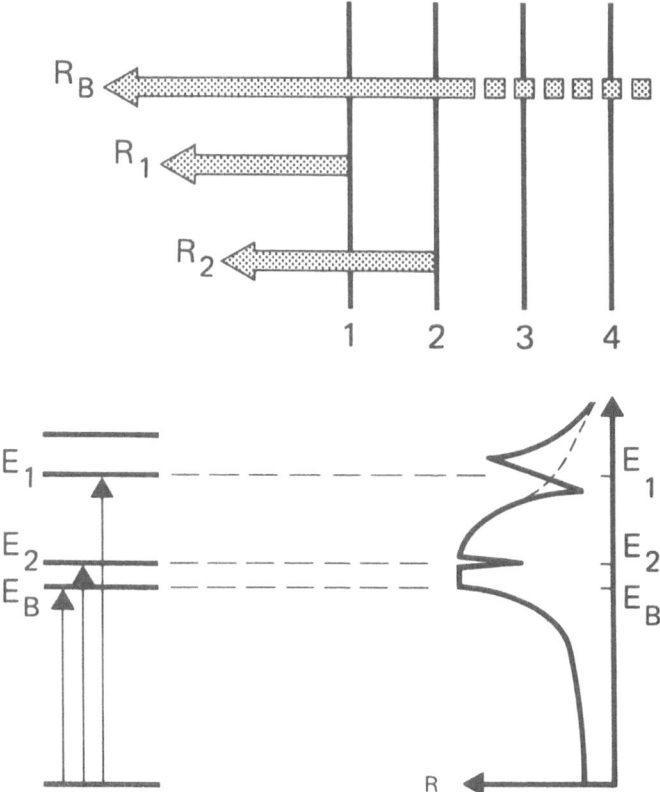

Figure 13.   Interferences in light reflected at frequencies near
a surface exciton transition.

molecules. In terms of an effective polarizability α_S and an
interaction v_S

$$n_S^2(\omega) = 1 + \frac{4\pi\alpha_S(\omega)}{1+4\pi\alpha_S(\omega)v_S} .$$
(4.8)

For (001) planes of anthracene, $\Omega = 0.015$ at a frequency
corresponding to a wavelength of 390 nm. The model with one surface
state gives a more complicated expression for the reflection power
than that obtained from Fresnel's equations. If the surface
molecules differ from those of the bulk by having energies higher
by an amount δ then

$$n_S(\omega) = n_B(\omega - \delta) \ . \tag{4.9}$$

The microscopic theory upon which (4.6) is based is general enough to allow for surface reconstruction (which would in general alter D_ℓ, I_o and the interplane spacing a), and to include the effect of gas absorbed on the surface during the mounting of the crystal on the sample holder. Thus the surface barrier heights $\delta_1, \delta_2, \ldots$ extracted from experimental data include the effects of surface contamination, reconstruction, strain, etc.

It has been demonstrated in earlier work [39,40] that the approximation $I_\ell = 0$ ($\ell \geq 1$) leads to the equations of classical optics where only one refractive index function is necessary to describe the optical properties of the solid. Therefore one suspects that the equations describing the reflection of light from thin films on a substrate can be manipulated into a form that is similar to Eqs. (4.6-7).

Consider the optical system consisting of three media with refractive indices n_o, n_1, and n_2 separated by two parallel planar surfaces that restrict medium 1 to a thickness a. The reflectivity amplitude for light incident normally from medium with refractive index n_o is [41]

$$\rho = \frac{\rho_1 + \rho_2 \exp(-i2n_1\Omega)}{1 + \rho_1\rho_2 \exp(-i2n_1\Omega)} \tag{4.10}$$

where $\Omega = \omega a/c$ as before and

$$\rho_1 = \frac{n_o - n_1}{n_o + n_1} \ , \tag{4.11}$$

$$\rho_2 = \frac{n_1 - n_2}{n_1 + n_2} \ . \tag{4.12}$$

For a very thin film such that $|n_1\Omega| \ll 1$, it is possible to manipulate the formula for ρ into

$$\rho = \frac{1 - n_B + i\Omega[(n_S)^2 + n_S - n_B - n_S n_B]}{1 + n_B - i\Omega[(n_S)^2 - n_S + n_B - n_S n_B]} \tag{4.13}$$

where for ease of comparison with (4.6) and (4.7) we have set $n_o = 1$, $n_1 = n_S$ and $n_2 = n_B$. Since $\Omega \ll 1$ we can neglect all terms except the first in the square brackets of the numerator and denominator, provided we assume that n_S and n_B peak at different frequencies. Under this condition, ρ yields a reflection power with the same form as Eqs. (4.6-7). Clearly one can extend these ideas to a substrate supporting a series of films, some of these films could represent frozen layers of absorbed gas or they could represent different surface regions of the same crystal.

Since equations (4.6-7) and (4.13) are compatible, the minima in reflection spectra are amenable to a phenomenological analysis. In crystals where spatial dispersion is negligible ($I_\ell=0, \ell \geq 1$), the equations of thin film optics can be used to fit the observed spectra. Such an analysis cannot show how the refractive indices of the films are related to the bulk crystal, one needs a microscopic theory to make this step. As mentioned above, the microscopic model is flexible enough to be able to describe the effects of gases absorbed during the cleaving and mounting, surface reconstruction and even distortions of surface planes caused by strains.

The existence of surface excitations with energies greater than the bulk transition implies that in a perfect crystal the bulk exciton is spatially trapped within the crystal, separated by several planes of less easily excited molecules. The surface states relax by radiative decay (which can be up to 10^4 times faster than the decay rate of the isolated molecule thereby making surface lasing a possibility at high excitation densities) or by phonon emission to the bulk states.

If the surface exciton hypothesis is correct, shifting the level by means of a condensed gas reveals the shape of the reflection band due to the bulk crystal transitions. One can therefore replace the reflectivity in the region of the reflection minima by values selected by comparison with coated crystals. By comparing a spectrum for an uncoated crystal with one coated with CH_4 we have derived a reflection band of the bulk transition alone. This bulk reflection spectrum has been used to calculate the optical properties of the bulk region using the Kramers-Kronig transformation.

Having obtained the bulk optical constants, we have used them in Eqs. (4.6), (4.7) and (4.9) to calculate theoretical reflection spectra with surface transitions corresponding to surface barrier heights of $\delta=10$, 150 and 200 cm^{-1}. The results are shown in Figure 14 and may be compared with the derived bulk crystal reflection spectrum which is shown in the upper left-hand panel. For $\delta=10$ cm^{-1} the calculated spectrum describes quite well the region around minimum II. Likewise for $\delta=200$ cm^{-1} the calculated spectrum shows striking agreement with the observed spectrum around minimum I. The intermediate case ($\delta=150$ cm^{-1}) closely resembles the spectrum

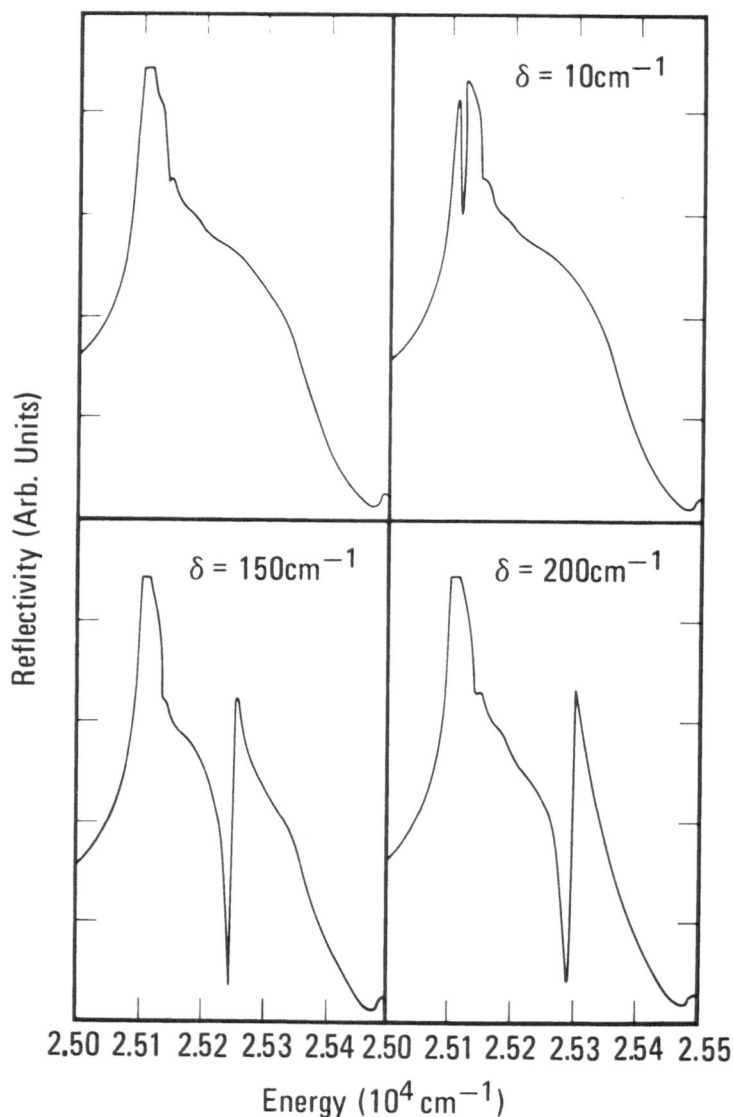

Figure 14.   The effect of a single surface transition on the bulk
crystal reflection spectrum (top left) calculated for surface
barrier heights of δ=10, 150 and 200 cm⁻¹. These calculated spectra
are qualitatively identical to the measured b-polarized reflection
spectra.

obtained when the crystal is coated with a frozen film of air.
These calculations prove that the equations used are capable of
describing the observed reflection spectrum and so represent a
self-consistent test of the theory.

On the basis of these calculations and the shift of minimum
I by frozen air films we assign the combined structure minimum I
and maximum 6 to a surface transition located at their midpoint
25299 cm^{-1}=E_{Bb}+205 cm^{-1}. The mean value of the minimum and maximum
is chosen because we recall that these structures occur through
an interference mechanism. The separation of the minimum and
maximum is 12 cm^{-1} and represents an approximate value for the
width of the surface state. The contribution of surface state II
is also well described by the simulated spectra in Figure 14.
When the crystal surface is disrupted by photo-oxidation [21] it
is erased from the reflection spectrum in the same way as minimum
I. Therefore, it too is assigned as surface transition located
at the midpoint of minimum II and maximum 3, at E_{II}=E_{Bb}+10 cm^{-1}.
There is no evidence presently available that minimum III is due
to a surface transition. Minimum III has not been observed to
shift after covering the (001) face with a frozen gas deposit.
This may be simply because the transition is localized too far
inside the crystal to be effected by frozen gas layers. Therefore
the only basis for assigning minimum III as a surface transition
is by analogy with the assignments for I and II. Very tentatively
a third surface transition is assigned to the mean of minimum III
and maximum 2 at E_{III}=E_{Bb}+2 cm^{-1}. Table IV summarizes the
assignments for the 0-0$_b$ reflection band.

Using Eqs. (4.4) and (4.5) extended to include L=3 for the
range of the D_ℓ interactions we obtain D_1=-195 cm^{-1}, D_2=-8 cm^{-1}
and D_3=-2 cm^{-1}. Note that these three numbers form a sequence
that drops off like $(\ell a)^{-4}$ as required for plane-sums of
van der Waals interactions (the individual interactions fall off as
the sixth inverse power of the separation). Note that if the
interactions fell off in strict accordance with an ℓ^{-4} power law
then the numbers would be -195, -12.2 and -2.4 cm^{-1}. Since the
surface transitions are associated in the theoretical model with
the first three crystal planes we see that E_I is a microscopic
surface exciton while E_{II} and E_{III} are subsurface states.

Glockner and Wolf [29] were the first to publish evidence of
fluorescence emission from levels lying higher than E_{Fb}. Lines
were observed at 25298 cm^{-1} (FWHM=16±4 cm^{-1}) and 25103 cm^{-1} (FWHM
~3 cm^{-1}). In later work an additional line was observed at 1.6K
situated halfway between E_{Fb} and 25103 cm^{-1} in crystals they
referred to as type A [32]. In type B crystals the first two
lines were shifted to longer wavelengths. The lines for type A
crystals agree sufficiently well with our estimates of the
positions and widths of I, II and III, that we have assigned the

Table IV. 0-0 Region of the b-polarized Reflection Spectrum of
          Anthracene (001) at 2K.

Label (See Fig. 9)	Position cm^{-1} (vacuum)	Description[a]	Difference from 25094	Assignment
1	25094	sm. max.	0	origin, E_{Bb}
III	25094	sm. min.	0 ⎫	surface state III $E_{III}=E_{Bb}+2$ cm^{-1}
2	25097	max.	3 ⎭	
II	25102	shp. min.	8 ⎫	surface state II $E_{II}=E_{Bb}+10$ cm^{-1}
3	25107	max.	13 ⎭	
4	25138	sm. shp. min.	44	phonon, $E_{Bb}+44$ cm^{-1}
5	25143	br. max.	49	
I	25293	dp. min.	199 ⎫	surface state I $E_{I}=E_{Bb}+205$ cm^{-1}
6	25305 (±4)	shp. max.	211 ⎭	

[a] sm.=small, shp.=sharp, max.=maximum, min.=minimum, dp.=deep.

lines observed by Glockner and Wolf to fluorescence from surface
exciton states. It should be noted however that except for state
I our absolute accuracy is insufficient to assign the fluorescence
lines to either minima, maxima or their midpoint. We have assumed
that the fluorescence peak corresponds to the mean of the
associated minimum and maximum because this structure arises by
the interference mechanism described earlier.

    Fluorescence from coherent exciton states of monomolecular
layers should have a radiative width larger than the free molecule
emission by the multiplicative factor $(\lambda/a)^2 \approx 10^4$, where a is the
lattice constant and λ is the vacuum wavelength divide by 2π. The
surface planes of anthracene behave like monolayers on a substrate
with refractive index n_B and it can be shown that the radiative

width includes $(n_B+1)^{-1}$ as an additional factor. This inverse
dependence on n_B causes the radiative width to decrease and
therefore the radiative lifetime to increase as the barrier height
δ decreases. Thus part of the decrease in the FWHM with decrease
in δ is explained as a change in the radiative width.

Finally this section is concluded with a brief report on work
in progress. The reflection spectra of multifaceted crystals
(grown by a vapour transport in evacuated cells) have been measured
[42]. The faces sufficiently well developed to be examined by
normal incidence spectroscopy are (00$\dot{1}$), (20$\bar{1}$), ($\bar{1}$01), (110),
($\bar{1}\bar{1}$1) and ($\bar{1}\bar{1}$2). Only the (001) face shows minima within the O-O$_b$
reflection band. The absence of surface transitions on (20$\bar{1}$) and
($\bar{1}$01) is significant since both faces contain the b crystal axis.
The most attractive explanation assumes that the interplane exciton
transfer interactions $I_\ell(\ell \geq 1)$ are not negligibly small for these
faces and so the surface transitions are smeared out by the mixing
of volume and surface states. Fluorescence from (001) has also
been reexamined recently [43], and it has been conclusively
demonstrated that the emission from I is peaked between the maximum
and minimum in the reflection spectrum lending strong support to
the interference mechanism described earlier.

V. EXCITON SURFACE POLARITONS

Consider a system consisting of two semi-infinite dielectrics
joined at the plane z=0. Suppose for z<0 the first halfspace is
isotropic with a dielectric constant ε_A and for z>0 the other
halfspace is also isotropic but with a frequency dependent
dielectric function $\varepsilon(\omega)$ given by Eq. (2.3). Maxwell's equations
for this system admit two types of solution. In the first type
the electromagnetic field has an oscillating (photon like)
component out to infinity in both directions along the z axis
($\omega < \omega_T$ and $\omega > \omega_L$), or an oscillating part for z<0 and an
exponentially decaying part for z>0 ($\omega_T < \omega < \omega_L$). These solutions
correspond to an incident field with reflected and transmitted
components. The second type of solution does not have an
oscillating part at $\pm\infty$ but decays exponentially in directions
perpendicular to the surface plane z=0. This second type of
solution corresponds to an electromagnetic wave travelling along
the interface of the two media [10]. It has three main
characteristics: (i) It is p-polarized i.e., the E-field is in the
plane containing the surface wavevector κ and the normal z to
the surface, (ii) it occurs only for frequencies ω such that
$\varepsilon_A + \varepsilon(\omega) < 0$, (iii) the field amplitudes decay exponentially in
directions perpendicular to the surface.

Surface electromagnetic waves were first investigated by
Sommerfeld and Zenneck in connection with problems in radio

physics. At infrared and optical frequencies surface electromagnetic waves arise from the coupling of photons to surface
electric dipole excitations. The dipoles may arise from a plasma
oscillation of free electrons, or from electric polarization due
to excitons or vibrating ions. More recently they have been
extensively studied on the surfaces of metals, where they are
referred to as surface plasmons (or more accurately plasmon surface
polaritons), and the surfaces of ionic crystals at infrared
frequencies (phonon surface polaritons). In the last two years
exciton surface polaritons (ESPs) have been reported for several
inorganic solids [44-45] and one organic solid [46-47].

For an isotropic surface active medium the dispersion of ESPs
is given by the following equation

$$c^2 \kappa^2 / \omega^2 = \varepsilon_A \varepsilon(\omega) / [\varepsilon_A + \varepsilon(\omega)] \ . \tag{5.1}$$

The inequality $\varepsilon_A + \varepsilon(\omega) < 0$ restricts the surface wave to frequencies
between ω_T and a cutoff frequency ω_S given by

$$\omega_S^2 = \omega_T^2 + \omega_{po}^2 f / (\varepsilon_o + \varepsilon_A) \ . \tag{5.2}$$

It is a characteristic of surface polaritons of all types that
the cutoff frequency depends on the dielectric constant ε_A of the
surface inactive medium. The effect of increasing ε_A is to
compress the frequency range of the surface mode. Figure 15 shows

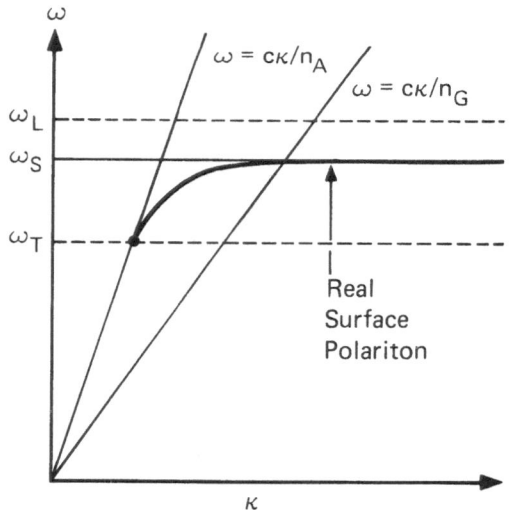

Figure 15. Dispersion of a real exciton surface polariton in a
molecular crystal.

the dispersion curve of an ESP and its position relative to the
stop-band, and the light-lines for the surface inactive medium
$\omega = c\kappa/n_A$ and a medium with refractive index $n_G > n_A$. Notice that
the ESP dispersion lies entirely on the right-hand side of the
light-line for the surface inactive medium. This means that
photons incident on the crystal face from the semi-infinite medium
A cannot excite the surface wave because it is not possible to
simultaneously match frequency and surface wavevector κ. If
another inactive medium with refractive index $n_G > n_A$ is brought
close to the surface (but not close enough to disturb the
dispersion of the ESP) then since the new light-line cuts the
surface polariton branch, it is possible to couple photons in
medium G to the ESP. This is the principle of detection of surface
waves by ATR spectroscopy.

Surface polaritons with a dispersion given by Eq. (5.1) are
defined for all values of the surface wavevector κ greater than
the minimum value $\kappa_T = \omega_T n_A/c$. This is clear from the curve in
Figure 15. These modes are called real surface polaritons because
there is no volume polariton with wavevector parallel to the
surface that can match phase and energy with the ESP. Put another
way, volume polaritons are excluded from the frequency range
$\omega_T < \omega < \omega_L$ because the stop-band is invariant under change in the
direction of the three dimensional wavevector k.

Organic solids are rarely isotropic, therefore a more complex
model of ESPs is needed. It is sufficient for present purposes
to assume that the surface active medium is orthorhombic with
dielectric axes X and Z parallel to κ and z respectively. It can
be shown that the dispersion relation for ESPs polarized in the
xz plane is given by

$$\kappa^2 c^2/\omega^2 = \frac{\varepsilon_A \varepsilon_Z (\varepsilon_A - \varepsilon_X)}{\varepsilon_A^2 - \varepsilon_X \varepsilon_Z} \quad . \tag{5.3}$$

The case of greatest practical interest occurs when ε_Z is
independent of frequency and $\varepsilon_X(\omega)$ is given by the right-hand side
of Eq. (2.3). This geometry approximates that of anthracene on
the (001) face with the b axis parallel to X, and TCNQ° on face
(010) with the direction of maximum reflectivity parallel to x.

It can be shown that there is no solution of Eq. (5.3) if $\varepsilon_Z < \varepsilon_A$,
and therefore there are no surface modes. When $\varepsilon_Z > \varepsilon_A$ there is a
volume polariton with wavevector k parallel to the x axis and E
vector parallel to z that intersects the ESP dispersion curve.
Thus the range of k is bounded and never exceeds $\omega_L n_Z/c$. Surface
modes of this type with bounded wavevectors are called virtual
ESPs to distinguish them from real ESPs. The intersection of the
volume and surface modes is shown schematically in Figure 16.

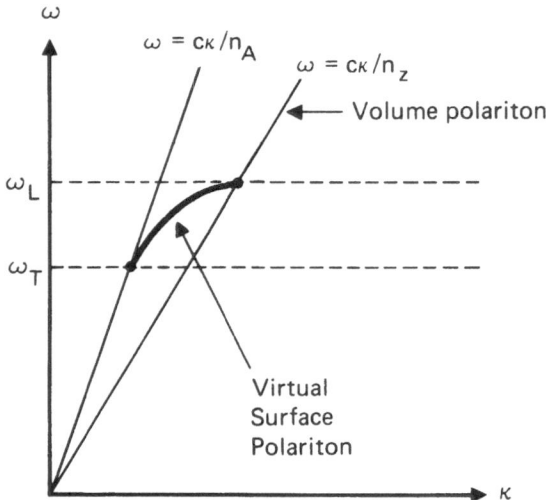

Figure 16. Dispersion of a virtual exciton surface polariton in a molecular crystal. The virtual mode stops at the point of intersection with the volume polariton $\omega = c\kappa/n_Z$.

Recently Tomioka, Sceats and Rice [46-47] reported the observation of an ESP on the (001) surface of anthracene using prism coupling to match phase and energy of the incident beam to the surface wave. Figure 17 shows an example of their results for a gap of 100 nm, a series of angles of incidence and a temperature of 16K. The ESP is visible in these spectra as a prominent peak that moves to higher energy as the angle of incidence increases. In these experiments the prism was made from IRG-2 glass with a refractive index $n_G=1.95$, and the surface inactive medium was a layer of the polymer PMMA with a refractive index $n_A=1.51$. The measurements utilized dual beams, one reflected from the sample giving signal R_p and the other displaced sufficiently to miss the crystal sample giving a signal R_0. The critical angle for the glass-PMMA interface in these experiments is $\theta_c=50.75°$.

The reflection minimum I is poorly resolved in these experiments due to thermal broadening and possible inhomogenous contact of the PMMA and anthracene. Also the reflection minimum does not appear to be shifted to the red, which would be expected if there were intimate molecular contact between the PMMA spacer layer and the crystal surface.

To investigate this problem further, theoretical ATR spectra have been calculated using the optical constants derived from the normal incidence reflection spectrum at 4K. No attempt was made to simulate the experimental ATR closely; instead it was decided

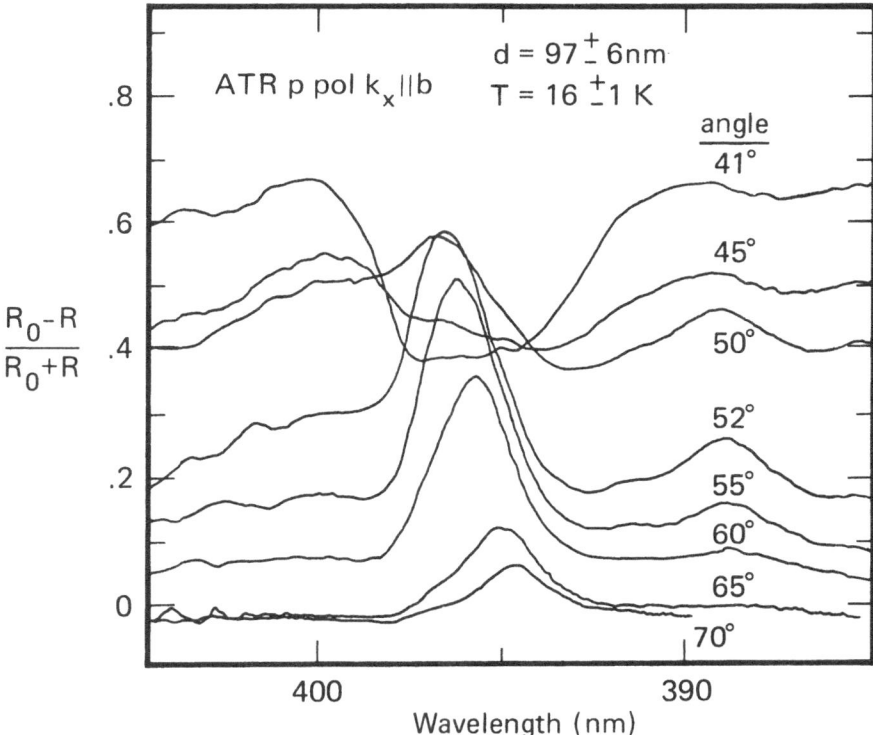

Figure 17. Experimental ATR spectra of an anthracene crystal at 16K off the (001) face. Peaks correspond to the exciton surface polariton resonance. Adapted from Tomioka, Sceats and Rice, J. Chem. Phys. 66, 2984 (1977).

to examine qualitatively the effect of surface exciton I on the ATR spectrum. Representative results are shown in Figure 18 using the bulk dielectric function (no surface exciton) and in Figure 19 with the combined bulk and surface contributions (surface exciton present). In all these calculations n_G=1.95, n_A=1.00 (vacuum gap) and the gap was 400 nm wide. The critical angle θ_c=30.85° in these calculations, which is smaller than the θ_c=50.75° in the experiment. The calculations predict that the anthracene ESP is observable as a large effect in the ATR spectrum. If surface exciton I is absent or only weakly present, then spectra like those of Figure 18 are expected. These qualitatively model the experimental ones of Tomioka, Sceats and Rice though they are shifted to higher frequencies, a result due in part to a smaller value of n_A. On the other hand, if surface exciton I is present then it occurs as strongly as the ESP and there is a shift in I to higher energies as θ increases. It is clear from a comparison of the theoretical and experimental ATR spectra that it is of some

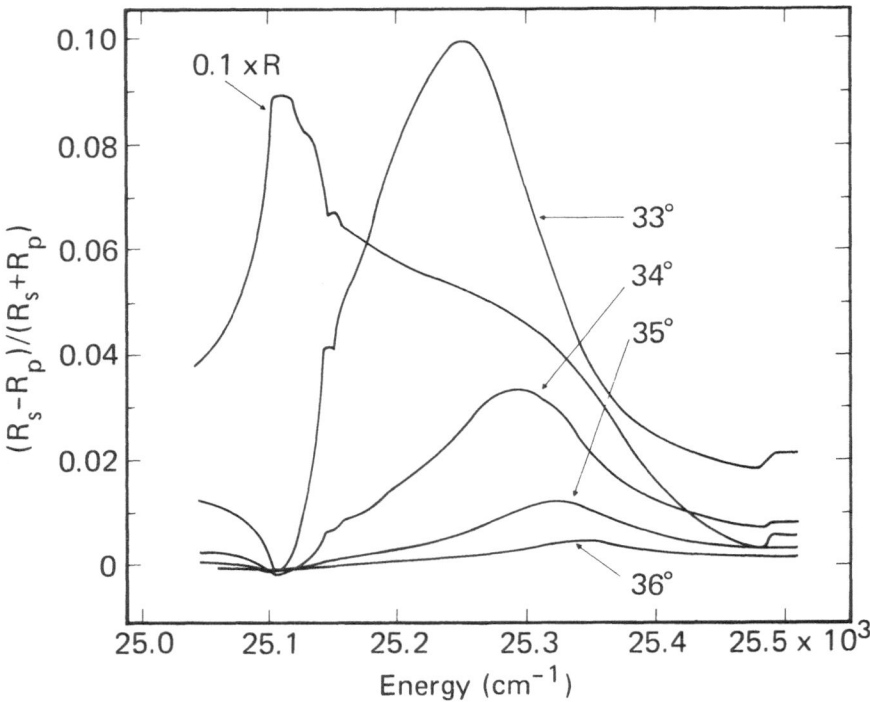

Figure 18.   Theoretical ATR spectra for an anthracene crystal off
the (001) face at 4K.   Surface exciton transitions are *not* included
in this calculation.

importance to repeat the experiments of Tomioka et al. at liquid
He temperatures, without the PMMA spacer layer.   These experiments
should yield detailed information about the dispersion of the ESP,
the electromagnetic components of the site shift surface excitons
and the interaction between these different types of surface wave.

VI.   PLASMON SURFACE POLARITON - MONOLAYER INTERACTIONS

The final topic is concerned with the effect on plasmon surface
polaritons (PSPs) of organic molecules laid down on the metal
surface by the Langmuir-Blodgett technique [48-49].   PSPs have
been studied on the surfaces of a number of metals, as has the
effect of inorganic overlayers on the resonances [6].   What has
not been done previously is the engineering of metal-insulator
interface by depositing a number of layers of known thickness.
By doping the monolayers with dye molecules the optical properties
of the organic film can be altered.   One long range goal of this

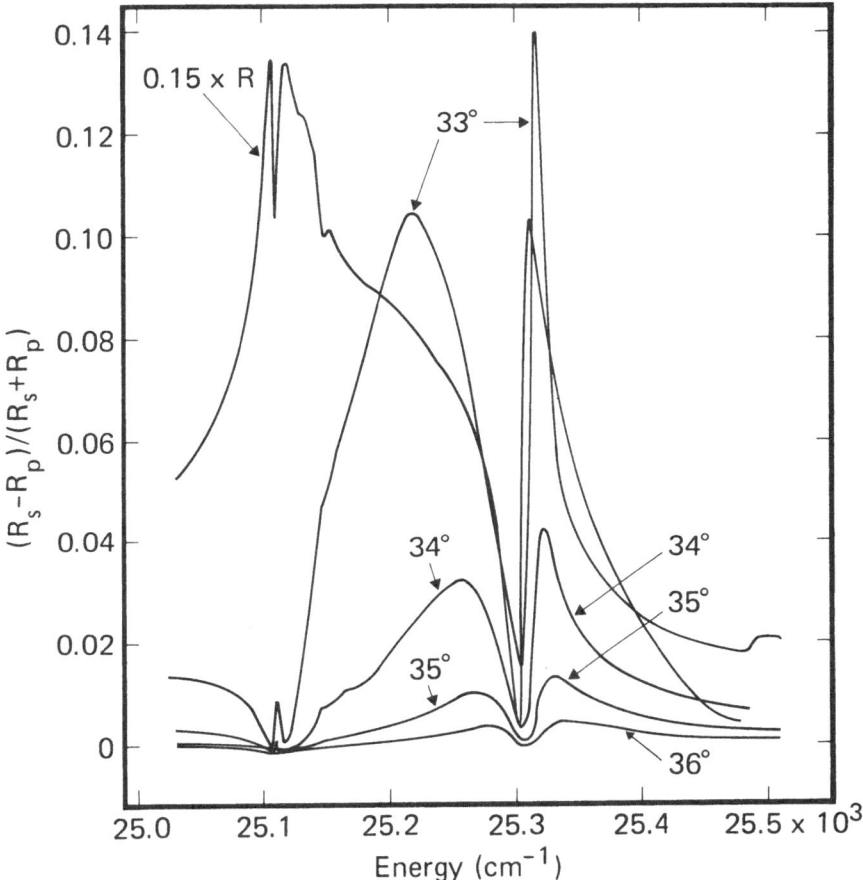

Figure 19. Theoretical ATR spectra for an anthracene crystal at 4K off the (001) face. Surface exciton transitions <u>are</u> included in this calculation.

work is the fabrication of an interface with exciton states that are coupled to plasmon surface polaritons.

The requirement for the observation of a narrow PSP resonance in an ATR experiment is the same as for exciton surface polaritons, namely that $\varepsilon(\omega)+\varepsilon_A<0$ and $|\mathrm{Re}\varepsilon(\omega)|>>\mathrm{Im}\varepsilon(\omega)$. For our purpose the properties of PSPs on metal films can be introduced by treating them as a special case of an exciton surface polariton. In the experiments to be described, a thin metal film is sandwiched between two insulators with different optical constants. The surface modes of such an asymmetrically placed metal slab can be understood easily if we consider first the surface of a metal halfspace and then the modes of a symmetrically sandwiched metal slab.

Consider first an insulator with isotropic dielectric tensor given by the right-hand side of Eq. (2.3). To obtain the optical response of free electrons in a metal we take the limit $\omega_T \to 0$ so that the dielectric function becomes

$$\varepsilon(\omega) = \varepsilon_o - \frac{\omega_p^2}{\omega^2 + i\omega\gamma} . \qquad (6.1)$$

Here we have induced a damping term $i\omega\gamma$ and set $\omega_p = \omega_{po}\sqrt{f}$. In the limit $\omega_T = 0$ the dispersion of the surface mode starts at $\omega = 0$ and $\kappa = 0$ in Figure 15, so that the PSP exists over a wide range of frequencies, formally all the way from the infrared to the ultraviolet. In practice, metals often have interband transitions that results in $|\text{Re}\varepsilon(\omega)| \leq \text{Im}\varepsilon(\omega)$ and a consequential severe restriction on the frequency window for detection of the PSP.

On the surface of a metal halfspace the PSP has an electromagnetic field that dies exponentially in directions perpendicular to the surface. The range of the field is typically the wavelength of the photon with the same energy for the light-like portion of the dispersion curve. At frequencies close to the cutoff frequency ω_S the range drops to a few nanometers. Figure 20a shows schematically the spatial extent and dispersion of a PSP on the surface of a metal halfspace. The dispersion relation is given by Eq. (5.1) with (6.1) substituted for $\varepsilon(\omega)$. The cutoff frequency

$$\omega_S = \omega_p(\varepsilon_o + \varepsilon_A)^{-1/2} \qquad (6.2)$$

has its maximum value $\omega_p/\sqrt{2}$ for a free electron gas $\varepsilon_o = 1$ adjacent to vacuum $\varepsilon_A = 1$.

In the symmetric case, a metal slab, infinite in lateral directions, has two PSPs, one associated with each surface, see Figure 20b. If the slab is thin enough, the two surface modes are coupled with the result that one rises above ω_S, and the other is depressed. At large κ both modes asymptotically approach ω_S since they become decoupled. Note that both PSPs lie on the right-hand side of the light-line $\omega = c\kappa/n_A$ and hence neither can be excited by a photon from the semi-infinite media A.

For an asymmetric slab e.g., a film metal film plated on the hypotenuse side of a right-angled prism, there are two cutoff frequencies ω_{SI} and ω_{SII} corresponding to the two different metal-insulator interfaces. The dispersion of the modes is shown schematically in Figure 20c. Note that mode I lies to the right

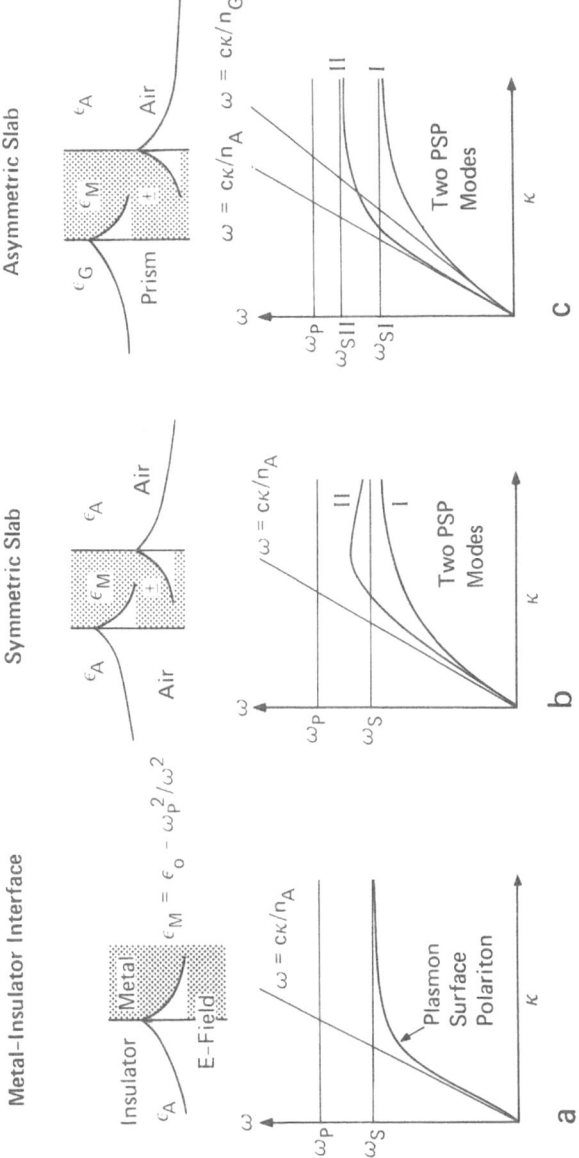

Figure 20. Schematic diagram depicting the spatial extent and ω vs. κ dispersion of plasmon surface polaritons on a single surface, symmetric slab and asymmetric slab surfaces.

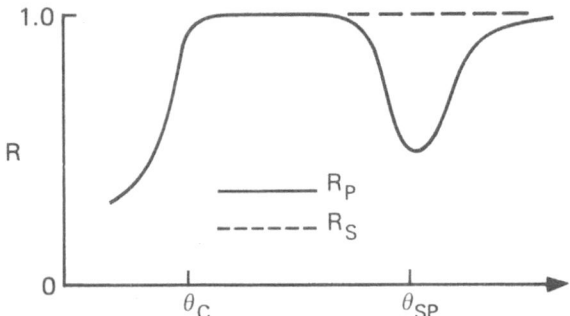

Figure 21. Schematic diagram depicting the detection of plasmon surface polaritons by sweeping angle of incidence θ whilst keeping the excitation frequency constant.

of the light lines of the prism and air; it is associated with the metal glass interface and cannot be detected by ATR from within the prism. The second mode, however, associated primarily with the metal-air interface, intersects the prism light line and can be detected in the ATR experiment. The dispersion relation for the PSP in the asymmetric slab configuration is

$$(k_1 + k_2)(k_2 + k_3) + (k_1 - k_2)(k_2 - k_3)e^{-2k_2 d} = 0 \qquad (6.3)$$

where d is the thickness of the metal slab and

$$k_i = \varepsilon_i^{-1}(\kappa^2 - \varepsilon_i \omega^2/c^2)^{1/2} \qquad (6.4)$$

Here ε_i is the dielectric function of glass (i=1), metal (i=2) and air (i=3).

The detection of PSPs on metal surfaces is generally done by bringing a prism close to the metal surface [50] so that there is a spacer gap of lower refractive index than the coupling prism,

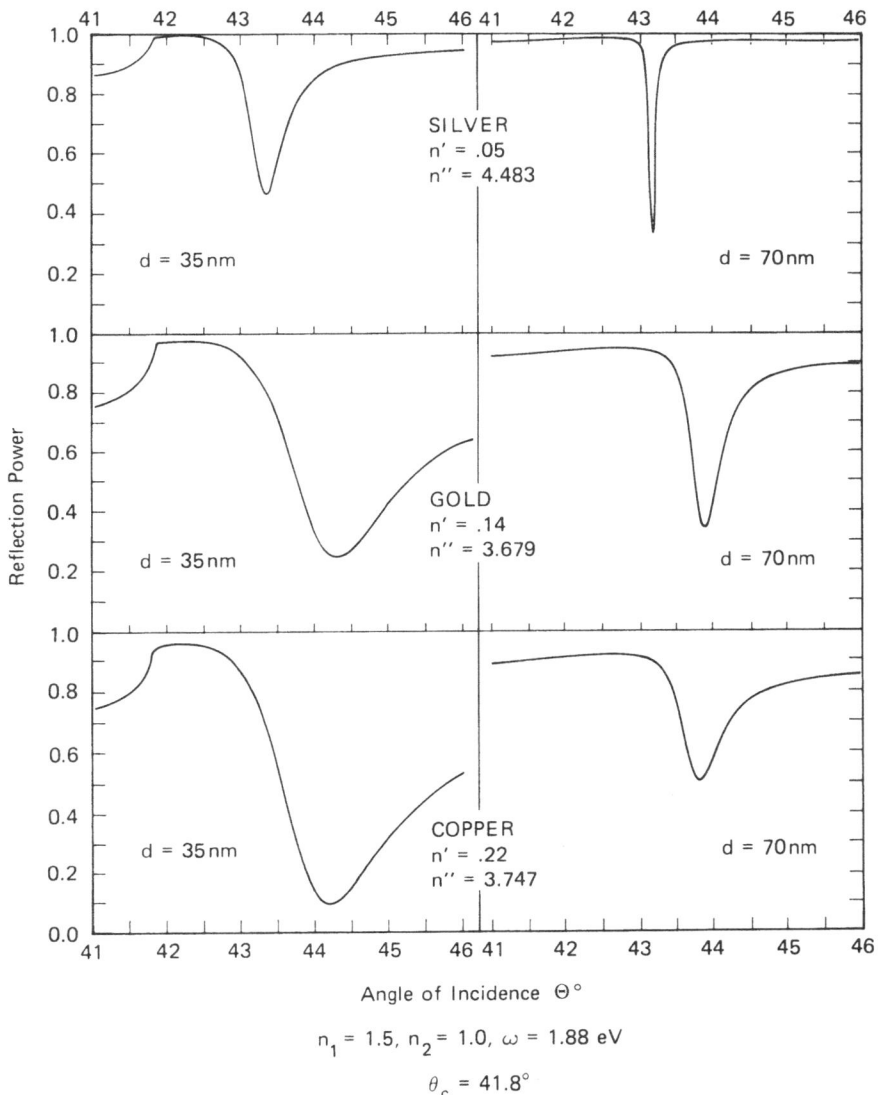

Figure 22. Theoretical ATR spectra of plasmon surface polaritons on Ag, Au and Cu films on a prism in the Kretschman configuration.

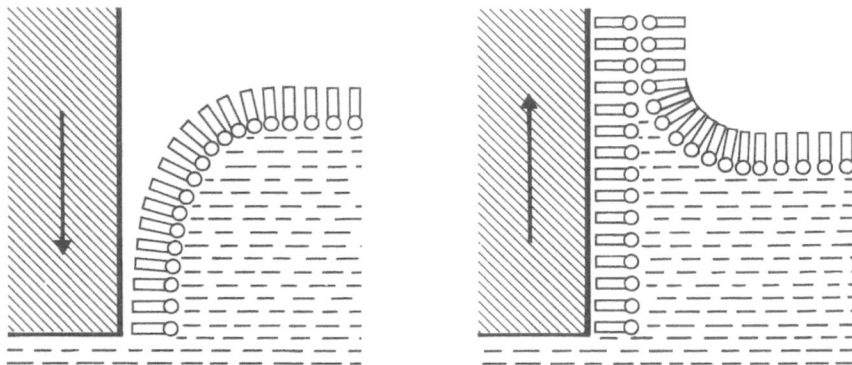

Figure 23. Deposition of a fatty acid bilayer on a gold film.
Each molecule is represented as a bar for the hydrocarbon portion
and an open circle for the carboxylate group and associated
counterions.

or by plating a thin film of metal onto a prism [51]. We shall
refer to these as the Otto and Kretschman techniques respectively.
In both configurations it is convenient to use a well columated
light beam like a laser and to rotate the prism so that the angle
of incidence is changed. The reflectivity becomes unity as the
angle of incidence passes the critical angle θ_c and then goes
through a minimum as the PSP resonance is passed. This is
schematically indicated in Figure 21. The results of a calculation
of theoretical ATR spectra are shown in Figure 22 for Ag, Au and
Cu. Notice the sensitivity to film thickness and optical
constants. The thickness is crucial since in very thin films the
PSP resonance is very broad due to radiation damping. The "window"
for PSP resonances is quite extensive for Ag, extending throughout
the visible. For Au and Cu the window is much narrower.

In an elegant series of experiments which marries two
techniques, deposition of monomolecular films and plasmon surface
polariton spectroscopy, Gordon and Swalen observed the shift of
the PSP resonance of a gold film due to organic films of increasing
thickness [5]. The monolayers were put down two at a time using
the so-called Y deposition technique illustrated in Figure 23. The
monolayers were made from cadmium arachidate
($(\text{n-CH}_3(\text{CH}_2)_{18}\text{COO}^-)_2\text{Cd}^{++}$) spread on a water surface from a
chloroform solution and held under a surface pressure of
approximately 30 dynes/cm. Each monolayer consists of a closely
packed parallel arrangement of rodlike arachidate ions with their
associated counterions and is 2.68 nm thick. As the gold surface
is hydrophobic, the hydrocarbon end of the first layer is adjacent
to the gold.

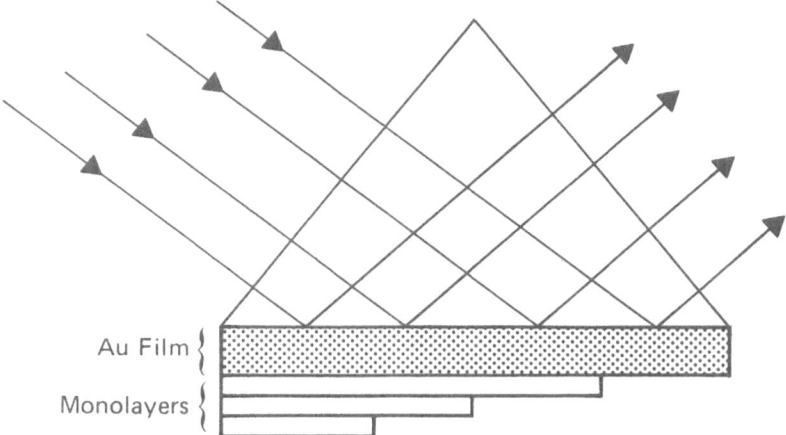

Figure 24. Schematic diagram illustrating the measurement of ATR spectra of cadmium arachidate monolayer on a gold film. Measurements on films of several different thicknesses are taken on the same gold coated prism.

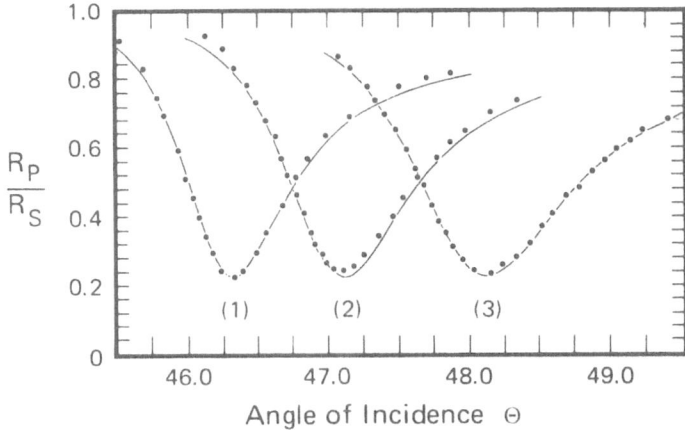

Figure 25. Reflectivity vs. angle of incidence at fixed frequency (632.8 nm) for the configuration of Figure 24, where θ is the angle of incidence. Curve 1 is the bare gold film. Curves 2 and 3 are for the same film covered with 2 and 4 monolayers, respectively, of cadmium arachidate. The dots are experimental points and the solid line is the calculated curve. Adapted from Ref. 5.

The experimental arrangement is shown schematically in Figure 24. The Kretschmann configuration was used because it is easier to control the thickness of a deposited metal film than to control that of an air gap only a few hundred nanometers thick. The thickness of the metal film is important as explained earlier. It must be sufficiently thin for the field to penetrate to the opposite surface, but still thick enough that there is not undue interaction between the two surface excitations, on the metal-air and on the metal-glass interfaces. The gold films in this work were approximately 70 nm thick and were e-beam evaporated onto microscope slides at 10^{-6} torr. The slide was brought into optical contact with a prism using an index matching liquid.

The experiments were performed using light of wavelength 632.8 nm, from a polarized He-Ne laser. The wavevector κ, was scanned by rotating the prism and varying the angle of incidence, θ. Since only p-polarized light can excite the surface plasmon, the spectra are conveniently reported as R_p/R_s, which avoids the measurement of absolute intensity. Figure 25 shows a typical set of ATR spectra. The PSP resonance shifts approximately 1° for each bilayer deposited. For a monolayer the shift of 0.5° corresponds to a change in κ of 900 cm^{-1}. Crudely this result is explained as a lowering of the cutoff frequency ω_s and a concomittant shift of the PSP dispersion curve to the right in Figure 20a.

The calculated reflectivities were obtained from the exact Fresnel equations for a four layer system. Each gold film was divided into three regions. One was left bare and the other two were coated with different thicknesses of cadmium arachidate. The reflectivity of the bare portion was used to determine the dielectric constant and thickness of the gold film. (The dielectric constants of the films differed about 15% from the accepted value for bulk gold, which is not unusual for evaporated films.) The gold parameters were held fixed and only the refractive index of the organic layer varied to generate the best fit to the reflectivity of the covered portions. The organic layer was assumed to be non-absorbing and its thickness was set at 2.68 times the number of cadmium arachidate monolayers.

The calculated isotropic refractive indices varied from 1.50 to two monolayers to 1.56 for six or more monolayers. The latter value agrees with that obtained on thicker layers with a more accurate waveguide technique [52], but differs from ellipsometrically determined value [53]. The variation within the first several layers does not seem unreasonable. It was also concluded in the ellipsometric study that the initial layers were different from the rest [53].

These experiments show that plasmon surface polariton spectroscopy, while using simpler equipment and requiring less effort, can provide results of comparable accuracy to ellipsometric measurements and that by employing ellipsometric techniques for detecting surface polariton absorption even greater accuracy and precision is attainable. It should be possible to detect organic layers only a few tenths of a nanometer in thickness by the present technique.

ACKNOWLEDGMENTS

All of the work described has been performed at the IBM Research Division Laboratory in San Jose. During the course of this work the author has benefited directly from discussions and active collaborations with J.-M. Turlet, K. Syassen, J. D. Swalen, J. Gordon II, P. M. Grant and I. Pockrand. The technical assistance of R. Santo and J. Duran cannot go unmentioned, as too the provision of purified materials by D. Burland, D. Haarer and T. Clarke.

REFERENCES

1. J. M. Adams, J. M. Thomas and M. J. Walters, J. Chem. Soc. Dalton Trans. 1459 (1975).

2. D. E. Aspnes, in B. O. Seraphin, Editor, Optical Properties of Solids, New Developments, Chap. 15, North Holland, Amsterdam (1976).

3. J. D. E. McIntyre, in B. O. Seraphin, Editor, Optical Properties of Solids, New Developments, Chap. 11, North Holland, Amsterdam (1976).

4. A. Otto, in B. O. Seraphin, Editor, Optical Properties of Solids, New Developments, Chap. 13, North Holland, Amsterdam (1976).

5. J. G. Gordon II and J. D. Swalen, "Effect of Thin Organic Films on the Surface Plasmon Resonance on Gold," Optics Comm., 22, 374 (1977).

6. H. Raether, "Surface Plasmon Oscillations and Their Applications," Physics of Thin Films 9, 145 (1977).

7. J. Frenkel, Phys. Rev. 37, 17 (1931).

8. R. S. Knox, Theory of Excitons, Academic Press, New York (1963).

9. J. J. Hopfield, Phys. Rev. 112, 1555 (1958).

10. D. L. Mills and E. Burstein, Rep. Prog. Phys 37, 817 (1974).

11. R. R. Pennelly and C. J. Eckhardt, Chem. Phys. 12, 89 (1976).

12. B. G. Annex and W. T. Simpson, Rev. Mod. Phys. 32, 466 (1960).

13. M. R. Philpott, J. Chem. Phys. 55, 2039 (1971).

14. a. K. Syassen and M. R. Philpott, unpublished results, in later experiments higher reflectivities were measured.
    b. D. P. Craig and R. D. Gordon, Proc. Roy. Soc. (London) A, 288, 69 (1965).

15. K. Syassen and M. R. Philpott, Chem. Phys. Lett., 50, 14 (1977).

16. K. Syassen and M. R. Philpott, unpublished result; See also, D. Bloor, D. J. Ando, F. H. Preston and G. C. Stevens, Chem. Phys. Lett. 24, 407 (1974).

17. M. R. Philpott, P. M. Grant, K. Syassen and J.-M. Turlet, J. Chem. Phys., 67, 4229 (1977).

18. S. V. Morisova, Ukr. Fiz. Zh. 12, 521 (1967).

19. M. S. Brodin, S. V. Morisova and S. A. Shturkhetskaya, Ukr. Fiz. Zh. 13, 353 (1968); Eng. trans. Ukr. Phys. J. 13, 249 (1968).

20. M. S. Brodin, M. A. Dudinskii and S. V. Morisova, Opt. Spektrosk. 34, 1120 (1973); Eng. trans. Opt. Spectrosc. 34, 651 (1973).

21. J.-M. Turlet and M. R. Philpott, J. Chem. Phys. 62, 2777 (1975).

22. J.-M. Turlet and M. R. Philpott, Chem. Phys. Lett. 35, 92 (1975).

23. J.-M. Turlet and M. R. Philpott, J. Chem. Phys. 62, 4260 (1975).

24. N. I. Ostapenko, M. P. Chernomorets and M. T. Shpak, Phys. Status Solidi B 72, K117 (1975).

25. A. Harmsen, E. E. Koch, V. Saile, M. Schwenter and M. Skibowski, in R. Haensel, E. E. Koch and C. Kunz, Editors, Vacuum

Ultraviolet Radiation Physics, Pergamon-Vieweg, New York, (1974).

26. J. Ferguson, Chem. Phys. Lett. 36 316 (1975).

27. J. Ferguson, Zeit. Phys. Chemie (Neue Folge) 101, S45-56 (1976).

28. V. Saile, P. Gürtler, E. E. Koch, A. Kozevnikov, M. Skibowski and W. Steinman, Phys. Rev. Lett. 37, 305 (1976).

29. E. Glockner and H. C. Wolf, Z. Naturforsch. 24a, 943 (1969).

30. M. S. Brodin, M. A. Dudinskii and S. V. Morisova, Opt. Spektrosk. 31, 749 (1971); Eng. trans. Opt. Spectrosc. 31, 401 (1971).

31. M. S. Brodin, M. A. Dudinskii and S. V. Morisova, Izv. Akad. Nauk SSSR 36, 1047 (1972).

32. E. Glockner and H. C. Wolf, Chem. Phys. Lett. 27, 161 (1974).

33. H. C. Wolf, Z. Naturforsch. 13a, 414 (1958).

34. A. Bree and L. E. Lyons, J. Chem. Soc. 1956, 2662 (1956).

35. J. P. Desvergne, H. Bouas-Laurent, E. V. Blackburn and R. Lapouyade, personal communication.

36. L. B. Clark and M. R. Philpott, J. Chem. Phys. 53, 3790 (1970).

37. V. I. Sugakov, Ukr. Fiz. Zh. 14, 1425 (1969).

38. V. I. Sugakov, Ukr. Fiz. Zh. 15, 2060 (1970).

39. M. R. Philpott, J. Chem. Phys. 60, 1410 (1974).

40. M. R. Philpott, J. Chem. Phys. 60, 2520 (1974).

41. H. Anders, Thin Films in Optics, The Focal Press, London, (1967).

42. K. Syassen and M. R. Philpott, "Reflection Spectra of Natural Faces of Crystalline Anthracene," J. Chem. Phys. in press, (1978).

43. J.-M. Turlet, J. Bernard and Ph. Kottis, "Experimental Evidence of Super-Radiant Fluorescence from Surface and Sub-Surface Exciton States in Anthracene Crystal," J. Chem. Phys. in press, 1978.

44. J. Lagois and B. Fischer, Phys. Rev. Lett. 36, 680 (1976).

45. I. Hiraboyoshi, T. Koda, Y. Tokura, J. Murata and Y. Kaneko, J. Phys. Soc. Japan 40, 471 (1976).

46. K. Tomioka, M. G. Sceats and S. A. Rice, J. Chem. Phys. 66, 2984 (1977).

47. M. G. Sceats, K. Tomioka and S. A. Rice, J. Chem. Phys. 66, 4486 (1977).

48. I. Langmuir, J. Am. Chem. Soc. 39, 1848 (1917); K. B. Blodgett, ibid., 57, 1007 (1935).

49. H. Kuhn, D. Möbius and H. Bücher, "Spectroscopy of Monolayer Assemblies," in A. Weissberger and B. W. Rossiter, Editors, Techniques of Chemistry, Chap. VII, Vol. 1, Physical Methods of Chemistry, Part III B, Wiley-Interscience, New York (1972).

50. A. Otto, Z. Phys. 216, 398 (1968).

51. E. Kretschmann, Z. Phys. 241, 313 (1971).

52. J. D. Swalen, K. Rieckhoff and M. Tacke, unpublished results.

53. R. Steiger, Helv. Chim. Acta. 54, 2645 (1971).

54. Abelès and Lopez-Rios have used ellipsometric detection of surface plasmons for several studies, including the growth of tarnish films on silver. F. Abelès, Surf. Sci. 56, 237 (1976), and F. Abelès and T. Lopez-Rios, in E. Burstein and F. De Martini, Editors, Polaritons, Proc. of the First Taormina Research Conf. on the Structure of Matter, p. 241, Pergamon, New York (1974).

Raman Spectroscopy at Surfaces

P. J. Hendra and M. Fleischmann

Department of Chemistry, University of Southampton

Southampton SO9 5NH, U.K.

I. INTRODUCTION

The study of species sorbed to surfaces is a daunting experimental problem. Numerous techniques have from time-to-time been developed which have provided information but few would suggest that our current state of knowledge is better than sparse. The classical techniques involve:

a) The measurement of vapour pressures of volatiles over surfaces as a function of temperature and mass of species sorbed and

b) Desorption (by pumping and heating) followed by chemical or physical analyses of the desorbed species.

Much more subtle methods have proliferated over the last few years including a wide variety of "electron" spectroscopies (e.g. secondary electron or Auger spectroscopy and ESCA), high resolution electron diffraction and L.E.E.D. and there is little doubt that their functions have contributed greatly to our understanding of chemisorption and some aspects of heterogeneous catalysis. Unfortunately, however, these methods rely on the use of highly evacuated systems and have almost inevitably led to an encouragement of the academic's tendency to *reductio ad absurdum*. Thus, it is intellectually attractive to study 'clean' single crystal surfaces and chemisorption thereto. The experimental methods listed above are ideal in this field and have therefore lead to an immense international investment in this type of research. It remains, however, that all catalytic processes of value, all corrosion or film problems in the real world, in fact surface chemistry is

concerned with species sorbed to dirty surfaces under relatively
high pressure conditions.

There is no doubt that this field is considerably more
complicated than that of the study of adsorption under high vacuum.
The systems themselves (e.g. oxide catalyst surfaces, polymer
layers or electrode systems) are difficult to define. Adsorption
normally occurs in the presence of other major components and the
chemistry which takes place on them frequently occurs at atypical
sites on the surfaces. Physical methods viable in this field are
scarce but infrared absorption spectroscopy has become a classic.
Thus, almost every surface laboratory of significance, all large-
scale catalyst producers and users find it essential to use this
method. The applications are manifold and two excellent books and
innumerable reviews have appeared surveying the field.

Infrared Absorption Spectroscopy of Surfaces

All materials absorb in the infrared region of the electromagnetic spectrum. They do so when, under appropriate conditions,
molecular vibrations resonate with the illuminating radiation.
Since molecular vibrational behaviour is itself complex, the infrared absorption spectrum constitutes an excellent fingerprint and
can be used as a valuable qualitative and quantitative analytical
method. See Figure 1. Further, the environment in which a molecule
finds itself subtly affects its vibrational behaviour and this too
"shows up" in the infrared spectrum. Thus, the attraction to the
surface chemist is obvious - sorption of a molecule on a surface
can lead to a spectral analysis identifying the sorbate, its
decomposition fragments and the creation of permanent or temporary
surface layers. In favourable cases, the spectrum will indicate
changes of the surface caused by sorption and/or the mode of
sorbate-sorbent interaction. Like all experimental methods,
infrared absorption has its limitations and it is these that have
lead to the research reported here.

The basic infrared absorption experiment consists of

$$I/I_o = e^{-\epsilon c l}$$

c = Conc. of absorber

ϵ = Extinction Coeff. of absorber

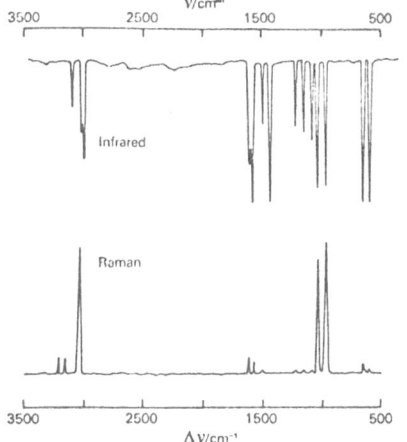

Fig. 1. I.R. and Raman of
         liquid Pyridene

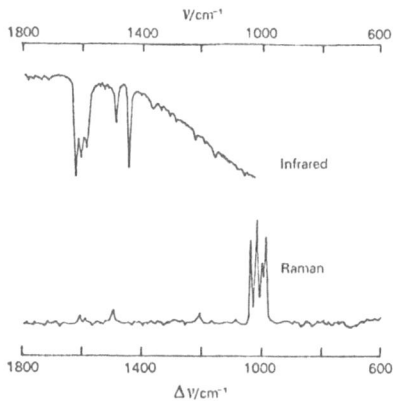

Fig. 2a. I.R. and Raman of
          Pyridene on Al_2O_3

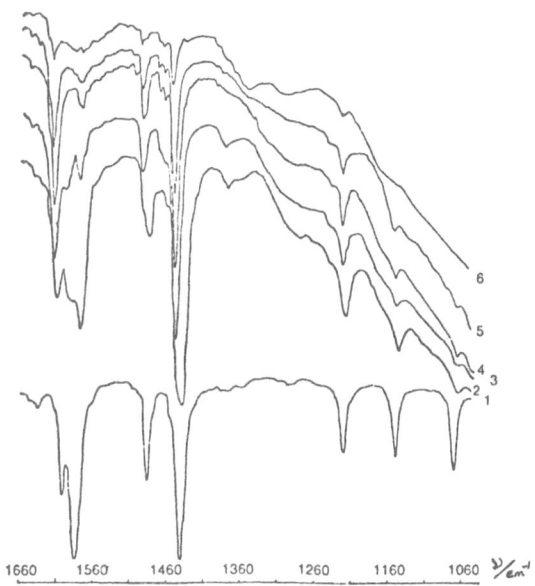

Fig. 2b. I.R. of Pyridene on Al_2O_3 (pretreated at 550 C).

1. liq. pyr.   2. Sorbed   3. After evac.

4. After heating to 150 C.   5. To 250 C   6. To 375 C.

from which it is clear that ALL absorbers present contribute to
the overall absorption spectrum. The result is that in many
experiments considerable zones of the analytically useful spectrum
are 'swamped' by unwanted absorption. Further, it is essential
that the amount of sorbate vs. sorbent be maximized. As a consequence, i.r. absorption is carried out on thin pressed discs of
adsorbent suspended in specially designed infrared cells. If the
adsorbent is opaque to infrared radiation (e.g. metals) little
can be achieved. One further limitation is that omnipresent
molecule, water! Water absorbs i.r. radiation very strongly and,
as a consequence, its presence usually renders the method
inappropriate.

    Considerable effort has been expended to overcome or reduce
these limitations, thus sophisticated reflection and multi-
reflection methods can be found in use whilst emission methods are
sometimes useful. Further, the most sophisticated spectrometric
equipment (e.g. computer interfaced spectrometers and interfero-
meters) is used when available, to maximize the sensitivity and
quantitative precision in infrared measurements.

    Thus, infrared absorption although invaluable has its limit-
ations. Vibrational spectra can be determined in an alternative
and quite distinct way - by light scattering utilising the Raman
effect.[†] The application of sophisticated laser Raman methods
to surface chemistry is the subject of this report.

II.   EXPERIMENTAL

    Laser Raman spectroscopy can be applied to the molecular
analysis of small or large samples in all phases. Over the last
ten years we have pioneered the application of the technique to
the study of species sorbed to catalyst surfaces. Immediately
prior to commencing the research described here, we were able to
show that the method was able to give data on molecules in the
electrode-electrolyte interface.

    Our efforts over the past $2\frac{1}{2}$ years have been concentrated
on

[†] Molecular vibrations are accompanied by changes in the dipole
    and/or polarizability of the molecular system. The former give
    rise to infrared absorption, the latter to Raman scattering.
    A brief account of the process will be found in ref. 5.

(a) The development of technique aimed at improving the versatility of our methods and

(b) A continuous application of the expertise to systems accessible at the moment.

Our approach has been primarily aimed at the electrochemical system but we have paralleled this with a lower level effort on dry catalyst surfaces.

Developments in Techniques

The Raman experiment consists of:

Laser Beam
$$\xrightarrow[\nu_o]{} \text{Sample} \xrightarrow[\substack{\text{Scattered} \\ \text{light}}]{\nu_o \pm \nu_{vibr.}} \text{Spectrometer} \rightarrow \text{Detector} \rightarrow \left\{\begin{array}{l}\text{Data} \\ \text{Processing} \\ \text{System}\end{array}\right.$$

The limitations on sensitivity are provided by:

(a) The brightness of the Raman scattered radiation and

(b) The efficiency of the total spectrometric system.

Progress has been made in both fields as follows:

The Raman sample. The amount of Raman scattered radiation available to the spectrometer is governed by:

(a) Power of the laser and the efficiency with which one can illuminate that part of the sample viewed by the spectrometer.

(b) The absorption of the sample (basically a function of the electrode or catalyst colour).

(c) The number of scatterers and their fundamental efficiency in the Raman process and

(d) The efficiency with which the light can be gathered.

It transpires that if the electrode or catalyst (particularly if dark) are not to be over-heated by the laser (a) is a serious

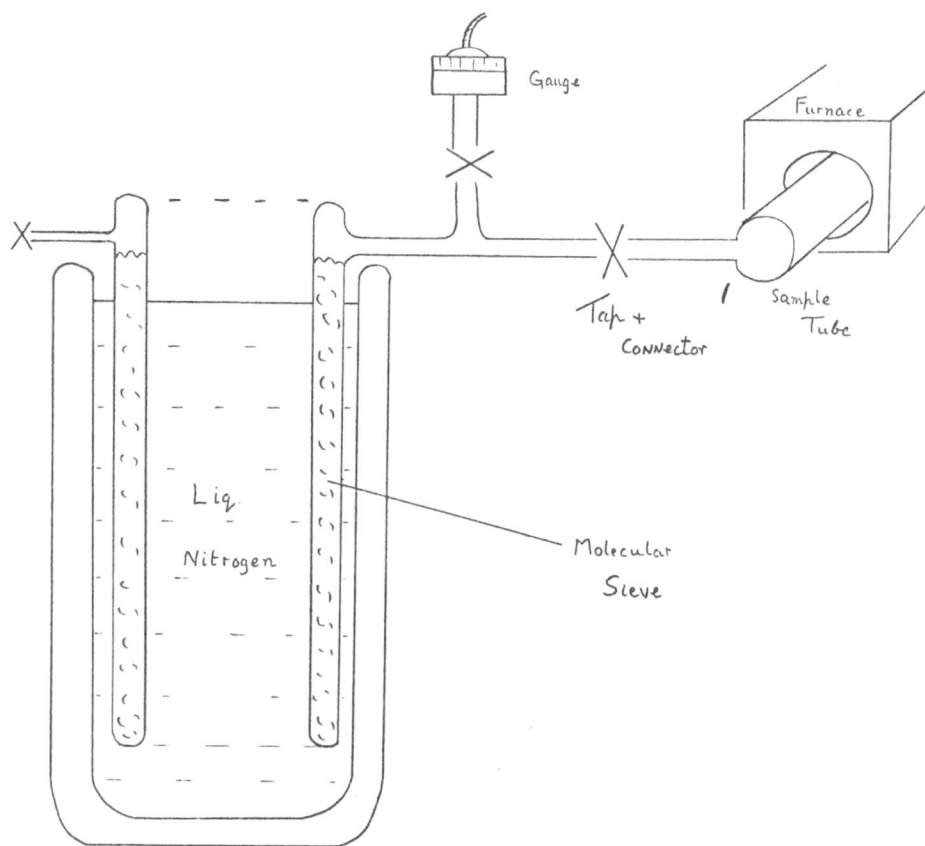

Fig. 3.   System for activating oxides in a grease free environment
          Apparatus pre-pumped with a water pump then seive cooled.
          Terminal pressure after activation $\sim 10^{-5}$ torr.

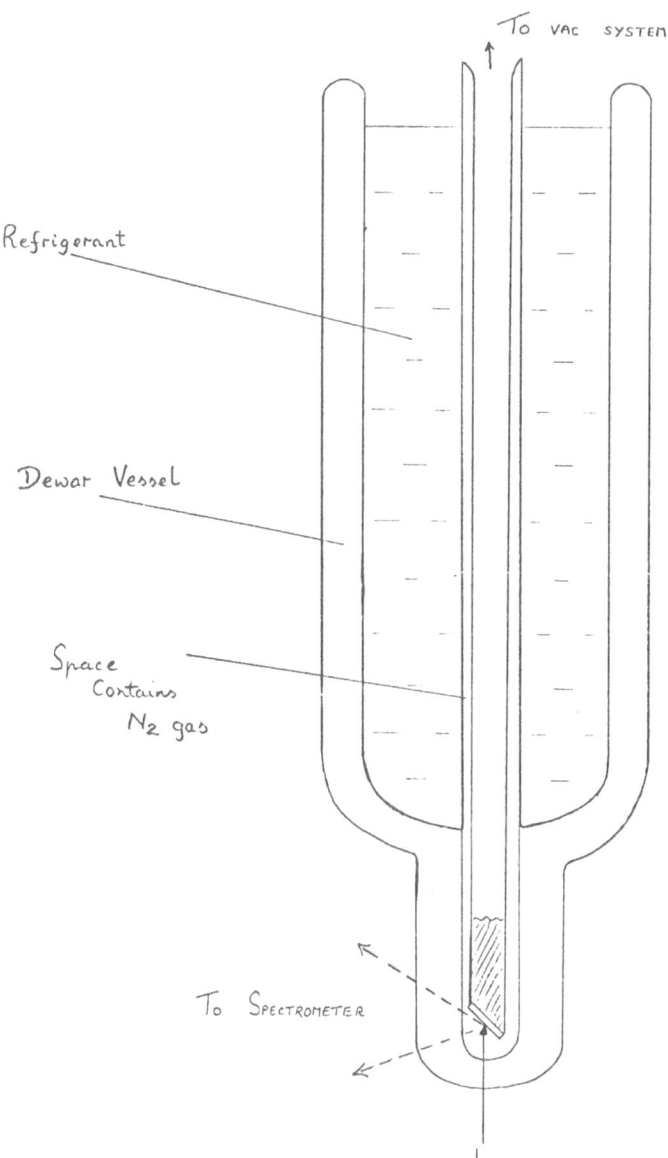

To VAC SYSTEM

Refrigerant

Dewar Vessel

Space
Contains
N_2 gas

To Spectrometer

L

Fig. 4.   Low Temperature Cell

Cell coupled to vacuum system in the same way as
conventional cells.

restriction. The number of scatterers can only be kept high at
low surface coverage if the surface area is kept at a high value.
This tends in metals to darken the surface.

Currently we are able to study the following surfaces thanks
to a technique developed at Southampton.

Electrodes. (a) Mercury droplets but not a smooth surface

(b) Roughened copper or silver. Neither of
these dark in colour

(c) Lead

(d) of most potential, a grade of platinum
electrode where the roughening is controlled to produce a bronze-coloured
surface. A few other electrodes look
promising e.g. gold.

Catalyst surfaces. White or pale-coloured high area
catalysts in general. These include silica, alumina, silica/
alumina, zeolites, magnesia, titania and others. We are currently
developing a system for studying dark coloured materials but
cannot yet report on its performance.

One of the major limitations in this field has been the
occurrence of persistent fluorescence in activated acidic
catalysts. The problem arises through breakdown of vacuum grease
on the catalyst surface during the activation routine. We have
developed a method of controlling this problem vis the use of a
simple grease-free sample preparing system. The system is shown
in figure 3 and its operation described in the caption. The new
technique has been used successfully on alumina activated over a
wide range of temperatures, a report on which will be found in
Section 3. We have also developed a highly efficient low
temperature sampling system. See Figure 4.

Progress in instrumental performance. The Raman spectrometer
we use in our work is described schematically in figure 5. It
incorporates a number of unusual features including

(a) a very high dispersion triple monochromator. The high
dispersion enables us to illuminate and view a very
large patch of the sample, hence minimizing laser
brightness (and hence sample heating) yet maximizing
the number of scatterers present.

(b) A cam driven repetitive scanning system enabling us to
accumulate very large numbers of scans

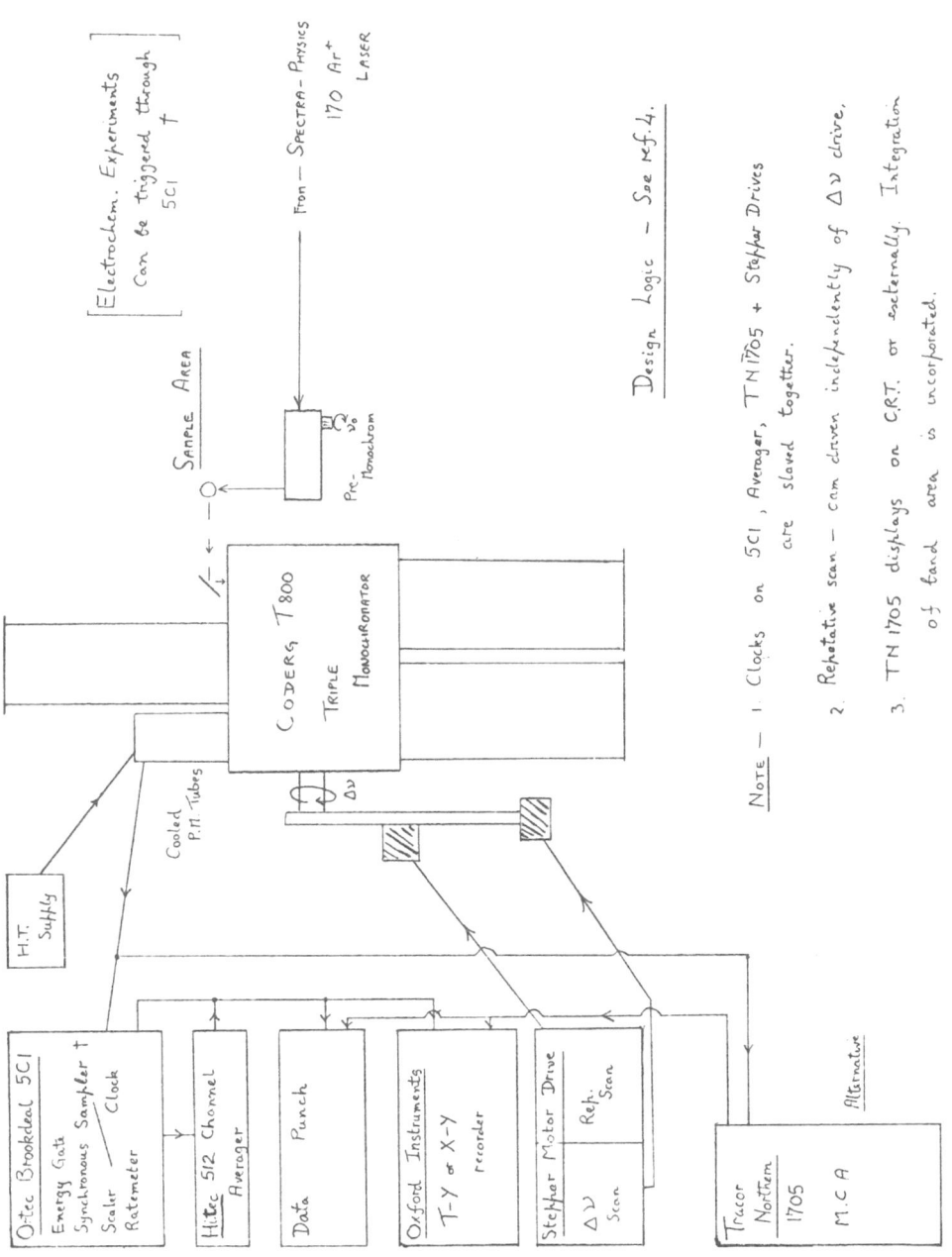

Fig. 5.   Coderg based Raman spectrometer at Southampton.

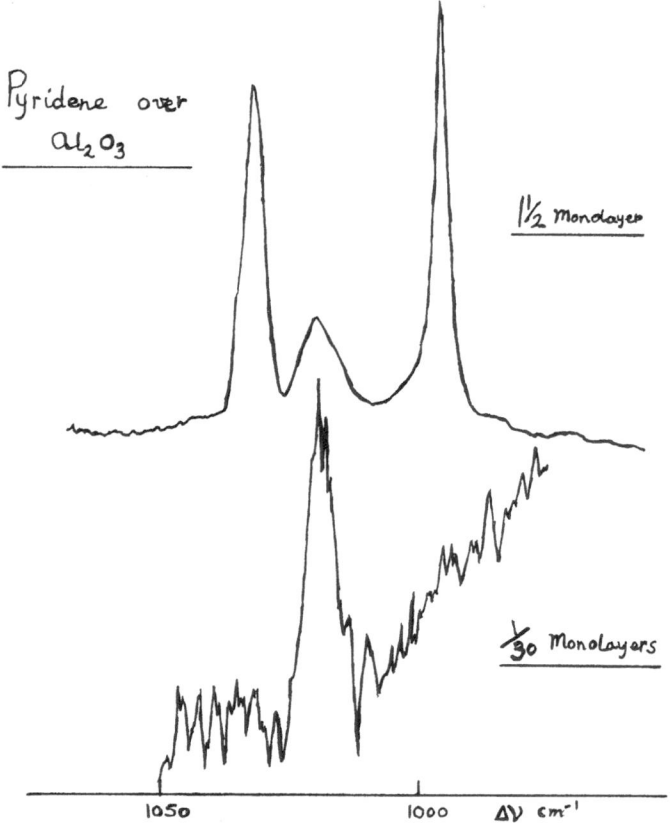

Fig. 6. Raman spectra at two different coverages of pyridene
          sorbed to γ Al₂O₃. The sensitivity used in the second
          spectrum is much higher than that used in the first
          and the spectrum is the summed result of 64 scans.
          Spectra recorded on a Coderg T800 based spectrometer
          at Southampton.

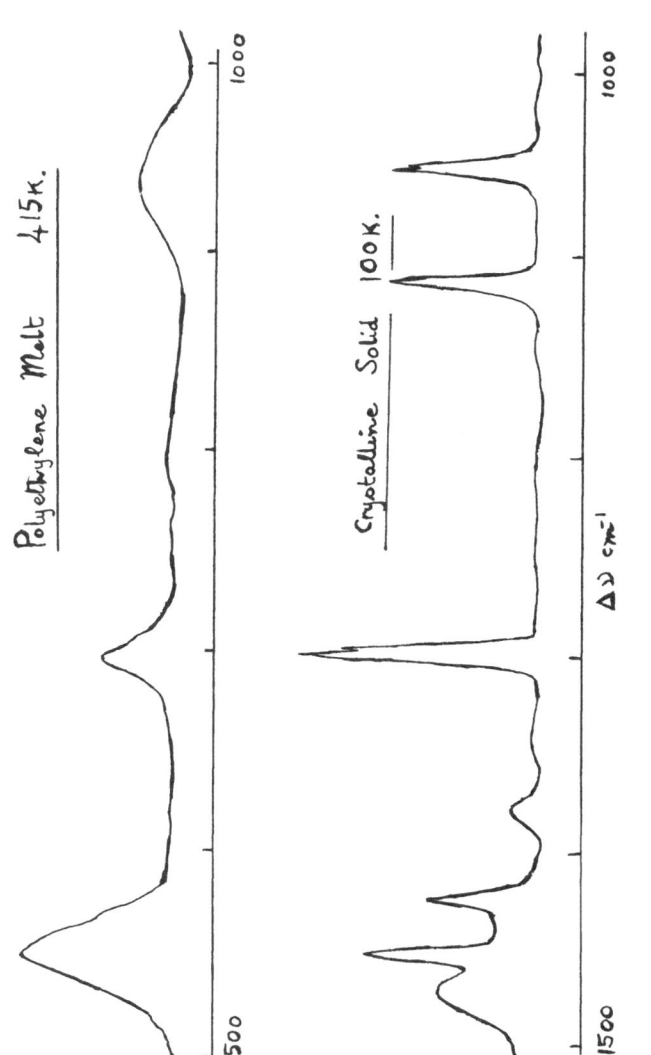

Fig. 7. An example of the effect of phase change on vibrational characteristics –
         Raman spectra of molten and crystalline linear polyethylene.

(c)   Scalar counting (rather a ratemeter analysis)

a newly developed device -

(d)   a sophisticated system for correcting for laser power
      fluctuations, vibrations etc. The device consists of a
      mirror and photomultiplier in the first monochromator
      which samples the total reflected and scattered light.
      The output drives a voltage to frequency convertor, the
      output of which over-rides the scaler and scanning
      system clocks. The result is that laser power variation,
      misallignment or vibration are all normalized out of the
      recorded data.

Proof of the machine's performance is provided in the spectrum of
pyridene sorbed to an alumina surface at a coverage of only about
4% monolayer shown in figure 6. We have found in the past that
coverages much below ½ monolayer are difficult to study, thus the
improvement is really significant. Over the last few months noise
has been further reduced and thus sensitivity has slightly improved.
In addition, we have recently carried out experiments using
Cassegranian high aperture collection optics. Also, an alternative
approach has been tried, a large area of sample is illuminated and
low collection apertures are used. In appropriate cases, one or
both of these systems will be attractive.

        The final outcome of this research should yield a sensitivity
in the Raman effect rivalling that obtainable with the best infrared methods.

        III.   RAMAN SPECTROSCOPY OF SURFACE SORBED SPECIES

        When a molecule's vibrational spectrum is determined in the
gas, liquid and solid phases, the spectra reveal

        (a)   A basic fingerprint arising for the gross molecular
              structure and identity and

        (b)   Subtle changes with phase arising from intermolecular
              interaction and/or symmetry charges due to crystallisation.
              An example is shown in figure 7.

Sorption of a molecule has a similar effect. The spectrum reveals
the nature of the molecule (which may have changed due to reaction
at the surface) and some information about its environment.

Summary of All Adsorbate-Adsorbent Systems Exhibiting Physisorption,
Studied using Laser-Raman Spectroscopy

Adsorbate	Adsorbent
CC1$_4$ Br$_2$ CS$_2$	silica gel
trans-CHCl=CHCl acetaldehyde	silica gel
trans-stilbene	cab-o-sil H5
acetonitrile	silica gel, aerosil 380 zeolites X and Y
benzene chlorobenzene aniline benzylamine dioxan acetone CH$_3$I	porous glass porous glass
CO$_2$	zeolites X and Y
acroiein	zeolite Y
nitrobenzene	porous glass, alumina, aerosil 380
styrene	silica gel, cab-o-sil, aerosil 380
ethylene	porous glass
propylene	porous glass zeolites A, X and Y
2-chloropyridine	silica gel
pyridine	magnesium oxide
thiophene	cab-o-sil

Physisorption on Oxide Surfaces

The various adsorbate-adsorbent systems listed in Table 1 have been investigated using laser Raman spectroscopy and all exhibit only weak interactions between the adsorbed molecules and the substrate. The Raman bands of the adsorbed molecules were found at almost the same frequencies as those of the pure compounds, even when the coverage was reduced to less than a monolayer. Physisorption reveals little about the nature of the surface, and Table 1 does not require much further discussion.

Egerton et al have found that the frequencies of the ring vibrations of aniline and benzylamine adsorbed on porous glass were shifted 5 cm^{-1} and 1 cm^{-1} compared with the liquid spectra. They attributed this to the amino group hydrogen bonding to surface OH groups, and since aniline has the amino group directly bonded to the benzene ring, the perturbation of the ring vibration would be expected to be larger than for benzylamine. The N-H stretch (expected $\sim$ 3400 cm^{-1}) could not be detected in this case due to the fall in photomultiplier sensitivity in the red.

Chemisorption

Chemisorption of pyridine on oxide surfaces. Raman spectroscopy has been used to identify pyridine chemisorbed on silica gel, porous silica glass, Cab-o-sil, Aerosil 380, silica gel, silica-alumina (13% Al), η- and γ-alumina, titanium dioxide, NH_4^+-mordenite, and Na Y zeolite. Pyridine can chemisorb to the surface by donation of the lone pair of electrons on the nitrogen atom to Lewis acid sites (a coordinate bond), by hydrogen bonding to surface OH groups, or by complete proton abstraction from surface OH groups to form the pyridinium cation (Bronsted acid sites). It is still debatable, however whether hydrogen bonded pyridine and the formation of a pyridinium cation should be considered as pyridine chemi-sorbed to the same type, or two distinctly different types, of surface sites.

The Raman bands of pyridine are sensitive to these different bonding environments, and we discuss below the frequencies of the C-H stretch and ring breathing vibrations of pyridine in some well defined environments.

Correlating the results of Hendra et al with those of Egerton et al it is apparent that hydrogen-bonded pyridine produces Raman bands in the ranges 994-1005, 1030-1040 and 3060-3072 cm^{-1}; pyridinium ion 1010-1015, 1025-1035, and 3090-3105 cm^{-1}; and Lewis coordinated pyridine 1018-1028, 1040-1050 (normally a weak band), and 3072-3087 cm^{-1}.

Hendra et al reported that pyridine adsorbed on various

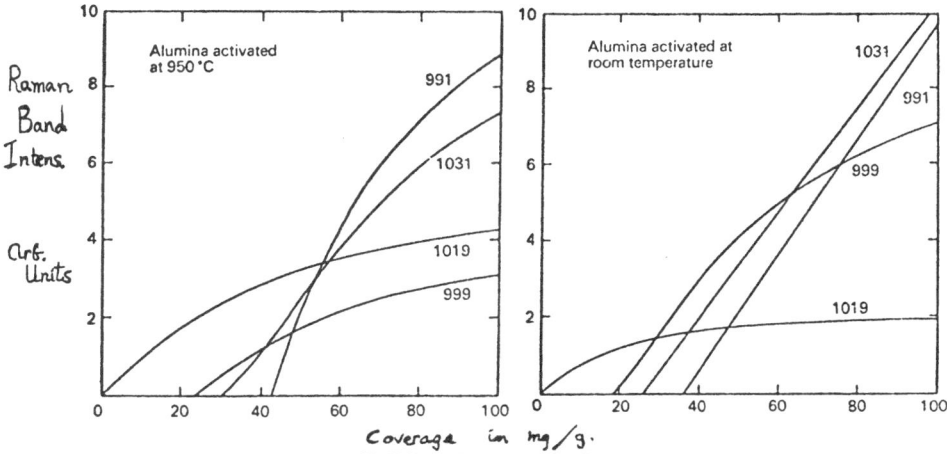

Fig. 8a. Raman intensities vs coverage for pyridene over γ Al₂O₃.

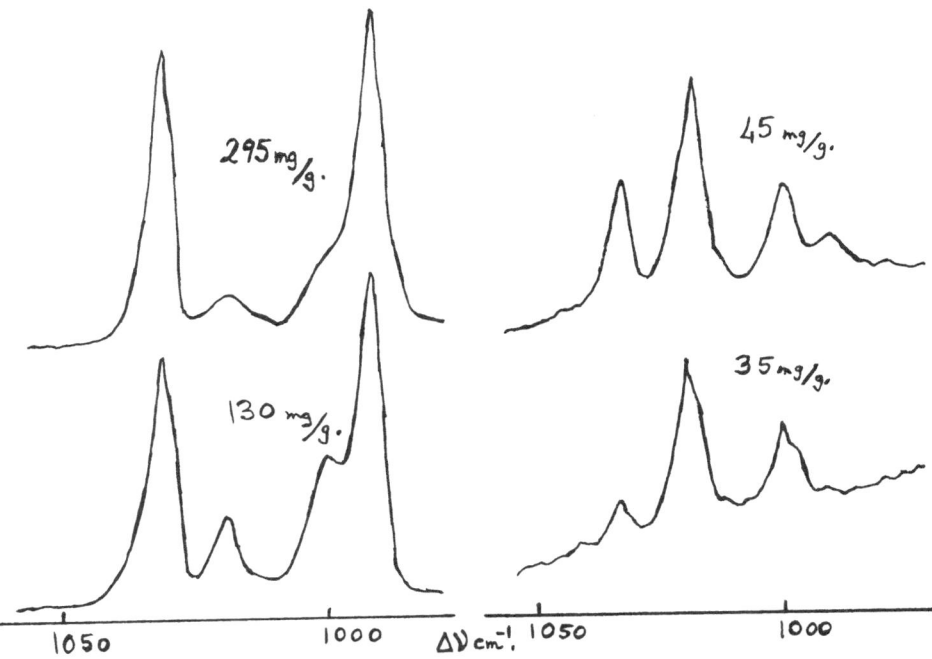

Fig. 8b. Part of the Raman spectra of pyridene sorbed to alumina pretreated to 950°C under vacuum. Spectra are recorded at sensitivities increasing as coverage is reduced. Normalised intensity values are used in fig. 8a.

physical forms of silica (e.g. silica gel, cab-o-sil, and aerosil)
that had previously been activated at elevated temperatures, was
mainly physically adsorbed at multilayer coverages (Raman lines at
991 and 1031 cm^{-1}) but with some chemisorbed pyridine (1007 cm^{-1}).
Reducing the coverage to approximately 0.1, monolayer simplified
the spectrum to two lines at 1007 and 1035 cm^{-1}, and this was
thought to indicate strong hydrogen bonding between pyridine and
surface silanol groups. These conclusions were supported later by
Egerton et al. It is interesting to note that 2-chloropyridine
only physisorbs to silica gel, due, it is thought by Kagel to
steric hinderance.

The 1000 cm^{-1} region of the Raman spectrum of pyridine
adsorbed on alumina changes markedly with coverage (Figure 8).
At multilayer coverages the dominating bands at 991 and 1031 cm^{-1}
are due to physisorbed material. As the coverage is reduced, these
bands fall sharply in intensity and additional bands at 999, 1019
and 3085 cm^{-1} become apparent. The bands at 1019 and 3085 cm^{-1} we
associated with pyridine Lewis coordinated to aluminium atoms at
the surface, while the band at 999 cm^{-1} is indicative of pyridine,
hydrogen bonded to surface OH groups. The areas of these Raman
bands as a function of coverage are markedly dependent on the
temperature at which the alumina is activated and this gives us an
indication of the relative distribution of the various types of
sites on the surface. (See figs. 8a and 8b). Alumina activated
at room temperature clearly contains relatively few Lewis acid
sites, and most of the 'chemisorbed' pyridine is hydrogen-bonded
to surface OH groups. Activating the alumina at 950°C removes a
large proportion of the aluminol groups, while at the same time
exposing more aluminium ions at the surface.

In view of the importance of Brönsted acid sites in catalysis,
Hendra et al have attempted a more rigorous characterisation of the
999 cm^{-1} band. Samples of alumina were pretreated at different
temperatures to vary the surface concentration of OH groups; the
alumina surface was deuterated prior to pyridine adsorption; a
surface known to have high Brönsted activity was prepared by
chlorinating a sample of alumina; and finally a competition
reaction for any available Brönsted sites and surface OH groups
capable of strong hydrogen bonding was attempted.

The technique used was to sorb pyridine to the surface and
then admit piperidine to the vacuum system. The Raman spectrum
clearly indicates the way the Lewis coordinated piperidene is
removed.

It is known that the frequency of the Raman band near 1015
cm^{-1} characteristic of Lewis coordinated pyridene is linearly
dependent upon the strength of the coordinate band. We therefore
expected to see a drift in the band head to higher frequencies as

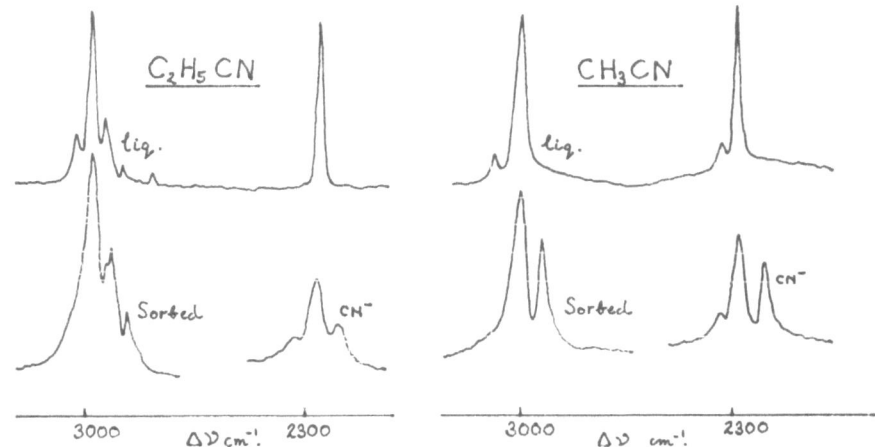

Fig. 9a. Raman spectra of acetonitrile and propionitrile as liquids
and when sorbed to Al_2O_3 pretreated at 950°C in vac.
Note the decomposition of the nitriles to CN⁻ anions.

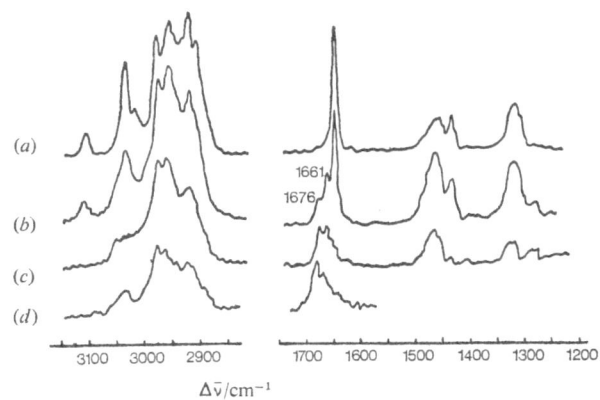

Fig. 9b. Raman spectra of hex-1-ene adsorbed to γ alumina.
Adsorbents pretreated at 950°C under vacuum.
Quantities of hex-1-ene adsorbed (mg g⁻¹ alumina)
(a) 450, (b) 90, (c) 40, (d) 30.

As the coverage is reduced the concentration of liquid
hex-1-ene reduces whilst that of isomers increases.

the coverage was reduced in fig. 6 but this does not appear to
take place. This may indicate that the most weakly retained
molecules are bonded as strongly as those which are very difficult
to remove: a point needing further study!

Silica - 13 per cent alumina exhibits from the Raman viewpoint strong Lewis acidity (bands at 1020 and 3078 cm^{-1}) and
strong hydrogen-bonding between pyridine and surface OH groups
(1007 cm^{-1}). The presence of weak Lewis acid sites on the surface
of titanium dioxide, which had been speculated by Hair from IR
evidence, has been confirmed by the Raman band at 1016 cm^{-1}.
No evidence for hydrogen-bonded pyridine was found in this case.

Egerton et al felt that the Raman bands at 1002 and 1036 cm^{-1}
of pyridine over NaY zeolite could not be ascribed to either
hydrogen bonding or pyridinium ion formation as their pretreatment
would have removed almost all the zeolitic water, thereby reducing
the surface concentration of OH groups to a level considerably
below that typical of silica, alumina, or silica-alumina. They
suggested that the upward shift in frequency was due to pyridine
coordinated to surface cations. Further studies indicated that
the Raman shifts do vary with the type of charge-balancing cation
present in the zeolite.

Olefin isomerization on oxide surfaces. The isomerization of
some mono-olefins and diolefins adsorbed on alumina and silica-
alumina have recently been investigated using laser Ramam spectroscopy. Isomerization of pentene-1 adsorbed on alumina (pretreated
at 950°C) was evident from the Raman bands at 1641, 1656 and 1671
cm^{-1}, which are characteristic of pentene-1,cis-2, and trans-2-
pentene respectively. Isomerization did not occur over silica, or
alumina that had only been activated at moderate temperatures
(< 450°C). Hexene-1 rapidly isomerizes over alumina (pretreated
at 950°C), producing trans-2-hexene and cis-2-hexene. As the
coverage is reduced, changes in intensities of the Raman bands in
the region 1640-1680 cm^{-1} are particularly noticeable, with the
cis form (1660 cm^{-1}) increasing in intensity relative to the trans
form (1674 cm^{-1}) and hexene-1 (1642 cm^{-1}). Isomerization has also
been observed with hexene-1 over silica-alumina (950°C) with the
cis form predominating as the coverage was reduced. No isomerisation
occurred over silica.

Diolefins adsorbed on alumina (960°C) showed much the same
behaviour as the mono-olefins. 1:4 pentadiene produced both the
cis and trans 1:3 pentadienes with evacuation gradually removing
the trans form to leave the cis isomer. When cis 1:3 pentadiene
was adsorbed on alumina (950°C) isomerization rapidly produced the
trans isomer (1656 cm^{-1}) but on evacuation the Raman bands began to
shift towards the frequency of the cis isomer (1650 cm^{-1}).
Isomerization of 1:5 hexadiene on alumina (950°C) resulted in many

different isomers of hexadiene and the Raman spectrum was difficult to interpret, only 2-cis-4-trans hexadiene being postively identified.

From the fact that isomerization over alumina and silica-alumina occurs only on samples that have been pretreated at high temperatures, it would appear that the process is dependent on surface aluminium ions that are exposed by the removal of hydroxyl groups. Hendra, Turner et al concluded that the mechanism of isomerization involved olefin adsorption to an exposed pair of aluminium ions with the high electrostatic field inducing a 1,2-hydride shift with the simultaneous formation of an allylic carbonium ion intermediate.

Spangenberg and Winde have reported that the frequency of the C-C stretching vibration of propylene (1717 cm^{-1}) is shifted upwards by 13 cm^{-1} when sorbed on A-zeolite. This was attributed to 6-7 propylene molecules becoming trapped in each void of the zeolite structure.

Nitrile adsorption on oxide surfaces. The Raman spectra of benzonitrile, propionitrile, and acetonitrile adsorbed on alumina pretreated at 950°C are significantly different from their liquid spectra, particularly in the frequency region typical of the $-C\equiv N$ stretching vibration. The large peaks at 2232, 2254 and 2257 cm^{-1} are due to physisorbed material, but the Raman bands on the high frequency side at 2275, 2305 and 2302 cm^{-1} are due to benzonitrile, propionitrile, and acetonitrile coordinatively bonded to Lewis acid sites. Propionitrile and acetonitrile also have a band (2178 and 2180 cm^{-1} respectively) on the low frequency side of the physisorbed material, which does not appear in the spectrum of adsorbed benzonitrile. These two bands have been attributed to the cyanide ion CN^- that is produced by the decomposition of the paraffinic nitrile.

This decomposition only occurs on alumina that has been dehydrated, i.e. in the presence of surface oxide ions. This surface presents sites of high Lewis acidity and Lewis basicity.

Although the surface of silica-alumina has sites of high Lewis acidity, the basic sites are weak in nature and thus the breakdown of nitriles does not occur.

Adsorption of Molecules Containing a Carbonyl Group

Winde reported a marked downward shift in the carbonyl stretching frequency of acetone adsorbed on alumina pretreated at 600°C and this was attributed to the carbonyl group hydrogen bonded to surface hydroxyl groups. Two additional bands (1559-1579 and 1612-1629 cm^{-1})

in the spectrum of adsorbed acetone were identified as arising from
Lewis coordinated acetone. Heating the alumina and adsorbed acetone to 200°C was found to give rise to Raman bands characteristic
of physisorbed mesityl oxide, a condensation product of acetone.

Hendra and Loader reported that the spectra of acetaldehyde
adsorbed on silica gel showed seven bands in the range 80-2000 cm^{-1},
which were not consistent with acetaldehyde. These bands were
subsequently shown to be due to physically adsorbed paraldehyde,

produced by the catalytic condensation of acetaldehyde.

The Raman spectrum of benzaldehyde adsorbed on alumina
(pretreated at 360°C) indicates both physisorption and chemisorption
with the physisorbed benzaldehyde readily removed by evacuation.
The chemisorbed material was identified as mainly aluminium benzoate
but a small proportion was benzyl alcohol, catalytically produced at
basic sites on the alumina surface.
We now survey recent progress by emphasising work completed in
the last 2-3 years at Southampton.

Recent Progress on Oxide Surfaces

We now report data on the sorption and polymerization of
acetylene over alumina and zeolites. The programme is not complete
but it appears that the polymerization is ion dependent in the case
of the zeolites and also activation temperature dependent. Lewis
sites would not appear to be involved in all cases. The product is
trans poly acetylene. The spectra are resonance enhanced since they
resemble those produced from degraded polyvinyl specimens.

Chlorine and bromine sorb to aluminas in an interesting manner,
the spectra indicating a wide variety of modes of sorption. It would
also seem that halogenation of the surface at low temperature is not
important.

Uranium hexafluoride sorption to alumina, metals and polymers
is also of interest. We feel that Lewis sites on oxides are
responsible for degradation to UOF$_4$ and related species.

In other laboratories, the quest for versatility continues
e.g. recent work on CO over supported nickel and it is now quite

clear that Raman spectroscopy is a really significant member of the
array of techniques available for surface analysis.

RAMAN SPECTROSCOPY AT ELECTRODE SURFACES

In recent years there has been a considerable increase in
research into the reactions which occur at electrode-electrolyte
interfaces. The quest for fuel cells and storage batteries with
high capacities and the control of metallic corrosion have added
urgency to this effort. Our knowledge of the reactants and final
products in an electrochemical process and of the electrical
parameters involved (current, voltage, capacitance, etc) is
excellent but the structure of the electrode-electrolyte interface
and the mechanisms of reactions at such interfaces are only poorly
understood.

As has been mentioned above, electron diffraction and related
techniques are not applicable under electrolytic conditions.
Optical methods of analysis at the electrode-electrolyte interface
are not normally feasible. Although some success has been achieved
with ultraviolet/visible absorption/reflection techniques and ellipsometry, however the data from these methods are not molecularly
specific. Infrared methods show promise but in aqueous electrolytic
cells are almost impossibly difficult due to the intensity of water
absorption. This limitation could vanish, however, with the
development of the tunable infrared laser. Infrared attenuated
total reflectance has been used on infrared transparent germanium
electrodes.

Raman on the other hand, would appear to be a particularly
attractive proposition. Since water is both a poor scatterer and
does not absorb in the visible region, its interference with
Raman spectra should be minimal. The first successful experiments
to be reported related to the mercury interface. The electrode,

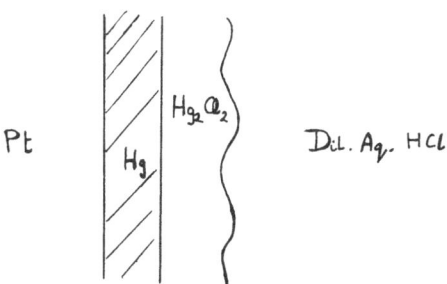

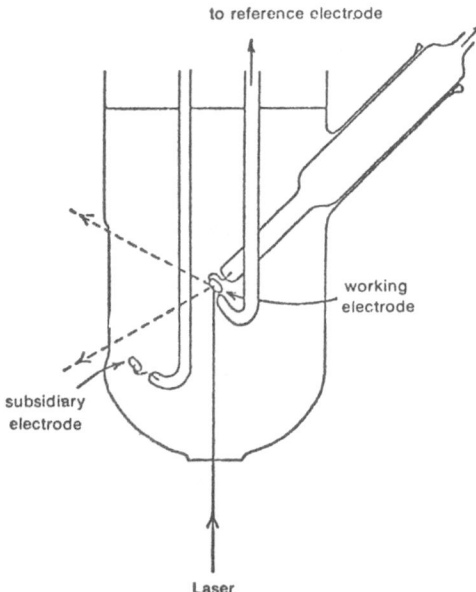

Fig. 10.   An electrochemical cell for Raman experiments. The work-
            ing electrode can be supported in a glass tube blown onto
            a modified hypodermic syringe or fitted into a machined
            Teflon sleeve.

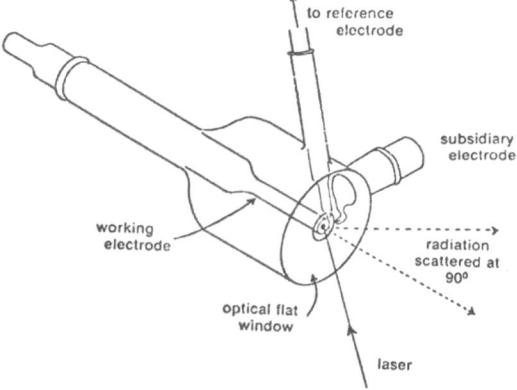

Fig. 11.   In this thin-layer cell a mercury drop only ∿ 1 mm in
            diameter is extruded through the hollow working electrode
            holder against a glass optical flat. The subsidiary
            electrode is a platinum wire ring surrounding the mercury
            drop.

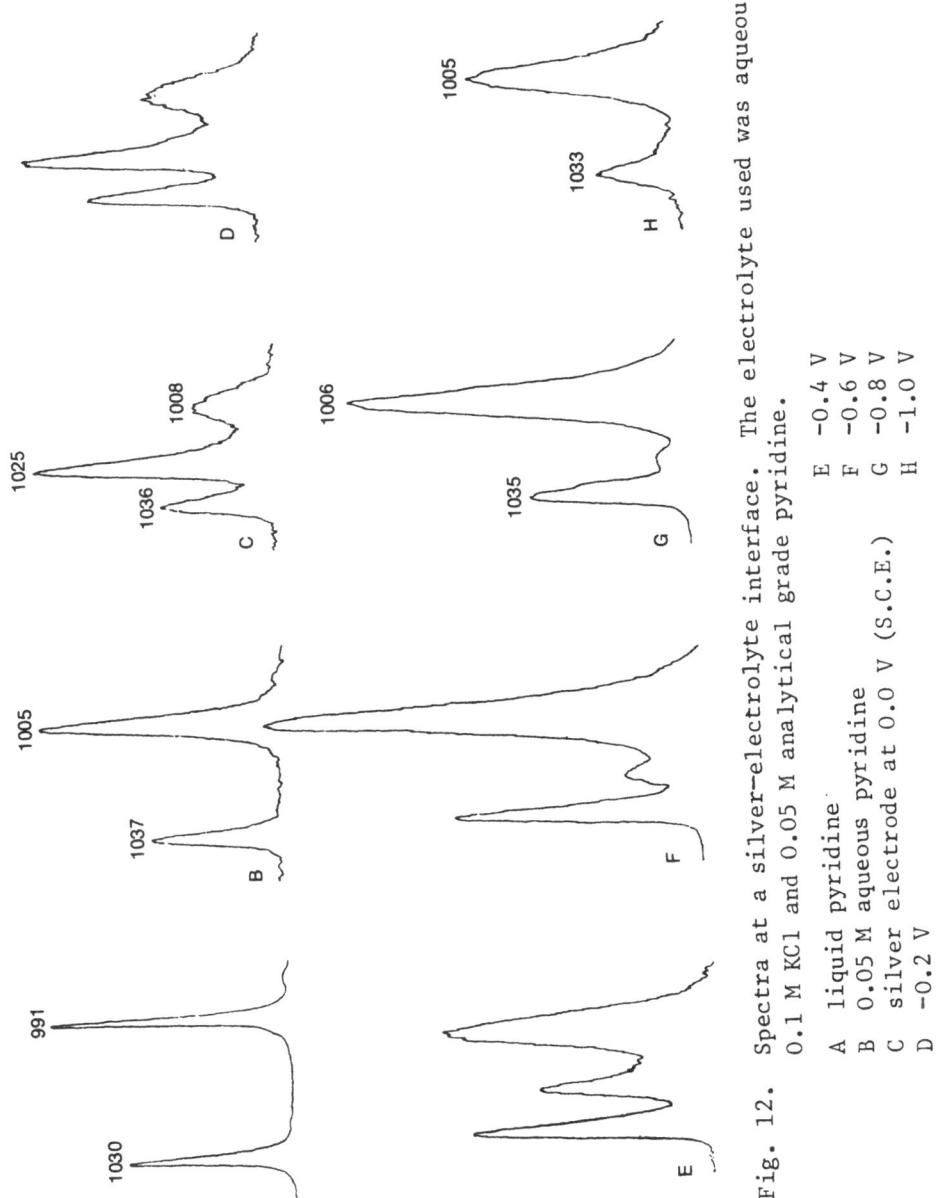

Fig. 12. Spectra at a silver—electrolyte interface. The electrolyte used was aqueous
0.1 M KCl and 0.05 M analytical grade pyridine.

A liquid pyridine
B 0.05 M aqueous pyridine
C silver electrode at 0.0 V (S.C.E.)
D −0.2 V

E −0.4 V
F −0.6 V
G −0.8 V
H −1.0 V

especially contrived to maximize its surface area (it consisted
of very finely divided droplets of mercury adhering to a platinum substrate) was placed in the Raman spectrometer with the
laser beam impinging at glancing incidence, thus illuminating
a considerable area of its surface. As the electrode potential
was cycled, the Raman-scattered spectrum of calomel (Hg_2Cl_2) could
be made to appear and vanish. Spectra of HgO and Hg_2Br_2 could be
found if KOH and hydrobromic acid were used as electrolyte. All
three species identified, HgO, Hg_2Cl_2 and Hg_2Br_2, are strong Raman
scatterers but the experiment did demonstrate the feasibility of
examining the electrode-electrolyte interface in this way. The cell
used is illustrated in figure 10.

Electrochemists have an almost insatiable attraction for the
plane mercury surface. Unfortunately this is far from ideal for
Raman work because it has little area and mercury surfaces wobble!
Confined in the cell drawn in Fig. 11, the surface can be kept still,
the thin film of electrolyte between the metal and the glass flat
being adequate to complete the cell. Using this device, admittedly
poor but recognizable spectra of calomel (Hg_2Cl_2) can be recorded
and, as before, their intensity is potential-sensitive thus demonstrating the feasibility of the Raman method in a situation really
well studied by the classical electrochemical, mercury polarographic
techniques. More recently, a new system has yielded significant
and much more detailed results. It is an electrochemically etched
silver electrode with a high surface-area.

The cell and the spectra produced as the electrode potential
is varied are shown in Fig. 11 and 12; the identity of the bands
given in the caption to Fig. 12. The appearance of pyridine molecules
Lewis-coordinated to silver atoms is not unexpected but the behaviour
of the hydrogen-bonded species is of considerable interest. As the
potential is made more cathodic, it would appear that sorption
maximizes near the "potential of zero charge". Further, the ring-
breathing mode frequency shifts. At anodic potentials, the frequency
is intermediate between that characteristic of hydrogen-bonded
pyridine in water and the pyridinium ion. At cathodic potentials
the frequency is close to that of the aqueous hydrogen-bonded hetero-
cycle. We have interpreted this as follows: at anodic potentials,
the water molecules are oriented with the oxygen atom towards the
electrode and the pyridine is bonded to the surface layer of water
molecules. At cathodic potentials the water molecules reverse in
orientation and at least one further water molecule must be
inserted between the metal and the pyridine (Fig. 13).

More recent research shows that this interpretation is a little
naïve. We now know from Raman experiments that in all our experiments carbon dioxide is absorbed and reduced. The reactions
involved are:

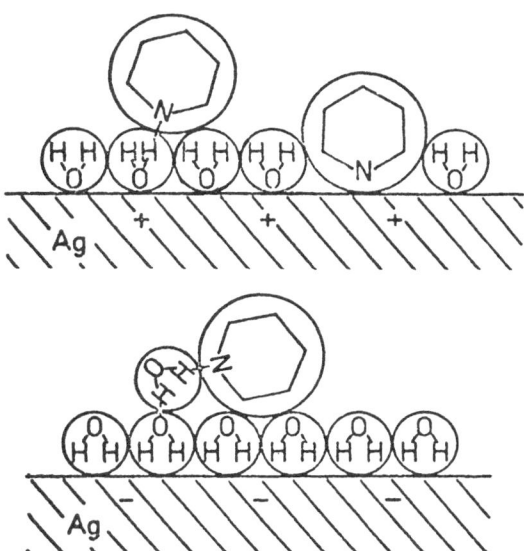

Fig. 13a. Possible models for pyridene sorbed to a silver electrode at anodic and cathodic potentials.

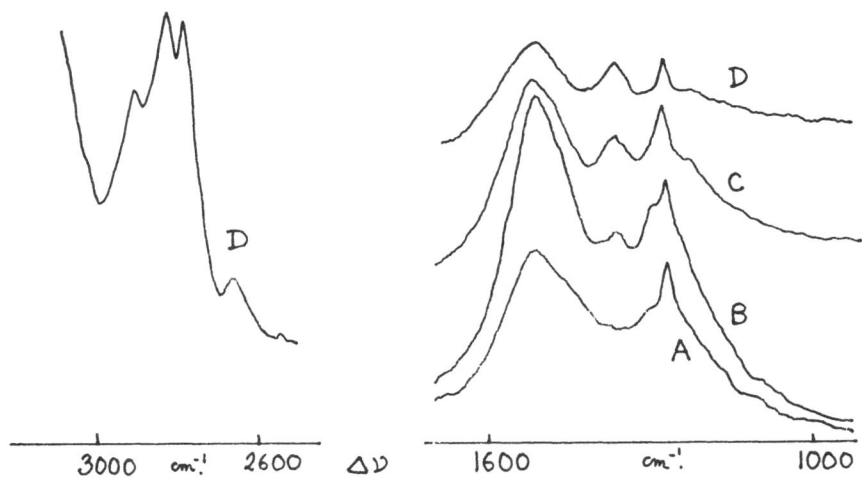

Fig. 13b. Raman spectra from a silver electrode at various potentials in 0.1 M potassium formate solution.
(A) 0.3 V (S.C.E.) (B) −0.3 V (C) −0.9 V (d) −1.2 V.

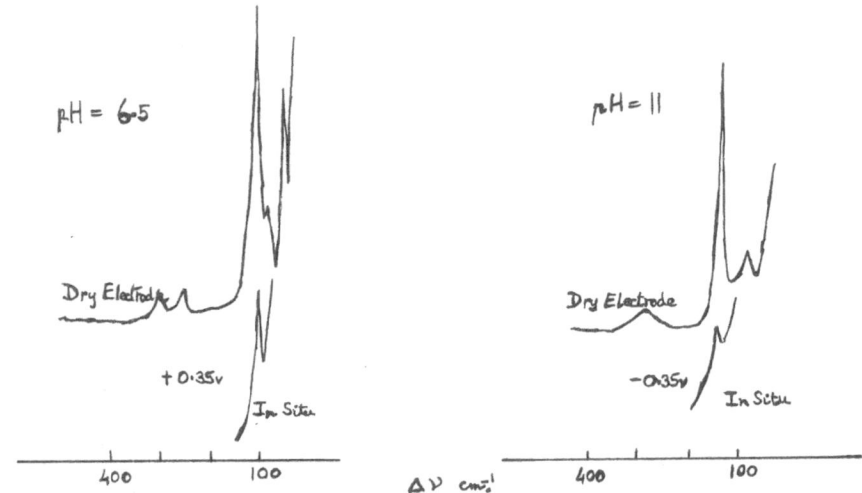

Fig. 14a.   Raman spectra of basic lead chloride corrosion layers
            on lead electrodes studied dry and in cells.

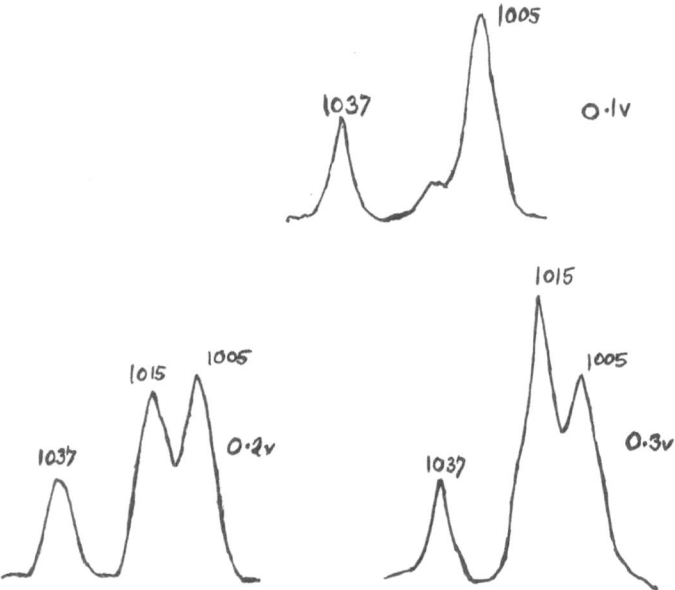

Fig. 14b.   Raman spectra at a copper electrode of aqueous
            0.05 M pyridene in 0.1 M KCl. Potentials refer
            to the S.C.E.

$$CO_2 \xrightarrow[\text{high pH}]{} \begin{array}{c}\text{Absorbed}\\\text{into}\\\text{solution}\end{array} \xrightarrow[\substack{\text{Potentials}\\\text{at Ag metal}}]{\text{Cathodic}} \begin{array}{c}HCOO^-\\\text{at}\\\text{surface}\end{array} \xrightarrow[\substack{\text{Cathodic}\\\text{potentials}}]{\text{More}} -(CH_2)_n COO^-$$

It appears that this series of reactions is difficult to
prevent - the carbon dioxide sorption is most persistent. It may
well be that this reduction could be of some significance
commercially.

Pyridene will adsorb to copper electrodes. This system is
very different in behaviour from that involving silver. It appears
from the Raman data that pyridine sorbs directly to the metal surface
which seems in most respects to be effectively chemically clean
(except of course, for the electrolyte!). See fig. 14a.

The lead electrode in aqueous chloride electrolyte is of some
interest. Raman spectroscopy shows that at the appropriate potential
a thin basic lead chloride film develops. The material is known in
bulk and its identity at the lead surface is clearly indicated in
the spectrum. See fig. 14b.

Even when the sorbed species is an excellent Raman scatterer
we usually find it necessary to investigate either thick (i.e. multilayer) films at smooth surfaces and/or thinner films on rough
surfaces. This results from limitations in instrumental sensitivity.
Recent progress in this respect has meant that versatility is
improving.

Platinum electrodes are of considerable interest in electrochemistry. Smooth surfaces coated with monolayers can not be used
except for very strongly scattering species such as those observed
for the polarisation of platinum in solutions containing iodide
ions. (fig. 15(b). With increased potential, the spectra due to
strongly sorbed iodine and triiodine ions are observed (0.4 V and
0.5 V) while at 0.5 V the spectrum of weakly sorbed iodine is also
seen. At the highest voltage only the spectra of very weakly
sorbed species are found.

In general, however, roughened platinum surfaces must be used.
The conventional platinum blacks are unattractive in view of their
colour but we have now developed a method of making such rough
platinum electrodes which are straw coloured. These have been used
to collect data for the sorption and oxidation of carbon monoxide.
The spectra data clearly indicate that the sorption of carbon
monoxide is terminal in nature i.e.

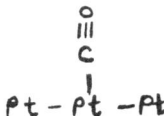

but its vibrational frequency i.e. $\nu_{C\equiv O}$ is unusually high
(compared with CO sorbed to Pt/Al$_2$O$_3$ catalysts. This caused some
concern until we learned from recent literature that the frequency
was typical of that due to CO sorbed to Pt 100 plane surfaces
derived from i.r. deflection data. It is this facet which pre-
dominates in microcrystalline rolled sheet. See fig. 15(a).

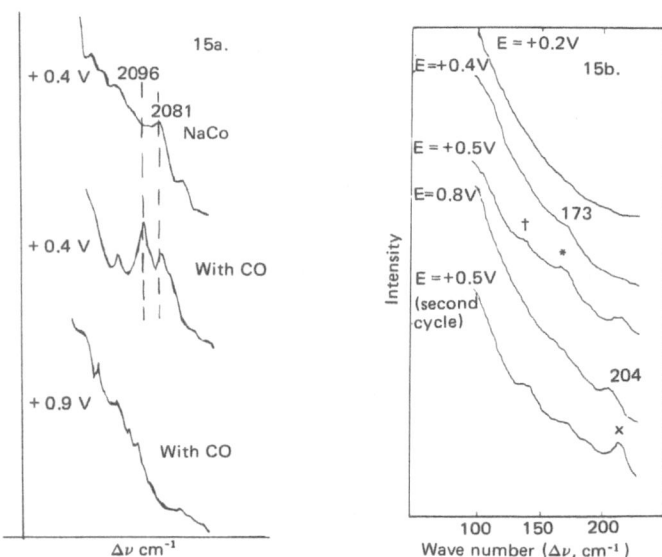

Figure 15.   a.   Pt(rough) + CO
              b.   Pt(smooth) + I$_2$

              * Strongly and X weakly sorbed iodine
              † Strongly sorbed I$_3^-$

Returning to the silver electrode we find that the silver/thio-
cyanate system is most informative, the spectrum clearly indicating
sorption via the silver atom at -0.7 and -0.6 V, while this is
changed to bridging coordination at less negative potentials and
eventually a solid phase of silver thiocyanate is formed, see
fig. 16.

All of the electrochemical data reported has been obtained
in aqueous electrolytes (primarily because the Raman scatter in
water is advantageously weak). Recently, we have succeeded in
recording spectra of good quality from the non-aqueous system:

        Pb/alkyl halides/dimethyl formamide
        (or Sn)
        We detect bands due to lead (methyl)$_n$ and lead (ethyl)$_n$ species.

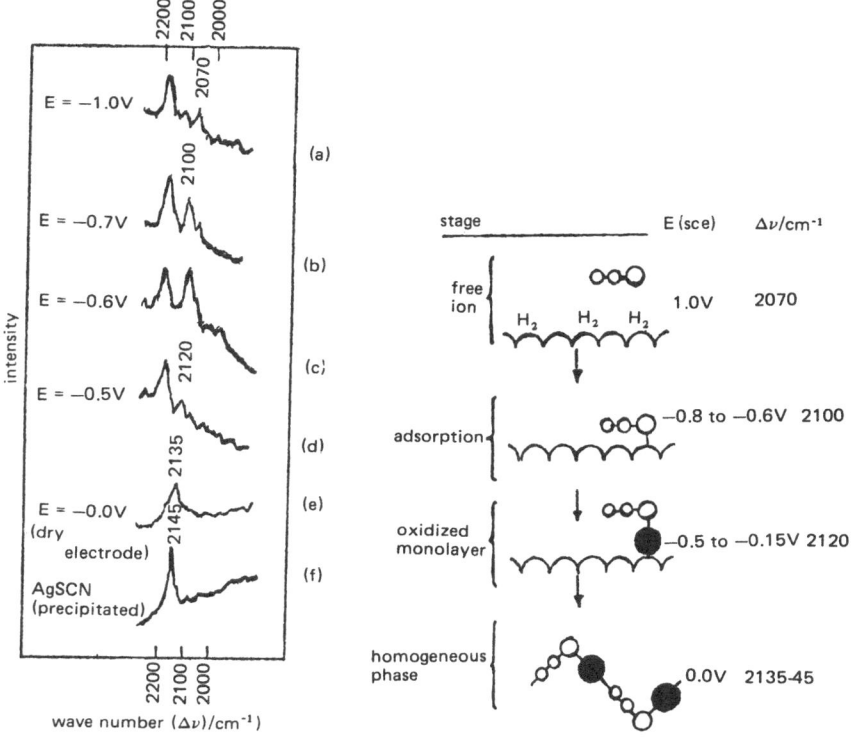

Fig. 16.   Raman spectra from the surface of an Ag electrode in
0.1 M NaClO₄ 0.01 M KSCN at pH 2.8.   (a-d) averaged
over 64 sweeps.   5145 Å at 80 mw for a-d and 20 mw for
e + f.

Schematic representation of our interpretation of
surface species is given beside the spectra.   The
3 atom group is SCN⁻. Lower semicircular pattern
is metal surface.   Black circles are oxidized Ag
atoms.

There is every prospect that this observation will prove to
be the first of many non-aqueous electrolytic systems.

CONCLUSION

We at Southampton were fortunate to pioneer the development
of Raman spectroscopy as a tool in the study of sorption and
reaction both at oxide and electrode surfaces.   Progress has been
steady and the field is opening up.   This comment is particularly
relevant to electrodes, where the data is quite unique.   As in

most branches of applied Raman spectroscopy, the rate of development of the subject is ultimately limited by the overall sensitivity of the spectrometric observation. This is inevitable in a field where the basic effect is weak. Significant further progress will necessitate considerable further improvements in sensitivity and we must look for ways of achieving these. Existing developments which may turn out to be significant are coherent anti-Stokes Raman spectroscopy (a most attractive technique for the study of gases at high sensitivity) and using conventional Raman processes, the utilization of the "multiplex advantage". The latter can crudely be stated to centre on the concept that instead of 'looking' with one's spectrometer at a single frequency for the time constant of the experiment and then moving to a new frequency - in the multiplex case one looks at all of the spectrum all of the time! Using conventional monochromators experiments are in-hand in a number of laboratories detecting the radiation with image intensifiers and t.v. cameras. The interferometer can be used too, and will be shortly, it is certain. Diffraction gratings have improved recently thanks to the development of the holographic production methods but very recently a double grating christened the 'griding' (rulings normal to one another on the same surface) has now been announced.

There seems an excellent chance that in a few years' time no first class surface chemical or electrochemical laboratory will be complete without its Raman facility.

BIBLIOGRAPHY

Rather than cite numerous references to original papers, readers are advised to consult:

Infrared spectra of adsorbed species

1.   L. H. Little, Infrared Spectra of Adsorbed Species, 1966, Academic Press, N.Y.

2.   Experimental developments: "Surface Adsorbed Samples" by N. D. Parkyns, Chapter in Lab. Methods in Infrared Spectroscopy", Ed. R. G. Miller and B. C. Stone, 2nd Edn. Heyden, London 1972.

Reviews on Raman Spectroscopy of Sorbed Species

3.   R. P. Cooney, G. Curthoys and N. T. Tam, Advances in Cat., 24, 293 (1975).

4.   R. L. Paul and P. J. Hendra, Minerals. Sci. Engng. 8, 171 (1976); and references cited therein.

INDEX